CARDIOVASCULAR PHARMACOLOGY:
ENDOTHELIAL CONTROL

CARDIOVASCULAR PHARMACOLOGY: ENDOTHELIAL CONTROL

Edited by
Paul M. Vanhoutte
Department of Pharmacology and Pharmacy
Li Ka Shing Faculty of Medicine
University of Hong Kong
Hong Kong, China

Serial Editor
S. J. Enna
Department of Molecular and Integrative Physiology
Department of Pharmacology, Toxicology and Therapeutics
University of Kansas Medical Center
Kansas City, Kansas, USA

ADVANCES IN
PHARMACOLOGY

VOLUME 60

AMSTERDAM • BOSTON • HEIDELBERG • LONDON
NEW YORK • OXFORD • PARIS • SAN DIEGO
SAN FRANCISCO • SINGAPORE • SYDNEY • TOKYO
Academic Press is an imprint of Elsevier

Academic Press is an imprint of Elsevier
525 B Street, Suite 1900, San Diego, CA 92101-4495, USA
30 Corporate Drive, Suite 400, Burlington, MA 01803, USA
32 Jamestown Road, London NW1 7BY, UK
Radarweg 29, PO Box 211, 1000 AE Amsterdam, The Netherlands

First edition 2010

British Library Cataloguing in Publication Data
A catalogue record for this book is available from the British Library

Library of Congress Cataloging-in-Publication Data
A catalog record for this book is available from the Library of Congress

ISBN: 978-0-12-385061-4

For information on all Academic Press publications visit our website at elsevierdirect.com

Printed and bound in USA
10 11 12 10 9 8 7 6 5 4 3 2 1

Contents

Contributors

Numbers in parentheses indicate the pages on which the authors' contributions begin.

Cyril Auger (133) Laboratoire de Biophotonique et Pharmacologie, UMR 7213 CNRS, Université de Strasbourg, Faculté de Pharmacie, Illkirch, France

William B. Campbell (27) Department of Pharmacology and Toxicology, Medical College of Wisconsin, Milwaukee, Wisconsin, USA

Thierry Chataigneau (133) Laboratoire de Biophotonique et Pharmacologie, UMR 7213 CNRS, Université de Strasbourg, Faculté de Pharmacie, Illkirch, France

Wei Chen (107) Department of Medicine and Division of Cardiology, Emory University School of Medicine and Atlanta Veterans Administration Medical Center, Decatur, Georgia, USA

Martine Clozel (1) Actelion Pharmaceuticals Ltd., Allschwil, Switzerland

Richard A. Cohen (85) Vascular Biology Unit, Whitaker Cardiovascular Institute, Boston University School of Medicine, Boston, Massachusetts, USA

Livius V. d'Uscio (257) Departments of Anesthesiology and Molecular Pharmacology and Experimental Therapeutics, Mayo Clinic College of Medicine, Rochester, Minnesota, USA

Andreas Daiber (177) II. Medizinische Klinik, Labor für Molekulare Kardiologie und Abteilung für Kardiologie und Angiologie Universitätsmedizin der Johannes-Gutenberg-Universität, Mainz, Germany

Sergey Dikalov (107) Department of Medicine and Division of Cardiology, Emory University School of Medicine and Atlanta Veterans Administration Medical Center, Decatur, Georgia, USA

Nelly Étienne-Selloum (133) Laboratoire de Biophotonique et Pharmacologie, UMR 7213 CNRS, Université de Strasbourg, Faculté de Pharmacie, Illkirch, France

Michel Félétou (85) Department of Angiology, Institut de Recherches Servier, Suresnes, France

Kathryn M. Gauthier (27) Department of Pharmacology and Toxicology, Medical College of Wisconsin, Milwaukee, Wisconsin, USA

Tommaso Gori (177) II. Medizinische Klinik, Labor für Molekulare Kardiologie und Abteilung für Kardiologie und Angiologie Universitätsmedizin der Johannes-Gutenberg-Universität, Mainz, Germany

David G. Harrison (107) Department of Medicine and Division of Cardiology, Emory University School of Medicine and Atlanta Veterans Administration Medical Center, Decatur, Georgia, USA

Yu Huang (61) Institute of Vascular Medicine, Li Ka Shing Institute of Health Sciences, and School of Biomedical Sciences, Chinese University of Hong Kong, Hong Kong SAR, China

Zvonimir S. Katusic (257) Departments of Anesthesiology and Molecular Pharmacology and Experimental Therapeutics, Mayo Clinic College of Medicine, Rochester, Minnesota, USA

Karen S. L. Lam (229) Department of Medicine; Research Centre of Heart, Brain, Hormone and Healthy Aging, University of Hong Kong, Hong Kong, China

Chi Wai Lau (61) Institute of Vascular Medicine, Li Ka Shing Institute of Health Sciences, and School of Biomedical Sciences, Chinese University of Hong Kong, Hong Kong SAR, China

Li Li (107) Department of Medicine and Division of Cardiology, Emory University School of Medicine and Atlanta Veterans Administration Medical Center, Decatur, Georgia, USA

Thomas Münzel (177) II. Medizinische Klinik, Labor für Molekulare Kardiologie und Abteilung für Kardiologie und Angiologie Universitätsmedizin der Johannes-Gutenberg-Universität, Mainz, Germany

Sandra L. Pfister (27) Department of Pharmacology and Toxicology, Medical College of Wisconsin, Milwaukee, Wisconsin, USA

Anantha Vijay R. Santhanam (257) Departments of Anesthesiology and Molecular Pharmacology and Experimental Therapeutics, Mayo Clinic College of Medicine, Rochester, Minnesota, USA

Valérie B. Schini-Kerth (133) Laboratoire de Biophotonique et Pharmacologie, UMR 7213 CNRS, Université de Strasbourg, Faculté de Pharmacie, Illkirch, France

Eric Thorin (1) Department of Surgery, Montreal Heart Institute, Research Center, Université de Montréal, Montreal, Quebec, Canada

Xiao Yu Tian (61) Institute of Vascular Medicine, Li Ka Shing Institute of Health Sciences, and School of Biomedical Sciences, Chinese University of Hong Kong, Hong Kong SAR, China

Paul M. Vanhoutte (85, 229) Department of Pharmacology and Pharmacy, Li Ka Shing Faculty Medicine; Research Centre of Heart, Brain, Hormone and Healthy Aging, University of Hong Kong, Hong Kong, China; Department BIN Fusion Technology, Chonbuk National University, Jeonju, Korea

Tony J. Verbeuren (85) Department of Angiology, Institut de Recherches Servier, Suresnes, France

Yu Wang (229) Research Centre of Heart, Brain, Hormone and Healthy Aging; Department of Pharmacology and Pharmacy, University of Hong Kong, Hong Kong, China

Siu Ling Wong (61) Institute of Vascular Medicine, Li Ka Shing Institute of Health Sciences, and School of Biomedical Sciences, Chinese University of Hong Kong, Hong Kong SAR, China

Wing Tak Wong (61) Institute of Vascular Medicine, Li Ka Shing Institute of Health Sciences, and School of Biomedical Sciences, Chinese University of Hong Kong, Hong Kong SAR, China

Aimin Xu (229) Department of Medicine; Research Centre of Heart, Brain, Hormone and Healthy Aging; Department of Pharmacology and Pharmacy, University of Hong Kong, Hong Kong, China

Foreword

The twentieth century has witnessed immense progress in preventing and treating cardiovascular disease. This has been fostered by improvements in life-style [exercise and diet] and the introduction of new therapeutics [antihypertensive and lipid lowering drugs]. Nonetheless, cardiovascular disease remains a major cause of death and disability in developed countries and, increasingly so, in the developing world. This is driven in part by demographics and the increase in longevity, and the obesity-metabolic syndrome-diabetes-atherosclerosis continuum that is reaching pandemic proportions. The hopes to address these issues using gene-therapy have faded over the past decade. Stem cell therapy, while scientifically exciting, is still in its infancy, so will be, in the near-term, a treatment for the privileged. Given its cost, this approach is likely to be inaccessible to most patients with cardiovascular disease, in particular those in the emerging countries. Accordingly, the discovery of novel targets involved in cardiovascular disease and the design of small molecules or biologics that interact with these sites still holds the greatest promise for treating large numbers of individuals afflicted with these conditions.

Presented in this second volume of *Cardiovascular Pharmacology* are some further promising possibilities in that regard that focus on the pivotal role played by the endothelium in the genesis of vascular disease. There is a discussion of the renewed interest in the role of endothelin-1 in cardiovascular function, an area that has been ignored of late because of the disappointing initial clinical results with endothelin antagonists.

Chapters are also devoted to describing the mounting evidence that arachidonic acid metabolites play a crucial role in the control of vascular function and its dysregulation. There is also coverage of studies of nitric oxide, the major endothelium-derived relaxing factor, with particular emphasis on the involvement of tetrahydrobiopterin as a cofactor in the release of this gas, the potential of natural products to stimulate its production and augment its bioavailability, and the continuing problem of tolerance to nitrates.

Contributors also consider new concepts relating to the impact of diabetes on vascular function. In addition, there is a summary of the complex actions of erythropoietin on the cardiovascular system that elegantly illustrates Paracelsus's observation that the dose makes the poison.

I thank the contributors, all of whom are internationally recognized experts in the field, for their efforts in making this volume possible. Together with them, I sincerely hope these reports will be a source of inspiration, instruction, and ideas for graduate students, cardiovascular scientists, and physicians interested in the function and dysfunction of the heart and the blood vessel wall. Not only will attainment of these goals be personally satisfying for us, but it will hopefully provide a stimulus for further advances in this important area.

Paul M. Vanhoutte, M.D., Ph.D.
Department of Pharmacology and Pharmacy
Li Ka Shing Faculty of Medicine
The University of Hong Kong
Hong Kong, China

Eric Thorin* and Martine Clozel[†]

*Department of Surgery, Montreal Heart Institute, Research Center, Université de Montréal, Montreal, Quebec, Canada

[†]Actelion Pharmaceuticals Ltd., Allschwil, Switzerland

The Cardiovascular Physiology and Pharmacology of Endothelin-1

Abstract

One year after the discovery in 1980 that the endothelium was obligatory for acetylcholine to relax isolated arteries, it was clearly shown that the endothelium could also promote contraction. In 1988, Dr Yanagisawa's group identified endothelin-1 (ET-1) as the first endothelium-derived contracting factor. The circulating levels of this short (21 amino acids) peptide were quickly determined in humans and it was reported that in most cardiovascular diseases, circulating levels of ET-1 were increased and ET-1 was then recognized as a likely mediator of pathological vasoconstriction in human. The discovery of two receptor subtypes in 1990, ET_A and ET_B, permitted optimization of

Advances in Pharmacology, Volume 60

1054-3589/10 $35.00
10.1016/S1054-3589(10)60009-9

bosentan, which entered clinical development in 1993, and was offered to patients with pulmonary arterial hypertension in 2001. In this report, we discuss the physiological and pathophysiological role of endothelium-derived ET-1, the pharmacology of its two receptors, focusing on the regulation of the vascular tone and as much as possible in humans. The coronary bed will be used as a running example, but references to the pulmonary, cerebral, and renal circulation will also be made. Many of the cardiovascular complications associated with aging and cardiovascular risk factors are initially attributable, at least in part, to endothelial dysfunction, particularly dysregulation of the vascular function associated with an imbalance in the close interdependence of NO and ET-1, in which the implication of the ET_B receptor may be central.

I. Introduction

The endothelium is an extraordinary organ that protects the arterial wall through the release of nitric oxide (NO) and prostacyclin (PGI_2) among other factors (Furchgott & Zawadzki, 1980; Palmer et al., 1987). Before 1980, it was merely considered an inert barrier (Aird, 2007). The presence of an endothelium-derived constricting factor (EDCF) was hypothesized 1 year after the revelation of the relaxant properties of the endothelium (De Mey & Vanhoutte, 1982; Vanhoutte et al., 1986). But it was only 8 years later that Yanagisawa and colleagues identified the peptide endothelin-1 (ET-1) has this long-lasting EDCF (Yanagisawa et al., 1988a, 1988b). Two receptors for ET-1 were identified 2 years later (Arai et al., 1990; Sakurai et al., 1990). Then, shortly after the discovery of ET_A and ET_B receptors, Martine Clozel and colleagues presented in 1993 the first orally active ET-1 receptor antagonist, Ro 46-2005 (Clozel et al., 1993), and the same team made a structurally modified analog, bosentan (Tracleer), available to patients with pulmonary arterial hypertension (PAH) at the end of 2001. In less than 15 years, a new factor was identified, its receptors were cloned and their pharmacology characterized, a pathology associated with the abnormal function of the ET-1 system, and an effective treatment offered to patients in need. Today, other ET receptor antagonists have been synthesized and are in development, all this being well reviewed recently (Kirkby et al., 2008; Motte et al., 2006). There is still much to be discovered on the role and the mechanisms of action of ET-1. We focus in this chapter on the pharmacology of ET-1 and review the role of ET-1 on the vasculature with as much as possible references to the human pathophysiology.

II. Cardiovascular Physiology of ET-1

ET-1 is one of the most potent vasoconstrictors identified so far (Yanagisawa et al., 1988b) inducing prolonged contraction of isolated canine and nonhuman primate coronary arteries with a half maximal effective concentration ($-\log[EC_{50}]$) of 8. The potency of ET-1 is unequaled, with the exception of urotensin II ($-\log[EC_{50}]=9.5$; Douglas et al., 2000). ET-1 elicits its effects through two receptors (http://www.iuphar.org/): ET_A receptors, located in vascular smooth muscle cells (VSMC) and cardiomyocytes, mediate contraction, whereas ET_B receptors, located on vascular endothelial cells (EC), mediate dilation and ET-1 uptake, and regulate ET-1 production (Arai et al., 1990; Barton & Yanagisawa, 2008; Brunner et al., 2006; Callera et al., 2007; Dupuis et al., 1997; Farhat et al., 2008; Komukai et al., 2010; Rubanyi & Polokoff, 1994; Sakurai et al., 1990; Sanchez et al., 2002). Additionally, ET_B receptors can also be expressed on VSMC and elicit contractions (Sanchez et al., 2002; Teerlink et al., 1994). It is a known fact that ET-1 is released continuously, mostly from EC, by a constitutive pathway and contributes to the regulation of the vascular tone in general (Brunner et al., 2006; Callera et al., 2007; Rubanyi & Polokoff, 1994). NO, however, strongly inhibits the release of ET-1 from the native endothelium (Boulanger & Luscher, 1990; Luscher et al., 1990); for this reason, it has been suggested that NO and ET-1 regulate each other through an autocrine feedback loop (Alonso & Radomski, 2003; Luscher et al., 1990). In addition to EC, ET-1 is also produced by VSMC, cardiomyocytes, leukocytes, macrophages, various neurons, and other cells (Kedzierski & Yanagisawa, 2001). This peptide is also proinflammatory and promotes VSMC proliferation (Anggrahini et al., 2009; Dashwood et al., 1998a; Davenport & Maguire, 2001; Ihling et al., 2001; Ivey et al., 2008; Ruschitzka et al., 2000). Thus, ET-1 contributes to the cardiovascular homeostasis by regulating basal vascular tone and remodeling (Brunner et al., 2006; Kedzierski & Yanagisawa, 2001).

A. The ET-1 System

1. Endothelins and Their Receptors

There are three isoforms of endothelin produced in humans, ET-1, ET-2, and ET-3 (Inoue et al., 1989a; Saida et al., 1989). They are encoded on chromosomes 6, 1, and 20, respectively (Inoue et al., 1989a). ET-1 binds ET_A and ET_B receptors with equal affinity, while ET_A receptors have approximately 100-fold less affinity for ET-3 than ET_B receptors (Davenport, 2002). In addition, snake venom toxins called “sarafotoxins” have been identified by homology. Sarafotoxin 6c (S6c) is a selective ET_B receptor agonist (Rubanyi & Polokoff, 1994). ET_A and ET_B receptors belong to the 7-transmembrane domain (7-TM) family and are encoded by distinct genes on chromosomes 4 and 13, respectively (Sakurai et al., 1990).

ET-1 is produced within the cell in two proteolytic steps from the preproET-1, a large precursor peptide of approximately 200 amino acid residues. First, a furin-like neutral endopeptidase cleaves the preproET-1 to bigET-1, an inactive peptide of 41 amino acid residues (Denault et al., 1995; Laporte et al., 1993). Second, bigET-1 is cleaved by the endothelin-converting enzymes (ECE-1 and ECE-2) to the biologically active ET-1, a 21-amino acid residue peptide enclosed by disulfur bonds (Takahashi et al., 1993) mainly by EC (Inoue et al., 1989b). Some alternative pathways to ECE for the synthesis of ET-1 have been reported: a chymostatin-sensitive enzyme, such as chymase, and the matrix metalloproteinase 2 are able to convert bigET-1 to mature ET-1 in human blood vessels (Maguire & Davenport, 2004; Maguire et al., 2001).

The clearance of ET-1 from the circulation after an intravenous injection of radiolabeled ET-1 in rats is rapid (half-life of 40 s; Sirvio et al., 1990), while its pressor effect is long lasting ($\approx$1 h at the doses administered) in man (Sirvio et al., 1990; Vierhapper et al., 1990). The majority of ET-1 is retained by the lungs and cleared from the circulation via binding to ET_B receptors (Dupuis et al., 1996a, 1996b).

2. Endothelin Receptor Ligands and Pharmacology: Emerging Concepts

There is no selective agonist for the ET_A receptor. ET-1 [1–31] has been shown to have more selectivity for ET_A compared to ET_B receptors (Rossi et al., 2002), but the 31-amino acid peptide has no direct pharmacological effects if not converted via a neutral endopeptidase-dependent mechanism to ET-1 [1–21] (Fecteau et al., 2005). Selective antagonists for the ET_A receptor include ZD4054, atrasentan, darusentan, macitentan, ambrisentan, and sitaxsentan (Motte et al., 2006).

In contrast to ET_A receptors, selective ET_B receptor agonists are available such as S6c and IRL-1620. Selective antagonists of the ET_B receptors include BQ788, A192621, RES7011, and IRL2500 (Alexander et al., 2009).

Because ET receptors are 7-TM receptors, their signal transduction (Fig. 1) was first interpreted as a sequential series of events initiated by the binding of the agonist on its receptor. This simplistic view had to be revised with the evidence that activation of a 7-TM receptor can activate simultaneously multiple pathways (Watts, 2010) including some G-protein-independent pathways (Galandrin et al., 2007; Kenakin, 2007; Violin & Lefkowitz, 2007). One well-known example is the activation of the angiotensin II (ANG II) receptor (AT_1): this receptor activates both G-protein-dependent pathways (PLC, PKC, channels, etc.) and β-arrestin-dependent pathways (independent of G proteins, i.e., the src/extracellular signal-regulated kinase/mitogen-activated protein kinase pathway). ANG II activates both signaling pathways; however, the substituted ANG II peptide Sar^1, Ile^4, Ile^8-ANG II (SII) almost exclusively activates the β-arrestin-dependent pathways (Violin & Lefkowitz, 2007). SII is called a "biased agonist" since it activates a preferential signal transduction pathway. Other biased ligands to several 7-TM receptors have been discovered (Gesty-Palmer et al., 2009; Rajagopal et al., 2010; Violin & Lefkowitz, 2007) including the β_2-adrenergic receptor:

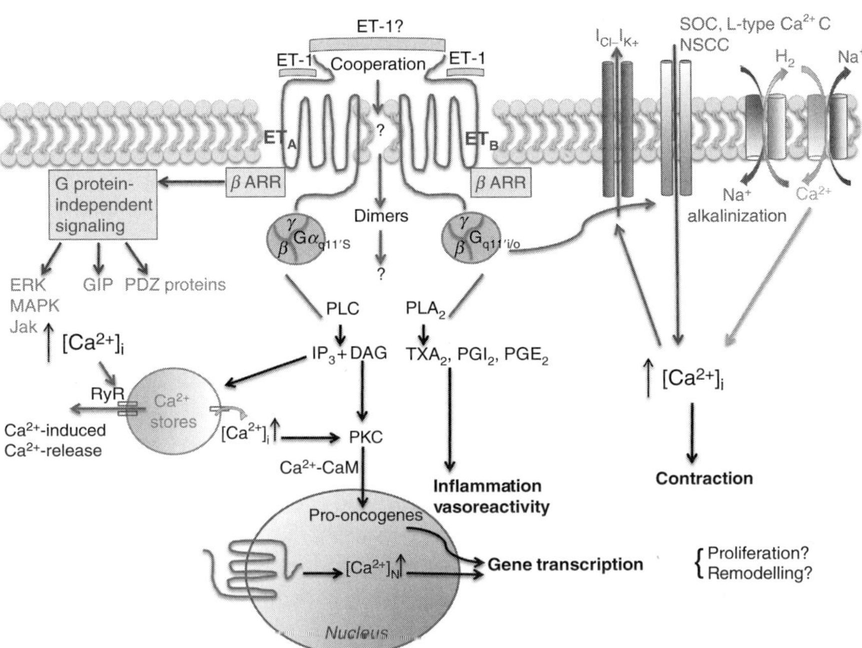

FIGURE 1 *The pleiotropic nature of ET-1 receptor signaling*. Schematic representation of the multiple pathways activated by 7-transmembrane ET_A and ET_B receptors either directly dependent on G protein activation or independent of G protein activation such as through direct interaction with β-arrestin or PDZ-domain-containing proteins that can act as scaffolds. In addition, ET-1 may act as a bivalent ligand leading to both ET_A and ET_B receptor activation and possibly dimerization, although this remains to be demonstrated. The signaling pathways activated by either a bivalent ET-1 or a dimerization remain unexplored. $G\alpha_{s/i/o/q/11}$, heterotrimeric G protein α subunits of different classes; β arr, β-arrestin; PKC, protein kinase C; Jak, Janus kinase; GIP, other GPCR interacting proteins; ERK, extracellular signal-regulated kinase; MAPK, mitogen-activated protein kinase.

carvedilol, a β-adrenergic receptor antagonist, is able to stabilize a receptor conformation, which, although uncoupled from G_s, is nonetheless able to stimulate β-arrestin-mediated signaling (Drake et al., 2008; Wisler et al., 2007). Through the activation of ET_A receptors, ET-1 stimulates both G-protein-dependent and independent pathways (Rosano et al., 2009; Spinella et al., 2009). It is therefore almost obvious that ET-1 acts as a biased ligand. This is also true for ET_B receptors, since ET-1 induces internalization of ET_B receptors and G-protein-dependent pathways (Farhat et al., 2008; Spinella et al., 2009). Are there specific conditions necessary to reveal the biased activation of ET_A by ET-1? What would be the conditions for ET-1 to act like carvedilol does on the β-adrenergic receptor, that is, solely activate the β-arrestin-mediated signaling? Are there pathological conditions that may affect ligand binding and signal transduction? This is an extremely important question because disease states or even aging alone could be responsible for changes in the microdomain such as lipid composition, influencing ligand binding, receptor dimerization (see later), and the subsequent signal transduction. It is known, for example, that oxidized low-density protein can change the microviscosity of the endothelium plasma membrane and alters signal transduction (Hamilton et al., 1994; Thorin et al., 1995). Much, however, needs to be understood from vascular primary cell cultures, isolated vessels, and *in vivo* preparations.

Another recent change in the pharmacological concepts of receptor signal transduction has been introduced by the evidence that ET receptors can form heterodimers, forcing us to change the way we interpreted pharmacological signals (Dai & Galligan, 2006; Evans & Walker, 2008a, 2008b; Gregan et al., 2004; Sauvageau et al., 2006). This observation, however, remains to be translated into physiological significance. So far, heterodimer formation has been reported in heterologous cell preparations expressing ET_A and ET_B receptors. To the best of our knowledge, we reported the only evidence of potential heterodimers expressed in rat pulmonary arteries (Sauvageau et al., 2006); in these vessels, the pharmacology is complex and difficult to interpret using the classical pharmacological concept of sequential events since for the least, cooperation between the two receptor subtypes exists. In addition, the pharmacology of ET receptors changes in pathological conditions such as experimental pulmonary hypertension in rats (Sauvageau et al., 2009). Although heterodimer formation of ET receptors is likely, the challenge will be to characterize their functions. Another level of complexity has been recently reached with the report that ET_B and dopamine D_3 receptor heterodimerization (Yu et al., 2009; Zeng et al., 2008): the authors reported aberrant interactions between these two receptors in cultured renal proximal tubule cells with basal D_3/ET_B receptor coimmunoprecipitation three times greater in Wistar Kyoto rats (WKY) than in spontaneously hypertensive rats (SHR). *In vivo*, the D_3 receptor agonist PD128907 caused natriuresis in WKY, which was partially blocked by ET_B receptor antagonism. In contrast, PD128907 blunted sodium excretion in SHR. The authors therefore speculated that there was interaction between the two receptors and that these heterodimers could be partly responsible for hypertension in SHR. If these results can be confirmed in other settings, it will only confirm the complexity of the ET-1 system and offer alternative explanations to unexplained effects of ET-1 in the various systems tested.

When trying to conceive a biological path linking two receptors as dimers, one cannot exclude the possibility that ET-1 is a bivalent ligand, capable of stimulating both receptors simultaneously and thus promoting dimerization of the receptors. This possibility was first proposed by Himeno and collaborators (Himeno et al., 1998), and it was based on the following observation: selective ET_B receptor ligands such as S6c, IRL1620, and BQ-788 competitively inhibited ^{125}I-ET-1 binding only when BQ-123 (selective ET_A receptor antagonist) was present in the incubation buffer. This therefore suggests that the ET_B

receptor is capable of binding ET-1 when the ET_A receptor is being occupied by BQ-123. A collaboration mechanism between the ET_A and the ET_B receptors may function in the recognition of ET-1, which is the qualification of a typical "bivalent" ligand. This could be at the basis of the formation of heterodimers at the surface of VSMC (Harada et al., 2002). Unfortunately, no other studies are available that could support this possibility.

Finally, ET-1 can bind with high affinity a newly identified atypical rat receptor, the dual ET-1/ANG II receptor (DEAR) and induces a rise in intracellular Ca^{2+} as efficiently as does ANG II (Ruiz-Opazo et al., 1998). The *Dear* gene maps to rat chromosome 2 and cosegregates with blood pressure in female F2(normotensive × hypertensive and salt-sensitive [R × S]) intercross rats with highly significant linkage (LOD 3.61) accounting for 14% of blood pressure variance. In $Dear^{-/-}$ mice, angiogenesis is impaired, the neuroepithelial development dysregulated, and is lethal by embryonic day 12.5 (Herrera et al., 2005). Interestingly, mouse DEAR does not bind ANG II as the rat DEAR does, but binds ET-1 and the vascular endothelial growth factor (VEGF) signal peptide (VEGFsp) with equal affinities (Herrera et al., 2005). The hypertension susceptibility in female F2(R × S) intercross rats was validated in humans, in a cohort from northern Sardinia (Glorioso et al., 2007). In the latter, the α_1N,K-ATPase (*ATP1A1*) polymorphism was also tested and concordant with the rat data, and associated with *Dear* gene polymorphism, albeit in men (and not women). It is interesting that *ATP1A1* and *Dear* are coexpressed in both renal tubular cells and vascular endothelium: it strongly suggests a role in the regulation of blood pressure for these two genes. Altered *ATP1A1* and *Dear* functions in the endothelium could contribute, in combination, to endothelial dysfunction through a putative imbalance of endothelial repair to turnover, because *ATP1A1* is implicated in cell proliferation and *Dear* in angiogenesis. Likewise, *ATP1A1* and *Dear* in renal tubular epithelial cells could affect sodium homeostasis, because ET-1 decreases renal Na, K-ATPase activity. Based on this observation, a net decrease in ET-1/*Dear* activation could result in greater renal Na^+, K^+ ATPase activity and increased Na^+ reabsorption given the same sodium load, hence salt sensitivity. All these data come from one group of scientists, and the physiology and pharmacology of DEAR has not been studied in depth. We therefore do not know if ET-1 binding site is sensitive to the classical small molecule ET receptor antagonists.

Altogether, these data demonstrate that ET-1 effects are more complex than predicted so far: ET_A and ET_B receptor cooperation, heterodimerization, the newly discovered DEAR, and a possible bivalent ligand (Fig. 1) are possibilities that have not been fully explored. Based on our data in rat resistance pulmonary arteries (Sauvageau et al., 2006, 2007), we propose that ET_A and ET_B receptor heterodimerization is an important component in the pharmacological effects of ET-1, although no technique is yet available to evaluate the type of interactions that take place between the two receptors *in vivo*.

B. Cardiovascular Effects of ET-1

In the systemic and pulmonary circulation, ET_A receptors are expressed in the VSMC (Hosoda et al., 1991), while both ET_A and ET_B receptors are expressed on the surface of VSMC (Ogawa et al., 1991) and EC (Davenport et al., 1993). Both receptors in VSMC induce contraction and cell proliferation in the presence of ET-1 (Clozel et al., 1992; Docherty & MacLean, 1998; LaDouceur et al., 1993; MacLean et al., 1994; Shetty et al., 1993; Sumner et al., 1992). In EC, activation of ET_B receptors activates the release of vasodilators and antiproliferative factors such as NO and PGI_2 (Clozel et al., 1992; de Nucci et al., 1988; Haynes & Webb, 1993; Muramatsu et al., 1999; Sato et al., 1995). The highest density of $ET_{A/B}$ receptors is found in the lungs and the heart (Simonson & Dunn, 1990).

ET-1 rapidly increases intracellular Ca^{2+} via activation of the phospholipase C that hydrolyzes phosphatidyl inositol trisphosphate (IP_3) and the neutral diacylglycerol (DAG) (Resink et al., 1988). This is followed by a sustained phase of Ca^{2+} influx associated with activation of secondary multiple intracellular events at the basis of ET-1-induced contraction, relaxation, and secretion (Fig. 1). The rise in IP_3 induces a fast and transient increase in $[Ca^{2+}]_i$ released from the reticulum, which is at the basis of the activation of membrane-bound channels leading to a sustained increase in $[Ca^{2+}]_i$ (Chen & Wagoner, 1991). This leads to numerous signals associated with Ca^{2+}-dependent pathways and Ca^{2+}/calmodulin-dependent pathways (Fig. 1): this includes activation of chloride channels (Haynes & Webb, 1993), the Na^+/H^+ exchanger resulting in cellular alkalinization and Ca^{2+} influx through the Na^+/Ca^{2+} exchanger (Grinstein & Rothstein, 1986; Koh et al., 1990), activates Ca^{2+}-induced Ca^{2+} release from the reticulum via ryanodine receptors and Ca^{2+}-activated K^+ channels (Bialecki et al., 1989; Nelson et al., 1995; Simpson & Ashley, 1989). In addition, DAG activates PKC, which leads to numerous intracellular events (Fig. 1) including membrane translocation and activation of phosphokinases (Newton & Keranen, 1994), and damping of the Ca^{2+} signal (Clerk et al., 1994).

Endothelin receptors are also expressed on the nuclear membrane of VSMC and cardiomyocytes, increasing nuclear Ca^{2+} concentration, and endogenous nuclear protein kinase activities (Bkaily et al., 2000; Boivin et al., 2003).

1. The Endothelium-Dependent Responses to ET-1

Although ET-1 is known as a potent vasoconstrictor, in healthy animals, in which low levels of blood ET-1 are measured, intracoronary injections of low doses of ET-1 induce a decrease in vascular resistance (Fig. 2): in anesthetized dogs, for example, an intracoronary bolus injection of S6c, a selective ET_B receptor agonist, induces a decrease in coronary resistance for doses lower than 1 μg (Teerlink et al., 1994). Likewise, the injection of ET-1 in isolated rat hearts leads to a drop in coronary perfusion pressure at low

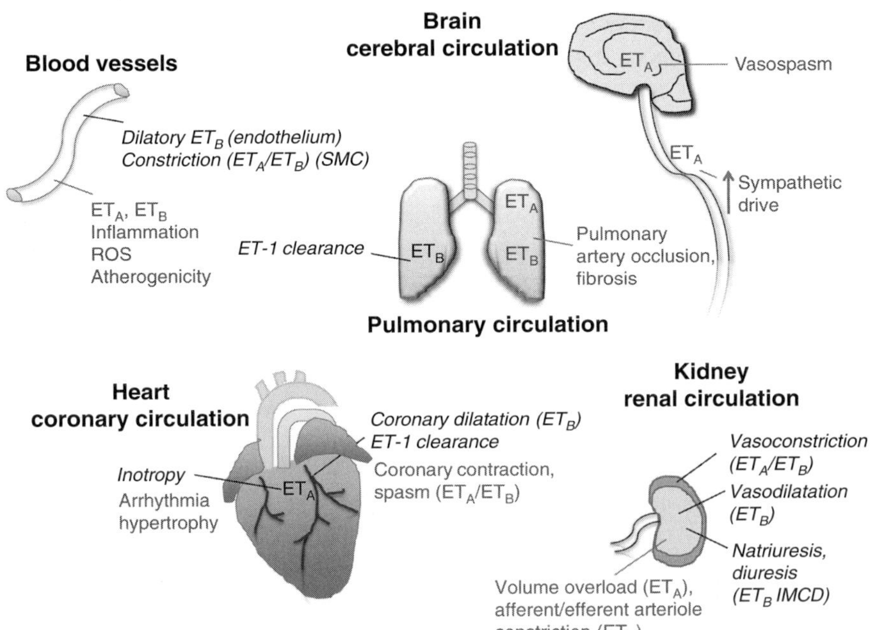

FIGURE 2 *Multiple effects of ET-1 in the cardiovascular system.* Physiological responses are presented in italic and black while the responses associated to pathological conditions are in light grey (and in red in the online version). (For interpretation of the references to color in this figure legend, the reader is referred to the Web version of this chapter.)

concentrations of ET-1 (Brunner et al., 2006). In coronary arterial rings isolated from young and healthy pigs, the activation of endothelial ET_B receptors induces a significant relaxation (Climent et al., 2005) through the release of NO and PGI_2 (Callera et al., 2007; Rubanyi & Polokoff, 1994). In addition, we know that in the human forearm circulation, the increase in blood flow induced by ET_A receptor blockade is blunted by ET_B receptor antagonism and NOS inhibition (Verhaar et al., 1998), suggesting that endogenous ET-1 exerts a dilatory tone by stimulating endothelial ET_B receptors. Nonetheless, the dilatory role of ET_B receptors is not significant in isolated human coronary arterial rings (Pierre & Davenport, 1998). However, such *ex vivo* studies are performed in human coronary vessels isolated from explanted hearts not of healthy subjects, but of patients undergoing cardiac transplantation for ischemic heart disease, vessels in which the expression of endothelial ET_B receptors is limited, except in the neovascularization of the atherosclerotic plaque (Bacon et al., 1996), and with a pronounced endothelial dysfunction (Thorin, 2001). This suggests that in pathology the vasodilating effect of endothelial ET_B receptor stimulation may be lost. Further data even suggest that in pathology, stimulation of endothelial ET_B receptors might be detrimental by inducing effects such as cell adhesion or contraction (Bergdahl et al., 2001; Schneider et al., 2007; Sen et al., 2009).

One final argument supporting a dilatory effect of ET-1 on normal coronary arteries is that the basal production of ET-1 is five times greater toward the lumen than in the interstitial space (Brunner, 1995), which would favor ET_B receptor stimulation on the endothelium, although other studies suggest, a polarized secretion of ET-1 toward the underlying VSMCs (Haynes & Webb, 1994; Unoki et al., 1999). Data from the study of Brunner demonstrated that the concentration of free ET-1 in the cardiac interstitial fluid never goes higher than 1 pg/ml (0.4 pM) in healthy animals, which is below the coronary constricting tone, supporting a vasodilatory tone associated with the stimulation of the endothelial ET_B autoreceptors in the heart, in physiological conditions.

One should also not underestimate the importance of the concentration of ET-1 in determining the dilatory versus constricting coronary response, because of the heterogeneous distribution of ET-1 receptors as illustrated in cerebral versus pulmonary arteries (Sauvageau et al., 2009). Saturation experiments using iodinated ligands, competition experiments, and reactivity studies using ET-1 receptor antagonists and autoradiography revealed that the expression of ET_A receptors is dominant compared to that of ET_B receptors in the coronaries of explanted ischemic heart (Bacon & Davenport, 1996; Pierre & Davenport, 1998). The overall effects of ET-1 on vascular tone *in vivo* are the clear result of the balance between the contraction mediated by VSMC ET_A and ET_B receptors and the dilation mediated by endothelial ET_B receptors (Callera et al., 2007; Thorin et al., 1999). This may explain that dual ET receptor antagonists such as bosentan or macitentan cause no vasodilation in healthy subjects but become vasodilators in pathological vascular beds.

2. Smooth Muscle Contraction, Inflammation, and Vascular Diseases

In rats, injection of bosentan, a dual $ET_{A/B}$ receptor antagonist, does not reduce blood pressure; after blockade of NO production, however, bosentan reduces blood pressure (Richard et al., 1995). This therefore suggests that NO inhibits ET-1-dependent activity *in vivo*. Bosentan and BQ123 dilate isolated and pressurized rabbit mesenteric arteries in no-flow conditions whether or not NO synthase is blocked (Nguyen et al., 1999). In these latter conditions, however, one might expect a lower influence of NO on the regulation of the vascular diameter since shear stress is nil, therefore favoring the effects of endogenous constrictors such as ET-1. In rat isolated pulmonary arteries, the contraction induced by ET-1 is both ET_A and ET_B receptor dependent, while in rat cerebral arteries, it is mostly ET_A receptor mediated, in agreement with the receptor expression profile in these arteries (Sauvageau et al., 2007, 2009). Coronary vessels, because of the ET-1/NO interdependence and due to their unique hemodynamic features (see later), are very susceptible to higher circulating and locally produced ET-1. The coronary endothelium is prone to dysfunction and highly sensitive to damage, which, with time, accumulates faster in the coronary than in other vascular beds. In addition, coronary endothelial dysfunction is associated with a decline in the contribution of NO in favor of a growing influence of ET-1 (Alonso & Radomski, 2003). Is the increased influence of ET-1 with time in the coronary bed only secondary to the loss of NO or is it due to a change in the ratio of ET_A and ET_B receptor expression? In any case, this supports the proinflammatory and proconstricting role of ET-1 via its predominant activation on smooth muscle receptors (Griendling et al., 2000; Marsden & Brenner, 1992; Sprague & Khalil, 2009), and could become the basis for the use of ET receptor antagonists to treat coronary artery diseases (CAD).

In cultured cells, it has been shown that ET-1 mRNA is upregulated by inflammatory factors such as TGF-β, TNF-α, interleukins, insulin, and ANG II, and downregulated by NO, PGI_2, and shear stress (Boulanger & Luscher, 1990; Brunner et al., 1995; Kohno et al., 1992; Kourembanas et al., 1993; Maemura et al., 1992; Prins et al., 1994). When considering these regulatory mechanisms within the coronary circulation, shear stress is a key element and NO is the key effector (Liu & Gutterman, 2009). In contrast to other vascular beds, wall shear stress in coronary arteries is uneven during the cardiac cycle (Heusch, 2008) and mechanical stress is therefore greatest in the coronary circulation (Thorin & Thorin-Trescases, 2009). In turn, it is not surprising that the coronary circulation is the prime site for endothelial dysfunction. It has been reported that because of these unique physiological hemodynamic features, coronary arteries display an unusual gene pattern when compared to the aorta: a fivefold lower eNOS and a 2.5-fold higher ET-1 mRNA expression (Dancu & Tarbell, 2007). Such a pattern predisposes coronary arteries to endothelial dysfunction and atherosclerosis. Therefore, based only on its physiological characteristics, the coronary circulation should be prone to an increased influence of ET-1 with age: the accumulation of age-related damages would favor ET-1 expression in contrast to that of eNOS and exacerbate endothelial dysfunction (Fig. 1).

Patients with atherosclerosis have elevated plasma levels of ET-1, and an upregulation of ET-1 and its receptors has been described in atherosclerotic arteries and plaques (Barton & Yanagisawa, 2008; Dagassan et al., 1996; Fan et al., 2000; Lerman et al., 1991). BigET-1 and ET-1 immunoreactivity has been found in atherosclerotic regions (Dashwood et al., 1998b; Hasdai et al., 1997). These observations have led to the hypothesis that ET-1 may be associated with the pathogenesis of atherosclerosis (Dashwood & Tsui, 2002; Ivey et al., 2008). In 1998, an important preclinical study (Barton et al., 1998) demonstrated that chronic ET_A receptor inhibition improved aortic endothelial dysfunction and reduced the development of atherosclerosis in ApoE knockout mice. Several studies have subsequently demonstrated the beneficial effects of acute intracoronary infusion of the ET_A receptor antagonist BQ123 on coronary diameter and coronary flow in patients with CAD. When narrowing the analysis of the dilatory effects of BQ123 to angiographically normal vessels, vessels with plaques and at stenosis, a higher dilation was observed after intracoronary infusion of BQ123 (40 nmol/min, for 60 min) in patients with CAD (Kinlay et al., 2001); in this study, compared with the dilation to nitroglycerin, ET-1 contributed to 39% of coronary vasomotor tone in healthy and angiographically clean vessels, 74% of tone in CAD arteries, and 106% of tone at stenosis. The contribution of NO was, however, not determined: one would assume that the more severe the disease condition, the less NO would be produced, and the more ET-1 would contribute to tone. This hypothesis was tested in 44 patients with CAD in a study published that same year: Halcox et al. provided the first evidence that ET-1, via ET_A receptors, contributed to the reduction of endothelial dilatory function (Halcox et al., 2001). The greatest improvement associated with the intracoronary infusion of the ET_A receptor antagonist was observed in patients with the greatest endothelial dysfunction as determined in the presence of acetylcholine (Halcox et al., 2001), suggesting that ET-1 contributes to the acute inactivation of NO. However, the tachyphylaxis of NO-dependent dilation occurring following systemic injections of ET-1 could also explain the apparent inactivation of NO by ET-1 (Le Monnier de Gouville et al., 1990); alternatively, we observed that chronic infusion of LU-135252 increased VSMC-sensitivity to NO, suggesting that ET-1 may regulate negatively the sensitivity of the soluble guanylate cyclase (Thorin et al., 2000). Both the selective ET_A receptor (BQ123) antagonist and the combination of selective ET_A (BQ123) and ET_B receptor (BQ788) antagonists improved endothelium-dependent dilation in the coronary arteries of patients with CAD (Bohm et al., 2008). In agreement with these data, using isolated human coronary arteries from idiopathic and atherosclerotic cardiomyopathic hearts, we demonstrated that ET-1-dependent constrictions became more pronounced when the endothelial function was altered (Thorin et al., 1999). Recently, a work by Dr Lerman's group (Reriani et al., 2010) showed that a chronic (6 months) treatment of patients with premature atherosclerosis with the ETA receptor antagonist atrasentan (10 mg/day) improves coronary endothelial function. Taken together, these data strongly suggest that ET-1 contributes to inactivate the dilatory function of the endothelium in the coronary arterial bed. Hence, the functional contribution of ET-1 is precocious and appears to rise with the severity of CAD.

ET-1, at low concentrations, potentiates coronary contractile responses to other vasoconstrictor substances such as norepinephrine and serotonin (Garcia-Villalon et al., 2008; Rubanyi & Polokoff, 1994; Thorin et al., 1998). In human cerebral arteries, a reduction in endothelium-derived ET-1 accounts for the dilatory effects of endothelial α_2-adrenergic receptor stimulation (Thorin et al., 1998). Consequently, even subthreshold concentrations of ET-1 may regulate vascular tone and reactivity in conditions where NO production is reduced, that is, with age and in patients presenting with risk factors for cardiovascular diseases.

The potential physiological roles of ET_B receptors, in addition to acting as clearance receptors for ET-1 and stimulating NO release, remain poorly understood. This is likely due to the limited final effects of the stimulation of endothelial ET_B receptors on the *in vitro* vascular function and the possible change in the expression of ET-1 receptors during the development of pathologies as evidenced in pulmonary hypertension (Sauvageau et al., 2009). A change in receptor expression is likely to change the pharmacology of the system. The consequences of receptor inhibition in young and healthy subjects or in old and diseased patients should obviously be different. Since most clinical data have been collected in an elderly population most likely showing some degree of endothelial dysfunction, it is quite possible that our understanding of ET-1 function as a proconstrictor and proinflammatory factor is only a reflection of these data and experimental environment, and thus may not illustrate the effects of ET-1 in young and healthy subjects. In support of this statement, the induction of endothelial damage eliminates ET_B-receptor-dependent relaxation in pig coronary arteries (Climent et al., 2005). The seminal demonstration that acetylcholine induces a contraction of coronary arteries in patients with CAD, but a dilation otherwise (Ludmer et al., 1986), is a good example of such a case.

Therefore, based on the literature reviewed so far, one can infer that at physiological and low concentrations, ET-1 predominantly induces dilations, while at pharmacological concentrations it induces contractions. The impact of the inevitable endothelial dysfunction when using isolated arteries from explanted human hearts may have led researchers to underestimate the endothelial dilatory component of ET-1. The production of NO may be reduced, but an alternative explanation may be the loss of coupling between the ET_B receptor and the NO pathway without affecting the ability of NO to clear ET-1 from the circulation. For example, acetylcholine induces a contraction of coronary vessels isolated from patients with ischemic heart disease, but substance-P still produces near-maximal relaxation by stimulating NO production (Thorin, 2001). A change in the expression or coupling of the endothelial ET_B receptor cannot be excluded in an elderly population (>65 years of age) and in patients with CAD.

3. Pulmonary Circulation

The pulmonary circulation is highly susceptible to elevated levels of ET-1 which have been associated with PAH (Stewart et al., 1991). Circulating levels of ET-1 are a good marker of disease severity (pulmonary vascular resistance, right atrial pressure, and pulmonary artery oxygen saturation) and predict poor prognosis (Stewart et al., 1991). Upregulation of ET-1 production by the lungs and changes in ET receptor expression could be at the basis of the dysregulation and PAH (Sauvageau et al., 2009; Takahashi et al., 2001). In animal models of PAH, both dual antagonists of ET_A and ET_B receptors (bosentan) and selective ET_A receptor antagonists (sitaxsentan, atrasentan, TBC-3711) are effective in reducing pulmonary artery resistance and inhibiting vascular remodeling. In humans, both types of antagonists are used (Kirkby et al., 2008; Motte et al., 2006).

Bosentan (Tracleer) was approved for the treatment of PAH in 2001 based on two clinical trials, "Study 351" with 32 class III patients with idiopathic PAH or associated with systemic sclerosis (Channick et al., 2001) and the important BREATHE-1 study that included 150 patients with idiopathic PAH, 47 with systemic sclerosis–associated PAH, and 16 with systemic lupus erythematosus–associated PAH (Rubin et al., 2002). In the latter study, bosentan improved exercise capacity, the functional class, and increased the time to clinical worsening. Sitaxsentan was approved for treatment of PAH in 2006 based on the STRIDE-1 results (Barst et al., 2004). Recently, ambrisentan has been approved for the treatment of PAH (Galie et al., 2008). No clinical advantages have been demonstrated between dual ET receptor antagonist and selective ET_A receptor antagonists.

4. Cardiac Myocyte Function and Heart Failure

ET-1 has positive cardiac inotropic effects in healthy animals and humans (Kang & Walker, 2006; Katoh et al., 1998; Kelly et al., 1990; Li et al., 1991; Pieske et al., 1999), but not in failing human hearts (MacCarthy et al., 2000; Pieske et al., 1999). ET-1 enhances myocyte contractility by activating ET_A receptor-phospholipase Cβ-PKCε signaling complexes preferentially localized in cardiac T-tubules (Robu et al., 2003). It has been shown that ET-1 is devoid of any significant effects on basal L-type Ca^{2+} channel activity, but exerts a potent inhibitory effect against isoprenaline-enhanced L-type Ca^{2+} channel current (He et al., 2000; Watanabe & Endoh, 2000). This effect is mediated through ET_A receptors coupled to pertussis toxin–sensitive G proteins (He et al., 2000; Thomas et al., 1997).

As mentioned earlier, ET-1 has growth-promoting effects (Inada et al., 1999). Because there is a correlation between ventricular mass and ET-1 concentration in the blood, it is possible that ET-1 contributes to the ventricular hypertrophy in patients with ischemic heart failure and dilated cardiomyopathy (Tsutamoto et al., 2000) and in rats following coronary artery ligation (Loennechen et al., 2001). ET-1 also promotes sympathetic tone, especially in heart failure as demonstrated in rabbits (Liu et al., 2001) and dogs (McConnell et al., 2000). This may partly explain the proarrhythmic effects of ET-1 (Burrell et al., 2000; Yorikane & Koike, 1990; Yorikane et al., 1990), while circulating levels of ET-1 have been associated with arrhythmia in patients with decompensated heart failure (Aronson & Burger, 2003a, 2003b; Aronson et al., 2001).

5. Renal Effects of ET-1

The effects of ET-1 on the kidney are complex. Exogenous administration of ET-1 induces a vasoconstriction in the renal cortex and a vasodilatation in the medulla (Rubinstein et al., 1995). The latter is mediated by ET_B receptors while the former is dependent on both ET_A and ET_B receptor activation (Dhaun et al., 2006). Acutely, selective ET_A receptor antagonism with BQ123 reduced blood pressure, proteinuria, and pulse wave velocity on top of standard treatment in patients with nondiabetic chronic kidney diseases (Dhaun et al., 2009). In diabetic patients with chronic kidney disease, however, chronic ET_A receptor antagonism with avosentan is deleterious due to fluid overload and congestive heart failure (Mann et al., 2010). In the inner-medullary-collecting duct of mice, ET-1 induces an autocrine natriuretic and diuretic effect which seems mediated by ET_B receptors, since specific inner-medullary-collecting duct deletion of ET_B but not ET_A receptors leads to salt-sensitive hypertension (Bagnall et al., 2006; Ge et al., 2006). However, the renal effects of ET_B receptor antagonism or deletion are inhibited by ET_A receptor antagonism, showing that it is the reactive increase in ET-1 acting on ET_A receptors, not the deletion of ET_B receptors *per se*, which is responsible for hypertension and tissue injury (Matsumura et al., 2000). Dual ET antagonists seem to give very low rates of fluid retention and edema in the clinical setting. In PAH clinical trials with Tracleer, combined adverse events of fluid retention or edema were reported in 1.7% (placebo-corrected) of patients (Tracleer US package insert, 2009). In a Phase II study in hypertensive patients, the novel dual ET antagonist macitentan did not cause peripheral edema (Press release Actelion Dec 2006). It is therefore possible that ET receptor antagonists can be used safely in patients with renal diseases, but this remains to be validated in a proper clinical trial.

III. Conclusion

Numerous clinical developments are ongoing with ET receptor antagonists (Aubert & Juillerat-Jeanneret, 2009). Our understanding of receptor pharmacology in general is changing with the appearance of new concepts including dimerization and G-protein-independent signaling. These changes apply to ET-1 and its receptors. One major weakness, however, which applies to many other pharmacological systems, is our lack of knowledge of the evolution of ET-1 and its receptors in the aging human and how this influences cardiovascular function in combination with risk factors for cardiovascular diseases. The critical role for ET-1 in controlling cardiovascular function is evident by the fact that its clinical importance was established within a few years after its discovery. It is likely that work in this area will continue to yield novel therapies for the treatment of cardiovascular disease.

Acknowledgments

The laboratory of Dr. Eric Thorin is supported by the Canadian Institutes for Health Research (MOP 14496 and 89733), the Heart and Stroke Foundation of Quebec, and the Foundation of the Montreal Heart Institute.

Conflict of Interest: Dr. Martine Clozel is Chief Scientific Officer at Actelion Pharmaceuticals.

Abbreviations

7-TM	7-transmembrane domain
ANG II	angiotensin II
EDCF	endothelium-derived constricting factor
ET-1	endothelin-1
GPCR	G protein-coupled receptor
PAH	pulmonary arterial hypertension
PGI_2	prostacyclin
S6c	sarafotoxin 6c

References

Aird, W. C. (2007). Phenotypic heterogeneity of the endothelium: I. Structure, function, and mechanisms. *Circulation Research*, *100*, 158–173.

Alexander, S. P., Mathie, A., & Peters, J. A. (2009). Guide to Receptors and Channels (GRAC), 4th edition. *British Journal of Pharmacology*, *158*(Suppl. 1), S42–S43.

Alonso, D., & Radomski, M. W. (2003). The nitric oxide-endothelin-1 connection. *Heart Failure Reviews*, *8*, 107–115.

Anggrahini, D. W., Emoto, N., Nakayama, K., Widyantoro, B., Adiarto, S., Iwasa, N., Nonaka, H., Rikitake, Y., Kisanuki, Y. Y., Yanagisawa, M., & Hirata, K. (2009). Vascular endothelial cell-derived endothelin-1 mediates vascular inflammation and neointima formation following blood flow cessation. *Cardiovascular Research*, *82*, 143–151.

Arai, H., Hori, S., Aramori, I., Ohkubo, H., & Nakanishi, S. (1990). Cloning and expression of a cDNA encoding an endothelin receptor. *Nature*, *348*, 730–732.

Aronson, D., & Burger, A. J. (2003a). Neurohormonal prediction of mortality following admission for decompensated heart failure. *The American Journal of Cardiology*, *91*, 245–248.

Aronson, D., & Burger, A. J. (2003b). Neurohumoral activation and ventricular arrhythmias in patients with decompensated congestive heart failure: Role of endothelin. *Pacing and Clinical Electrophysiology*, *26*, 703–710.

Aronson, D., Mittleman, M. A., & Burger, A. J. (2001). Role of endothelin in modulation of heart rate variability in patients with decompensated heart failure. *Pacing and Clinical Electrophysiology*, *24*, 1607–1615.

Aubert, J. D., & Juillerat-Jeanneret, L. (2009). Therapeutic potential of endothelin receptor modulators: Lessons from human clinical trials. *Expert Opinion on Therapeutic Targets*, *13*, 1069–1084.

Bacon, C. R., Cary, N. R., & Davenport, A. P. (1996). Endothelin peptide and receptors in human atherosclerotic coronary artery and aorta. *Circulation Research*, *79*, 794–801.

Bacon, C. R., & Davenport, A. P. (1996). Endothelin receptors in human coronary artery and aorta. *British Journal of Pharmacology*, *117*, 986–992.

Bagnall, A. J., Kelland, N. F., Gulliver-Sloan, F., Davenport, A. P., Gray, G. A., Yanagisawa, M., Webb, D. J., & Kotelevtsev, Y. V. (2006). Deletion of endothelial cell endothelin B receptors does not affect blood pressure or sensitivity to salt. *Hypertension*, *48*, 286–293.

Barst, R. J., Langleben, D., Frost, A., Horn, E. M., Oudiz, R., Shapiro, S., McLaughlin, V., Hill, N., Tapson, V. F., Robbins, I. M., Zwicke, D., Duncan, B., Dixon, R. A., & Frumkin, L. R. (2004). Sitaxsentan therapy for pulmonary arterial hypertension. *American Journal of Respiratory and Critical Care Medicine*, *169*, 441–447.

Barton, M., Haudenschild, C. C., d'Uscio, L. V., Shaw, S., Munter, K., & Luscher, T. F. (1998). Endothelin ETA receptor blockade restores NO-mediated endothelial function and inhibits atherosclerosis in apolipoprotein E-deficient mice. *Proceedings of the National Academy of Sciences of the United States of America*, *95*, 14367–14372.

Barton, M., & Yanagisawa, M. (2008). Endothelin: 20 years from discovery to therapy. *Canadian Journal of Physiology and Pharmacology*, *86*, 485–498.

Bergdahl, A., Valdemarsson, S., Adner, M., Sun, X. Y., Hedner, T., & Edvinsson, L. (2001). Enhanced endothelin-1-induced contractions in mesenteric arteries from rats with congestive heart failure: Role of ET(B) receptors. *European Journal of Heart Failure*, *3*, 293–299.

Bialecki, R. A., Izzo, N. J. Jr., & Colucci, W. S. (1989). Endothelin-1 increases intracellular calcium mobilization but not calcium uptake in rabbit vascular smooth muscle cells. *Biochemical and Biophysical Research Communications*, *164*, 474–479.

Bkaily, G., Choufani, S., Hassan, G., El-Bizri, N., Jacques, D., & D'Orleans-Juste, P. (2000). Presence of functional endothelin-1 receptors in nuclear membranes of human aortic vascular smooth muscle cells. *Journal of Cardiovascular Pharmacology*, *36*, S414–S417.

Bohm, F., Jensen, J., Svane, B., Settergren, M., & Pernow, J. (2008). Intracoronary endothelin receptor blockade improves endothelial function in patients with coronary artery disease. *Canadian Journal of Physiology and Pharmacology*, *86*, 745–751.

Boivin, B., Chevalier, D., Villeneuve, L. R., Rousseau, E., & Allen, B. G. (2003). Functional endothelin receptors are present on nuclei in cardiac ventricular myocytes. *The Journal of Biological Chemistry*, *278*, 29153–29163.

Boulanger, C., & Luscher, T. F. (1990). Release of endothelin from the porcine aorta. Inhibition by endothelium-derived nitric oxide. *The Journal of Clinical Investigation*, *85*, 587–590.

Brunner, F. (1995). Tissue endothelin-1 levels in perfused rat heart following stimulation with agonists and in ischaemia and reperfusion. *Journal of Molecular and Cellular Cardiology*, *27*, 1953–1963.

Brunner, F., Bras-Silva, C., Cerdeira, A. S., & Leite-Moreira, A. F. (2006). Cardiovascular endothelins: Essential regulators of cardiovascular homeostasis. *Pharmacology & Therapeutics*, *111*, 508–531.

Brunner, F., Stessel, H., & Kukovetz, W. R. (1995). Novel guanylyl cyclase inhibitor, ODQ reveals role of nitric oxide, but not of cyclic GMP in endothelin-1 secretion. *FEBS Letters*, *376*, 262–266.

Burrell, K. M., Molenaar, P., Dawson, P. J., & Kaumann, A. J. (2000). Contractile and arrhythmic effects of endothelin receptor agonists in human heart in vitro: Blockade with SB 209670. *The Journal of Pharmacology and Experimental Therapeutics*, *292*, 449–459.

Callera, G., Tostes, R., Savoia, C., Muscara, M. N., & Touyz, R. M. (2007). Vasoactive peptides in cardiovascular (patho)physiology. *Expert Review of Cardiovascular Therapy*, *5*, 531–552.

Channick, R. N., Simonneau, G., Sitbon, O., Robbins, I. M., Frost, A., Tapson, V. F., Badesch, D. B., Roux, S., Rainisio, M., Bodin, F., & Rubin, L. J. (2001). Effects of the dual endothelin-receptor antagonist bosentan in patients with pulmonary hypertension: A randomised placebo-controlled study. *Lancet*, *358*, 1119–1123.

Chen, C., & Wagoner, P. K. (1991). Endothelin induces a nonselective cation current in vascular smooth muscle cells. *Circulation Research*, *69*, 447–454.

Clerk, A., Bogoyevitch, M. A., Anderson, M. B., & Sugden, P. H. (1994). Differential activation of protein kinase C isoforms by endothelin-1 and phenylephrine and subsequent stimulation of p42 and p44 mitogen-activated protein kinases in ventricular myocytes cultured from neonatal rat hearts. *The Journal of Biological Chemistry*, *269*, 32848–32857.

Climent, B., Fernandez, N., Sanz, E., Sanchez, A., Monge, L., Garcia-Villalon, A. L., & Dieguez, G. (2005). Enhanced response of pig coronary arteries to endothelin-1 after ischemia-reperfusion. Role of endothelin receptors, nitric oxide and prostanoids. *European Journal of Pharmacology*, *524*, 102–110.

Clozel, M., Breu, V., Burri, K., Cassal, J. M., Fischli, W., Gray, G. A., Hirth, G., Loffler, B. M., Muller, M., Neidhart, W., et al. (1993). Pathophysiological role of endothelin revealed by the first orally active endothelin receptor antagonist. *Nature*, *365*, 759–761.

Clozel, M., Gray, G. A., Breu, V., Loffler, B. M., & Osterwalder, R. (1992). The endothelin ETB receptor mediates both vasodilation and vasoconstriction in vivo. *Biochemical and Biophysical Research Communications*, *186*, 867–873.

Dagassan, P. H., Breu, V., Clozel, M., Kunzli, A., Vogt, P., Turina, M., Kiowski, W., & Clozel, J. P. (1996). Up-regulation of endothelin-B receptors in atherosclerotic human coronary arteries. *Journal of Cardiovascular Pharmacology*, *27*, 147–153.

Dai, X., & Galligan, J. J. (2006). Differential trafficking and desensitization of human ET(A) and ET(B) receptors expressed in HEK 293 cells. *Experimental Biology and Medicine (Maywood)*, *231*, 746–751.

Dancu, M. B., & Tarbell, J. M. (2007). Coronary endothelium expresses a pathologic gene pattern compared to aortic endothelium: Correlation of asynchronous hemodynamics and pathology in vivo. *Atherosclerosis*, *192*, 9–14.

Dashwood, M. R., Mehta, D., Izzat, M. B., Timm, M., Bryan, A. J., Angelini, G. D., & Jeremy, J. Y. (1998a). Distribution of endothelin-1 (ET) receptors (ET(A) and ET(B)) and immunoreactive ET-1 in porcine saphenous vein-carotid artery interposition grafts. *Atherosclerosis*, *137*, 233–242.

Dashwood, M. R., Timm, M., Muddle, J. R., Ong, A. C., Tippins, J. R., Parker, R., McManus, D., Murday, A. J., Madden, B. P., & Kaski, J. C. (1998b). Regional variations in endothelin-1 and its receptor subtypes in human coronary vasculature: Pathophysiological implications in coronary disease. *Endothelium*, *6*, 61–70.

Dashwood, M. R., & Tsui, J. C. (2002). Endothelin-1 and atherosclerosis: Potential complications associated with endothelin-receptor blockade. *Atherosclerosis*, *160*, 297–304.

Davenport, A. P. (2002). International Union of Pharmacology. XXIX. Update on endothelin receptor nomenclature. *Pharmacological Reviews*, *54*, 219–226.

Davenport, A. P., & Maguire, J. J. (2001). The endothelin system in human saphenous vein graft disease. *Current Opinion in Pharmacology*, *1*, 176–182.

Davenport, A. P., O'Reilly, G., Molenaar, P., Maguire, J. J., Kuc, R. E., Sharkey, A., Bacon, C. R., & Ferro, A. (1993). Human endothelin receptors characterized using reverse transcriptase-polymerase chain reaction, in situ hybridization, and subtype-selective ligands BQ123 and BQ3020: Evidence for expression of ETB receptors in human vascular smooth muscle. *Journal of Cardiovascular Pharmacology*, *22*(Suppl 8), S22–S25.

De Mey, J. G., & Vanhoutte, P. M. (1982). Heterogeneous behavior of the canine arterial and venous wall. Importance of the endothelium. *Circulation Research*, *51*, 439–447.

de Nucci, G., Thomas, R., D'Orleans-Juste, P., Antunes, E., Walder, C., Warner, T. D., & Vane, J. R. (1988). Pressor effects of circulating endothelin are limited by its removal in the pulmonary circulation and by the release of prostacyclin and endothelium-derived relaxing factor. *Proceedings of the National Academy of Sciences of the United States of America*, *85*, 9797–9800.

Denault, J. B., Claing, A., D'Orleans-Juste, P., Sawamura, T., Kido, T., Masaki, T., & Leduc, R. (1995). Processing of proendothelin-1 by human furin convertase. *FEBS Letters*, *362*, 276–280.

Dhaun, N., Goddard, J., & Webb, D. J. (2006). The endothelin system and its antagonism in chronic kidney disease. *Journal of the American Society of Nephrology*, *17*, 943–955.

Dhaun, N., Macintyre, I. M., Melville, V., Lilitkarntakul, P., Johnston, N. R., Goddard, J., & Webb, D. J. (2009). Blood pressure-independent reduction in proteinuria and arterial stiffness after acute endothelin-a receptor antagonism in chronic kidney disease. *Hypertension*, *54*, 113–119.

Docherty, C. C., & MacLean, M. R. (1998). EndothelinB receptors in rabbit pulmonary resistance arteries: Effect of left ventricular dysfunction. *The Journal of Pharmacology and Experimental Therapeutics*, *284*, 895–903.

Douglas, S. A., Sulpizio, A. C., Piercy, V., Sarau, H. M., Ames, R. S., Aiyar, N. V., Ohlstein, E. H., & Willette, R. N. (2000). Differential vasoconstrictor activity of human urotensin-II in vascular tissue isolated from the rat, mouse, dog, pig, marmoset and cynomolgus monkey. *British Journal of Pharmacology*, *131*, 1262–1274.

Drake, M. T., Violin, J. D., Whalen, E. J., Wisler, J. W., Shenoy, S. K., & Lefkowitz, R. J. (2008). beta-arrestin-biased agonism at the beta2-adrenergic receptor. *The Journal of Biological Chemistry*, *283*, 5669–5676.

Dupuis, J., Goresky, C. A., & Fournier, A. (1996a). Pulmonary clearance of circulating endothelin-1 in dogs in vivo: Exclusive role of ETB receptors. *Journal of Applied Physiology*, *81*, 1510–1515.

Dupuis, J., Goresky, C. A., Rose, C. P., Stewart, D. J., Cernacek, P., Schwab, A. J., & Simard, A. (1997). Endothelin-1 myocardial clearance, production, and effect on capillary permeability in vivo. *The American Journal of Physiology*, *273*, H1239–H1245.

Dupuis, J., Stewart, D. J., Cernacek, P., & Gosselin, G. (1996b). Human pulmonary circulation is an important site for both clearance and production of endothelin-1. *Circulation*, *94*, 1578–1584.

Evans, N. J., & Walker, J. W. (2008a). Endothelin receptor dimers evaluated by FRET, ligand binding, and calcium mobilization. *Biophysical Journal*, *95*, 483–492.

Evans, N. J., & Walker, J. W. (2008b). Sustained Ca2+ signaling and delayed internalization associated with endothelin receptor heterodimers linked through a PDZ finger. *Canadian Journal of Physiology and Pharmacology*, *86*, 526–535.

Fan, J., Unoki, H., Iwasa, S., & Watanabe, T. (2000). Role of endothelin-1 in atherosclerosis. *Annals of New York Academy of Sciences*, *902*, 84–93, discussion 93–94.

Farhat, N., Matouk, C. C., Mamarbachi, A. M., Marsden, P. A., Allen, B. G., & Thorin, E. (2008). Activation of ETB receptors regulates the abundance of ET-1 mRNA in vascular endothelial cells. *British Journal of Pharmacology*, *153*, 1420–1431.

Fecteau, M. H., Honore, J. C., Plante, M., Labonte, J., Rae, G. A., & D'Orleans-Juste, P. (2005). Endothelin-1 (1-31) is an intermediate in the production of endothelin-1 after big endothelin-1 administration in vivo. *Hypertension*, *46*, 87–92.

Furchgott, R. F., & Zawadzki, J. V. (1980). The obligatory role of endothelial cells in the relaxation of arterial smooth muscle by acetylcholine. *Nature*, *288*, 373–376.

Galandrin, S., Oligny-Longpre, G., & Bouvier, M. (2007). The evasive nature of drug efficacy: Implications for drug discovery. *Trends in Pharmacological Sciences*, *28*, 423–430.

Galie, N., Olschewski, H., Oudiz, R. J., Torres, F., Frost, A., Ghofrani, H. A., Badesch, D. B., McGoon, M. D., McLaughlin, V. V., Roecker, E. B., Gerber, M. J., Dufton, C., Wiens, B. L., & Rubin, L. J. (2008). Ambrisentan for the treatment of pulmonary arterial hypertension: Results of the ambrisentan in pulmonary arterial hypertension, randomized, double-blind, placebo-controlled, multicenter, efficacy (ARIES) study 1 and 2. *Circulation*, *117*, 3010–3019.

Garcia-Villalon, A. L., Amezquita, Y. M., Monge, L., Fernandez, N., Salcedo, A., & Dieguez, G. (2008). Endothelin-1 potentiation of coronary artery contraction after ischemia-reperfusion. *Vascular Pharmacology*, *48*, 109–114.

Ge, Y., Bagnall, A., Stricklett, P. K., Strait, K., Webb, D. J., Kotelevtsev, Y., & Kohan, D. E. (2006). Collecting duct-specific knockout of the endothelin B receptor causes hypertension and sodium retention. *The American Journal of Physiology—Renal Physiology*, *291*, F1274–F1280.

Gesty-Palmer, D., Flannery, P., Yuan, L., Corsino, L., Spurney, R., Lefkowitz, R. J., & Luttrell, L. M. (2009). A beta-arrestin-biased agonist of the parathyroid hormone receptor (PTH1R) promotes bone formation independent of G protein activation. *Science Translational Medicine*, *1*, 1ra1.

Glorioso, N., Herrera, V. L., Bagamasbad, P., Filigheddu, F., Troffa, C., Argiolas, G., Bulla, E., Decano, J. L., & Ruiz-Opazo, N. (2007). Association of ATP1A1 and dear single-nucleotide polymorphism haplotypes with essential hypertension: Sex-specific and haplotype-specific effects. *Circulation Research*, *100*, 1522–1529.

Gregan, B., Jurgensen, J., Papsdorf, G., Furkert, J., Schaefer, M., Beyermann, M., Rosenthal, W., & Oksche, A. (2004). Ligand-dependent differences in the internalization of endothelin A and endothelin B receptor heterodimers. *The Journal of Biological Chemistry*, *279*, 27679–27687.

Griendling, K. K., Sorescu, D., Lassegue, B., & Ushio-Fukai, M. (2000). Modulation of protein kinase activity and gene expression by reactive oxygen species and their role in vascular physiology and pathophysiology. *Arteriosclerosis, Thrombosis, and Vascular Biology*, *20*, 2175–2183.

Grinstein, S., & Rothstein, A. (1986). Mechanisms of regulation of the Na+/H+ exchanger. *The Journal of Membrane Biology*, *90*, 1–12.

Halcox, J. P., Nour, K. R., Zalos, G., & Quyyumi, A. A. (2001). Coronary vasodilation and improvement in endothelial dysfunction with endothelin ET(A) receptor blockade. *Circulation Research*, *89*, 969–976.

Hamilton, C. A., Thorin, E., McCulloch, J., Dominiczak, M. H., & Reid, J. L. (1994). Chronic exposure of bovine aortic endothelial cells to native and oxidized LDL modifies phosphatidylinositol metabolism. *Atherosclerosis*, *107*, 55–63.

Harada, N., Himeno, A., Shigematsu, K., Sumikawa, K., & Niwa, M. (2002). Endothelin-1 binding to endothelin receptors in the rat anterior pituitary gland: Possible formation of an ETA-ETB receptor heterodimer. *Cellular and Molecular Neurobiology*, *22*, 207–226.

Hasdai, D., Mathew, V., Schwartz, R. S., Smith, L. A., Holmes, D. R. Jr., Katusic, Z. S., & Lerman, A. (1997). Enhanced endothelin-B-receptor-mediated vasoconstriction of small porcine coronary arteries in diet-induced hypercholesterolemia. *Arteriosclerosis, Thrombosis, and Vascular Biology*, *17*, 2737–2743.

Haynes, W. G., & Webb, D. J. (1993). Endothelium-dependent modulation of responses to endothelin-I in human veins. *Clinical Science (London)*, *84*, 427–433.

Haynes, W. G., & Webb, D. J. (1994). Contribution of endogenous generation of endothelin-1 to basal vascular tone. *Lancet*, *344*, 852–854.

He, J. Q., Pi, Y., Walker, J. W., & Kamp, T. J. (2000). Endothelin-1 and photoreleased diacylglycerol increase L-type Ca2+ current by activation of protein kinase C in rat ventricular myocytes. *Journal de Physiologie*, *524*(Pt 3), 807–820.

Herrera, V. L., Ponce, L. R., Bagamasbad, P. D., VanPelt, B. D., Didishvili, T., & Ruiz-Opazo, N. (2005). Embryonic lethality in Dear gene-deficient mice: New player in angiogenesis. *Physiological Genomics*, *23*, 257–268.

Heusch, G. (2008). Heart rate in the pathophysiology of coronary blood flow and myocardial ischaemia: Benefit from selective bradycardic agents. *British Journal of Pharmacology*, *153*, 1589–1601.

Himeno, A., Shigematsu, K., Taguchi, T., & Niwa, M. (1998). Endothelin-1 binding to endothelin receptors in the rat anterior pituitary gland: Interaction in the recognition of endothelin-1 between ETA and ETB receptors. *Cellular and Molecular Neurobiology*, *18*, 447–452.

Hosoda, K., Nakao, K., Hiroshi, A., Suga, S., Ogawa, Y., Mukoyama, M., Shirakami, G., Saito, Y., Nakanishi, S., & Imura, H. (1991). Cloning and expression of human endothelin-1 receptor cDNA. *FEBS Letters*, *287*, 23–26.

Ihling, C., Szombathy, T., Bohrmann, B., Brockhaus, M., Schaefer, H. E., & Loeffler, B. M. (2001). Coexpression of endothelin-converting enzyme-1 and endothelin-1 in different stages of human atherosclerosis. *Circulation*, *104*, 864–869.

Inada, T., Fujiwara, H., Hasegawa, K., Araki, M., Yamauchi-Kohno, R., Yabana, H., Fujiwara, T., Tanaka, M., & Sasayama, S. (1999). Upregulated expression of cardiac endothelin-1 participates in myocardial cell growth in Bio14.6 Syrian cardiomyopathic hamsters. *Journal of the American College of Cardiology*, *33*, 565–571.

Inoue, A., Yanagisawa, M., Kimura, S., Kasuya, Y., Miyauchi, T., Goto, K., & Masaki, T. (1989a). The human endothelin family: Three structurally and pharmacologically distinct isopeptides predicted by three separate genes. *Proceedings of the National Academy of Sciences of the United States of America*, *86*, 2863–2867.

Inoue, A., Yanagisawa, M., Takuwa, Y., Mitsui, Y., Kobayashi, M., & Masaki, T. (1989b). The human preproendothelin-1 gene. Complete nucleotide sequence and regulation of expression. *The Journal of Biological Chemistry*, *264*, 14954–14959.

Ivey, M. E., Osman, N., & Little, P. J. (2008). Endothelin-1 signalling in vascular smooth muscle: Pathways controlling cellular functions associated with atherosclerosis. *Atherosclerosis*, *199*, 237–247.

Kang, M., & Walker, J. W. (2006). Endothelin-1 and PKC induce positive inotropy without affecting pHi in ventricular myocytes. *Experimental Biology and Medicine (Maywood)*, *231*, 865–870.

Katoh, H., Terada, H., Iimuro, M., Sugiyama, S., Qing, K., Satoh, H., & Hayashi, H. (1998). Heterogeneity and underlying mechanism for inotropic action of endothelin-1 in rat ventricular myocytes. *British Journal of Pharmacology*, *123*, 1343–1350.

Kedzierski, R. M., & Yanagisawa, M. (2001). Endothelin system: The double-edged sword in health and disease. *Annual Review of Pharmacology and Toxicology*, *41*, 851–876.

Kelly, R. A., Eid, H., Kramer, B. K., O'Neill, M., Liang, B. T., Reers, M., & Smith, T. W. (1990). Endothelin enhances the contractile responsiveness of adult rat ventricular myocytes to calcium by a pertussis toxin-sensitive pathway. *The Journal of Clinical Investigation*, *86*, 1164–1171.

Kenakin, T. (2007). Collateral efficacy in drug discovery: Taking advantage of the good (allosteric) nature of 7TM receptors. *Trends in Pharmacological Sciences*, *28*, 407–415.

Kinlay, S., Behrendt, D., Wainstein, M., Beltrame, J., Fang, J. C., Creager, M. A., Selwyn, A. P., & Ganz, P. (2001). Role of endothelin-1 in the active constriction of human atherosclerotic coronary arteries. *Circulation*, *104*, 1114–1118.

Kirkby, N. S., Hadoke, P. W., Bagnall, A. J., & Webb, D. J. (2008). The endothelin system as a therapeutic target in cardiovascular disease: Great expectations or bleak house? *British Journal of Pharmacology*, *153*, 1105–1119.

Koh, E., Morimoto, S., Kim, S., Nabata, T., Miyashita, Y., & Ogihara, T. (1990). Endothelin stimulates Na+/H+ exchange in vascular smooth muscle cells. *Biochemistry International*, *20*, 375–380.

Kohno, M., Horio, T., Yokokawa, K., Kurihara, N., & Takeda, T. (1992). C-type natriuretic peptide inhibits thrombin- and angiotensin II-stimulated endothelin release via cyclic guanosine 3′,5′-monophosphate. *Hypertension*, *19*, 320–325.

Komukai, K., Jin, O. U., Morimoto, S., Kawai, M., Hongo, K., Yoshimura, M., & Kurihara, S. (2010). Role of Ca2+/calmodulin-dependent protein kinase II in the regulation of the cardiac L-type Ca2+ current during endothelin-1 stimulation. *American Journal of Physiology Heart and Circulatory Physiology*, *298*, H1902–H1907.

Kourembanas, S., McQuillan, L. P., Leung, G. K., & Faller, D. V. (1993). Nitric oxide regulates the expression of vasoconstrictors and growth factors by vascular endothelium under both normoxia and hypoxia. *The Journal of Clinical Investigation*, *92*, 99–104.

LaDouceur, D. M., Flynn, M. A., Keiser, J. A., Reynolds, E., & Haleen, S. J. (1993). ETA and ETB receptors coexist on rabbit pulmonary artery vascular smooth muscle mediating contraction. *Biochemical and Biophysical Research Communications*, *196*, 209–215.

Laporte, S., Denault, J. B., D'Orleans-Juste, P., & Leduc, R. (1993). Presence of furin mRNA in cultured bovine endothelial cells and possible involvement of furin in the processing of the endothelin precursor. *Journal of Cardiovascular Pharmacology*, *22*(Suppl. 8), S7–S10.

Le Monnier de Gouville, A. C., Lippton, H., Cohen, G., Cavero, I., & Hyman, A. (1990). Vasodilator activity of endothelin-1 and endothelin-3: Rapid development of cross-tachyphylaxis and dependence on the rate of endothelin administration. *The Journal of Pharmacology and Experimental Therapeutics*, *254*, 1024–1028.

Lerman, A., Edwards, B. S., Hallett, J. W., Heublein, D. M., Sandberg, S. M., & Burnett, J. C. Jr. (1991). Circulating and tissue endothelin immunoreactivity in advanced atherosclerosis. *The New England Journal of Medicine*, *325*, 997–1001.

Li, K., Stewart, D. J., & Rouleau, J. L. (1991). Myocardial contractile actions of endothelin-1 in rat and rabbit papillary muscles. Role of endocardial endothelium. *Circulation Research*, *69*, 301–312.

Liu, Y., & Gutterman, D. D. (2009). Vascular control in humans: Focus on the coronary microcirculation. *Basic Research in Cardiology*, *104*, 211–227.

Liu, J. L., Pliquett, R. U., Brewer, E., Cornish, K. G., Shen, Y. T., & Zucker, I. H. (2001). Chronic endothelin-1 blockade reduces sympathetic nerve activity in rabbits with heart failure. *American Journal of Physiology: Regulatory, Integrative and Comparative Physiology*, *280*, R1906–R1913.

Loennechen, J. P., Stoylen, A., Beisvag, V., Wisloff, U., & Ellingsen, O. (2001). Regional expression of endothelin-1, ANP, IGF-1, and LV wall stress in the infarcted rat heart. *American Journal of Physiology. Heart and Circulatory Physiology*, *280*, H2902–H2910.

Ludmer, P. L., Selwyn, A. P., Shook, T. L., Wayne, R. R., Mudge, G. H., Alexander, R. W., & Ganz, P. (1986). Paradoxical vasoconstriction induced by acetylcholine in atherosclerotic coronary arteries. *The New England Journal of Medicine*, *315*, 1046–1051.

Luscher, T. F., Yang, Z., Tschudi, M., von Segesser, L., Stulz, P., Boulanger, C., Siebenmann, R., Turina, M., & Buhler, F. R. (1990). Interaction between endothelin-1 and endothelium-derived relaxing factor in human arteries and veins. *Circulation Research*, *66*, 1088–1094.

MacCarthy, P. A., Grocott-Mason, R., Prendergast, B. D., & Shah, A. M. (2000). Contrasting inotropic effects of endogenous endothelin in the normal and failing human heart: Studies with an intracoronary ET(A) receptor antagonist. *Circulation*, *101*, 142–147.

MacLean, M. R., McCulloch, K. M., & Baird, M. (1994). Endothelin ETA- and ETB-receptor-mediated vasoconstriction in rat pulmonary arteries and arterioles. *Journal of Cardiovascular Pharmacology*, *23*, 838–845.

Maemura, K., Kurihara, H., Morita, T., Oh-hashi, Y., & Yazaki, Y. (1992). Production of endothelin-1 in vascular endothelial cells is regulated by factors associated with vascular injury. *Gerontology*, *38*(Suppl. 1), 29–35.

Maguire, J., & Davenport, A. P. (2004). Alternative pathway to endothelin-converting enzyme for the synthesis of endothelin in human blood vessels. *Journal of Cardiovascular Pharmacology*, *44*(Suppl 1), S27–S29.

Maguire, J. J., Kuc, R. E., & Davenport, A. P. (2001). Vasoconstrictor activity of novel endothelin peptide, ET-1(1–31), in human mammary and coronary arteries in vitro. *British Journal of Pharmacology*, *134*, 1360–1366.

Mann, J. F., Green, D., Jamerson, K., Ruilope, L. M., Kuranoff, S. J., Littke, T., & Viberti, G. (2010). Avosentan for overt diabetic nephropathy. *Journal of American Society of Nephrology*, *21*, 527–535.

Marsden, P. A., & Brenner, B. M. (1992). Transcriptional regulation of the endothelin-1 gene by TNF-alpha. *The American Journal of Physiology*, *262*, C854–C861.

Matsumura, Y., Kuro, T., Kobayashi, Y., Konishi, F., Takaoka, M., Wessale, J. L., Opgenorth, T. J., Gariepy, C. E., & Yanagisawa, M. (2000). Exaggerated vascular and renal pathology in endothelin-B receptor-deficient rats with deoxycorticosterone acetate-salt hypertension. *Circulation*, *102*, 2765–2773.

McConnell, P. I., Olson, C. E., Patel, K. P., Blank, D. U., Olivari, M. T., Gallagher, K. P., Quenby-Brown, E., & Zucker, I. H. (2000). Chronic endothelin blockade in dogs with pacing-induced heart failure: Possible modulation of sympathoexcitation. *Journal of Cardiac Failure*, *6*, 56–65.

Motte, S., McEntee, K., & Naeije, R. (2006). Endothelin receptor antagonists. *Pharmacology & Therapeutics*, *110*, 386–414.

Muramatsu, M., Oka, M., Morio, Y., Soma, S., Takahashi, H., & Fukuchi, Y. (1999). Chronic hypoxia augments endothelin-B receptor-mediated vasodilation in isolated perfused rat lungs. *The American Journal of Physiology*, *276*, L358–L364.

Nelson, M. T., Cheng, H., Rubart, M., Santana, L. F., Bonev, A. D., Knot, H. J., & Lederer, W. J. (1995). Relaxation of arterial smooth muscle by calcium sparks. *Science*, *270*, 633–637.

Newton, A. C., & Keranen, L. M. (1994). Phosphatidyl-L-serine is necessary for protein kinase C's high-affinity interaction with diacylglycerol-containing membranes. *Biochemistry*, *33*, 6651–6658.

Nguyen, T. D., Vequaud, P., & Thorin, E. (1999). Effects of endothelin receptor antagonists and nitric oxide on myogenic tone and alpha-adrenergic-dependent contractions of rabbit resistance arteries. *Cardiovascular Research*, *43*, 755–761.

Ogawa, Y., Nakao, K., Arai, H., Nakagawa, O., Hosoda, K., Suga, S., Nakanishi, S., & Imura, H. (1991). Molecular cloning of a non-isopeptide-selective human endothelin receptor. *Biochemical and Biophysical Research Communications*, *178*, 248–255.

Palmer, R. M., Ferrige, A. G., & Moncada, S. (1987). Nitric oxide release accounts for the biological activity of endothelium-derived relaxing factor. *Nature*, *327*, 524–526.

Pierre, L. N., & Davenport, A. P. (1998). Endothelin receptor subtypes and their functional relevance in human small coronary arteries. *British Journal of Pharmacology*, *124*, 499–506.

Pieske, B., Beyermann, B., Breu, V., Loffler, B. M., Schlotthauer, K., Maier, L. S., Schmidt-Schweda, S., Just, H., & Hasenfuss, G. (1999). Functional effects of endothelin and regulation of endothelin receptors in isolated human nonfailing and failing myocardium. *Circulation*, *99*, 1802–1809.

Prins, B. A., Hu, R. M., Nazario, B., Pedram, A., Frank, H. J., Weber, M. A., & Levin, E. R. (1994). Prostaglandin E2 and prostacyclin inhibit the production and secretion of endothelin from cultured endothelial cells. *The Journal of Biological Chemistry*, *269*, 11938–11944.

Rajagopal, S., Kim, J., Ahn, S., Craig, S., Lam, C. M., Gerard, N. P., Gerard, C., & Lefkowitz, R. J. (2010). Beta-arrestin- but not G protein-mediated signaling by the "decoy" receptor CXCR7. *Proceedings of the National Academy of Sciences of the United States of America*, *107*, 628–632.

Reriani, M., Raichlin, E., Prasad, A., Mathew, V., Pumper, G. M., Nelson, R. E., Lennon, R., Rihal, C., Lerman, L. O., & Lerman, A. (2010). Long-term administration of endothelin receptor antagonist improves coronary endothelial function in patients with early atherosclerosis. *Circulation*, *122*, 958–966.

Resink, T. J., Scott-Burden, T., & Buhler, F. R. (1988). Endothelin stimulates phospholipase C in cultured vascular smooth muscle cells. *Biochemical and Biophysical Research Communications*, *157*, 1360–1368.

Richard, V., Hogie, M., Clozel, M., Loffler, B. M., & Thuillez, C. (1995). In vivo evidence of an endothelin-induced vasopressor tone after inhibition of nitric oxide synthesis in rats. *Circulation*, *91*, 771–775.

Robu, V. G., Pfeiffer, E. S., Robia, S. L., Balijepalli, R. C., Pi, Y., Kamp, T. J., & Walker, J. W. (2003). Localization of functional endothelin receptor signaling complexes in cardiac transverse tubules. *The Journal of Biological Chemistry*, *278*, 48154–48161.

Rosano, L., Cianfrocca, R., Masi, S., Spinella, F., Di Castro, V., Biroccio, A., Salvati, E., Nicotra, M. R., Natali, P. G., & Bagnato, A. (2009). Beta-arrestin links endothelin A receptor to beta-catenin signaling to induce ovarian cancer cell invasion and metastasis. *Proceedings of the National Academy of Sciences of the United States of America*, *106*, 2806–2811.

Rossi, G. P., Andreis, P. G., Colonna, S., Albertin, G., Aragona, F., Belloni, A. S., & Nussdorfer, G. G. (2002). Endothelin-1[1-31]: A novel autocrine-paracrine regulator of human adrenal cortex secretion and growth. *The Journal of Clinical Endocrinology and Metabolism*, *87*, 322–328.

Rubanyi, G. M., & Polokoff, M. A. (1994). Endothelins: Molecular biology, biochemistry, pharmacology, physiology, and pathophysiology. *Pharmacological Reviews*, *46*, 325–415.

Rubin, L. J., Badesch, D. B., Barst, R. J., Galie, N., Black, C. M., Keogh, A., Pulido, T., Frost, A., Roux, S., Leconte, I., Landzberg, M., & Simonneau, G. (2002). Bosentan therapy for pulmonary arterial hypertension. *The New England Journal of Medicine*, *346*, 896–903.

Rubinstein, I., Gurbanov, K., Hoffman, A., Better, O. S., & Winaver, J. (1995). Differential effect of endothelin-1 on renal regional blood flow: Role of nitric oxide. *Journal of Cardiovascular Pharmacology*, *26*(Suppl. 3), S208–S210.

Ruiz-Opazo, N., Hirayama, K., Akimoto, K., & Herrera, V. L. (1998). Molecular characterization of a dual endothelin-1/Angiotensin II receptor. *Molecular Medicine*, *4*, 96–108.

Ruschitzka, F., Moehrlen, U., Quaschning, T., Lachat, M., Noll, G., Shaw, S., Yang, Z., Teupser, D., Subkowski, T., Turina, M. I., & Luscher, T. F. (2000). Tissue endothelin-converting enzyme activity correlates with cardiovascular risk factors in coronary artery disease. *Circulation*, *102*, 1086–1092.

Saida, K., Mitsui, Y., & Ishida, N. (1989). A novel peptide, vasoactive intestinal contractor, of a new (endothelin) peptide family. Molecular cloning, expression, and biological activity. *The Journal of Biological Chemistry*, *264*, 14613–14616.

Sakurai, T., Yanagisawa, M., Takuwa, Y., Miyazaki, H., Kimura, S., Goto, K., & Masaki, T. (1990). Cloning of a cDNA encoding a non-isopeptide-selective subtype of the endothelin receptor. *Nature*, *348*, 732–735.

Sanchez, R., MacKenzie, A., Farhat, N., Nguyen, T. D., Stewart, D. J., Mercier, I., Calderone, A., & Thorin, E. (2002). Endothelin B receptor-mediated regulation of endothelin-1 content and release in cultured porcine aorta endothelial cell. *Journal of Cardiovascular Pharmacology*, *39*, 652–659.

Sato, K., Oka, M., Hasunuma, K., Ohnishi, M., & Kira, S. (1995). Effects of separate and combined ETA and ETB blockade on ET-1-induced constriction in perfused rat lungs. *The American Journal of Physiology*, *269*, L668–L672.

Sauvageau, S., Thorin, E., Caron, A., & Dupuis, J. (2006). Evaluation of endothelin-1-induced pulmonary vasoconstriction following myocardial infarction. *Experimental Biology and Medicine (Maywood)*, *231*, 840–846.

Sauvageau, S., Thorin, E., Caron, A., & Dupuis, J. (2007). Endothelin-1-induced pulmonary vasoreactivity is regulated by ET(A) and ET(B) receptor interactions. *Journal of Vascular Research*, *44*, 375–381.

Sauvageau, S., Thorin, E., Villeneuve, L., & Dupuis, J. (2009). Change in pharmacological effect of endothelin receptor antagonists in rats with pulmonary hypertension: Role of ETB-receptor expression levels. *Pulmonary Pharmacology & Therapeutics*, *22*, 311–317.

Schneider, M. P., Boesen, E. I., & Pollock, D. M. (2007). Contrasting actions of endothelin ET (A) and ET(B) receptors in cardiovascular disease. *Annual Review of Pharmacology and Toxicology*, *47*, 731–759.

Sen, U., Tyagi, N., Patibandla, P. K., Dean, W. L., Tyagi, S. C., Roberts, A. M., & Lominadze, D. (2009). Fibrinogen-induced endothelin-1 production from endothelial cells. *American Journal of Physiology. Cell Physiology*, *296*, C840–C847.

Shetty, S. S., Okada, T., Webb, R. L., DelGrande, D., & Lappe, R. W. (1993). Functionally distinct endothelin B receptors in vascular endothelium and smooth muscle. *Biochemical and Biophysical Research Communications*, *191*, 459–464.

Simonson, M. S., & Dunn, M. J. (1990). Cellular signaling by peptides of the endothelin gene family. *The FASEB Journal*, *4*, 2989–3000.

Simpson, A. W., & Ashley, C. C. (1989). Endothelin evoked Ca2+ transients and oscillations in A10 vascular smooth muscle cells. *Biochemical and Biophysical Research Communications*, *163*, 1223–1229.

Sirvio, M. L., Metsarinne, K., Saijonmaa, O., & Fyhrquist, F. (1990). Tissue distribution and half-life of 125I-endothelin in the rat: Importance of pulmonary clearance. *Biochemical and Biophysical Research Communications*, *167*, 1191–1195.

Spinella, F., Garrafa, E., Di Castro, V., Rosano, L., Nicotra, M. R., Caruso, A., Natali, P. G., & Bagnato, A. (2009). Endothelin-1 stimulates lymphatic endothelial cells and lymphatic vessels to grow and invade. *Cancer Research*, *69*, 2669–2676.

Sprague, A. H., & Khalil, R. A. (2009). Inflammatory cytokines in vascular dysfunction and vascular disease. *Biochemical Pharmacology*, *78*, 539–552.

Stewart, D. J., Levy, R. D., Cernacek, P., & Langleben, D. (1991). Increased plasma endothelin-1 in pulmonary hypertension: Marker or mediator of disease? *Annals of Internal Medicine*, *114*, 464–469.

Sumner, M. J., Cannon, T. R., Mundin, J. W., White, D. G., & Watts, I. S. (1992). Endothelin ETA and ETB receptors mediate vascular smooth muscle contraction. *British Journal of Pharmacology*, *107*, 858–860.

Takahashi, M., Matsushita, Y., Iijima, Y., & Tanzawa, K. (1993). Purification and characterization of endothelin-converting enzyme from rat lung. *The Journal of Biological Chemistry*, *268*, 21394–21398.

Takahashi, H., Soma, S., Muramatsu, M., Oka, M., Ienaga, H., & Fukuchi, Y. (2001). Discrepant distribution of big endothelin (ET)-1 and ET receptors in the pulmonary artery. *The European Respiratory Journal*, *18*, 5–14.

Teerlink, J. R., Breu, V., Sprecher, U., Clozel, M., & Clozel, J. P. (1994). Potent vasoconstriction mediated by endothelin ETB receptors in canine coronary arteries. *Circulation Research*, *74*, 105–114.

Thomas, G. P., Sims, S. M., & Karmazyn, M. (1997). Differential effects of endothelin-1 on basal and isoprenaline-enhanced Ca2+ current in guinea-pig ventricular myocytes. *Journal de Physiologie*, *503*(Pt. 1), 55–65.

Thorin, E. (2001). Different contribution of endothelial nitric oxide in the relaxation of human coronary arteries of ischemic and dilated cardiomyopathic hearts. *Journal of Cardiovascular Pharmacology*, *37*, 227–232.

Thorin, E., Hamilton, C., Dominiczak, A. F., Dominiczak, M. H., & Reid, J. L. (1995). Oxidized-LDL induced changes in membrane physico-chemical properties and [Ca2+]i of bovine aortic endothelial cells. Influence of vitamin E. *Atherosclerosis*, *114*, 185–195.

Thorin, E., Lucas, M., Cernacek, P., & Dupuis, J. (2000). Role of ET(A) receptors in the regulation of vascular reactivity in rats with congestive heart failure. *American Journal of Physiology. Heart and Circulatory Physiology*, *279*, H844–H851.

Thorin, E., Nguyen, T. D., & Bouthillier, A. (1998). Control of vascular tone by endogenous endothelin-1 in human pial arteries. *Stroke*, *29*, 175–180.

Thorin, E., Parent, R., Ming, Z., & Lavallee, M. (1999). Contribution of endogenous endothelin to large epicardial coronary artery tone in dogs and humans. *The American Journal of Physiology*, *277*, H524–H532.

Thorin, E., & Thorin-Trescases, N. (2009). Vascular endothelial ageing, heartbeat after heartbeat. *Cardiovascular Research*, *84*, 24–32.

Tsutamoto, T., Wada, A., Maeda, K., Mabuchi, N., Hayashi, M., Tsutsui, T., Ohnishi, M., Sawaki, M., Fujii, M., Matsumoto, T., Horie, H., Sugimoto, Y., & Kinoshita, M. (2000). Transcardiac extraction of circulating endothelin-1 across the failing heart. *The American Journal of Cardiology*, *86*, 524–528.

Unoki, H., Fan, J., & Watanabe, T. (1999). Low-density lipoproteins modulate endothelial cells to secrete endothelin-1 in a polarized pattern: A study using a culture model system simulating arterial intima. *Cell and Tissue Research*, *295*, 89–99.

Vanhoutte, P. M., Rubanyi, G. M., Miller, V. M., & Houston, D. S. (1986). Modulation of vascular smooth muscle contraction by the endothelium. *Annual Review of Physiology*, *48*, 307–320.

Verhaar, M. C., Strachan, F. E., Newby, D. E., Cruden, N. L., Koomans, H. A., Rabelink, T. J., & Webb, D. J. (1998). Endothelin-A receptor antagonist-mediated vasodilatation is attenuated by inhibition of nitric oxide synthesis and by endothelin-B receptor blockade. *Circulation*, *97*, 752–756.

Vierhapper, H., Wagner, O., Nowotny, P., & Waldhausl, W. (1990). Effect of endothelin-1 in man. *Circulation*, *81*, 1415–1418.

Violin, J. D., & Lefkowitz, R. J. (2007). Beta-arrestin-biased ligands at seven-transmembrane receptors. *Trends in Pharmacological Sciences*, *28*, 416–422.

Watanabe, T., & Endoh, M. (2000). Antiadrenergic effects of endothelin-1 on the L-type Ca2+ current in dog ventricular myocytes. *Journal of Cardiovascular Pharmacology*, *36*, 344–350.

Watts, S. W. (2010). Endothelin receptors: What's new and what do we need to know? *American Journal of Physiology—Regulatory, Integrative and Comparative Physiology*, *298*, R254–R260.

Wisler, J. W., DeWire, S. M., Whalen, E. J., Violin, J. D., Drake, M. T., Ahn, S., Shenoy, S. K., & Lefkowitz, R. J. (2007). A unique mechanism of beta-blocker action: Carvedilol stimulates beta-arrestin signaling. *Proceedings of the National Academy of Sciences of the United States of America*, *104*, 16657–16662.

Yanagisawa, M., Inoue, A., Ishikawa, T., Kasuya, Y., Kimura, S., Kumagaye, S., Nakajima, K., Watanabe, T. X., Sakakibara, S., Goto, K., et al. (1988a). Primary structure, synthesis, and biological activity of rat endothelin, an endothelium-derived vasoconstrictor peptide. *Proceedings of the National Academy of Sciences of the United States of America*, *85*, 6964–6967.

Yanagisawa, M., Kurihara, H., Kimura, S., Tomobe, Y., Kobayashi, M., Mitsui, Y., Yazaki, Y., Goto, K., & Masaki, T. (1988b). A novel potent vasoconstrictor peptide produced by vascular endothelial cells. *Nature*, *332*, 411–415.

Yorikane, R., & Koike, H. (1990). The arrhythmogenic action of endothelin in rats. *Japanese Journal of Pharmacology*, *53*, 259–263.

Yorikane, R., Shiga, H., Miyake, S., & Koike, H. (1990). Evidence for direct arrhythmogenic action of endothelin. *Biochemical and Biophysical Research Communications*, *173*, 457–462.

Yu, C., Yang, Z., Ren, H., Zhang, Y., Han, Y., He, D., Lu, Q., Wang, X., Yang, C., Asico, L. D., Hopfer, U., Eisner, G. M., Jose, P. A., & Zeng, C. (2009). D3 dopamine receptor regulation of ETB receptors in renal proximal tubule cells from WKY and SHRs. *American Journal of Hypertension*, *22*, 877–883.

Zeng, C., Asico, L. D., Yu, C., Villar, V. A., Shi, W., Luo, Y., Wang, Z., He, D., Liu, Y., Huang, L., Yang, C., Wang, X., Hopfer, U., Eisner, G. M., & Jose, P. A. (2008). Renal D3 dopamine receptor stimulation induces natriuresis by endothelin B receptor interactions. *Kidney International*, *74*, 750–759.

Sandra L. Pfister, Kathryn M. Gauthier, and William B. Campbell

Department of Pharmacology and Toxicology,
Medical College of Wisconsin, Milwaukee, Wisconsin, USA

Vascular Pharmacology of Epoxyeicosatrienoic Acids

Abstract

Epoxyeicosatrienoic acids (EETs) are cytochrome P450 metabolites of arachidonic acid that are produced by the vascular endothelium in responses to various stimuli such as the agonists acetylcholine (ACH) or bradykinin or by shear stress which activates phospholipase A_2 to release arachidonic acid. EETs are important regulators of vascular tone and homeostasis. In the modulation of vascular tone, EETs function as endothelium-derived hyperpolarizing factors (EDHFs). In models of vascular inflammation, EETs attenuate inflammatory signaling pathways in both the endothelium and vascular smooth muscle. Likewise, EETs regulate blood vessel formation or angiogenesis by mechanisms

Advances in Pharmacology, Volume 60

1054-3589/10 $35.00
10.1016/S1054-3589(10)60008-7

that are still not completely understood. Soluble epoxide hydrolase (sEH) converts EETs to dihydroxyeicosatrienoic acids (DHETs) and this metabolism limits many of the biological actions of EETs. The recent development of inhibitors of sEH provides an emerging target for pharmacological manipulation of EETs. Additionally, EETs may initiate their biological effects by interacting with a cell surface protein that is a G protein-coupled receptor (GPCR). Since GPCRs represent a common target of most drugs, further characterization of the EET receptor and synthesis of specific EET agonists and antagonist can be used to exploit many of the beneficial effects of EETs in vascular diseases, such as hypertension and atherosclerosis. This review will focus on the current understanding of the contribution of EETs to the regulation of vascular tone, inflammation, and angiogenesis. Furthermore, the therapeutic potential of targeting the EET pathway in vascular disease will be highlighted.

I. Introduction

Epoxyeicosatrienoic acids (EETs) play a pivotal role in numerous cellular processes involved in vascular function, including vasodilation and inflammation. The multifunctional nature of EETs underlies the importance of these compounds in cardiovascular disease. This review will focus on the current understanding of the contribution of EETs to the regulation of vascular tone, inflammation, and angiogenesis. Furthermore, the therapeutic potential of targeting the EET pathway in vascular disease will be highlighted.

II. Biochemistry

In endothelial cells, arachidonic acid is metabolized by the cyclooxygenase (COX), lipoxygenase, and cytochrome P450 (CYP) epoxygenase pathways (Rosolowsky & Campbell, 1993, 1996). The epoxygenase pathway leads to the formation of four regioisomeric EETs, 14,15-EET, 11,12-EET, 8,9-EET, and 5,6-EET (Fig. 1; Capdevila et al., 1981; Zeldin, 2001). The EETs are released by endothelial cells in response to receptor agonists such as acetylcholine (ACH) or bradykinin to function as autocrine and paracrine hormones (Campbell et al., 1996a; Fisslthaler et al., 2001; Gauthier et al., 2005; Huang et al., 2005; Nithipatikom et al., 2000). The relative abundance of each EET regioisomer differs among vascular beds depending on which CYP isoforms are expressed, as each CYP isoform generates its own unique profile of EET regioisomers. While many CYP isozymes have been identified in blood vessels, endothelial CYP2C8/2C9 and CYP2J2 function in humans to produce mainly 14,15-EET, with lesser amounts of 11,12-EET

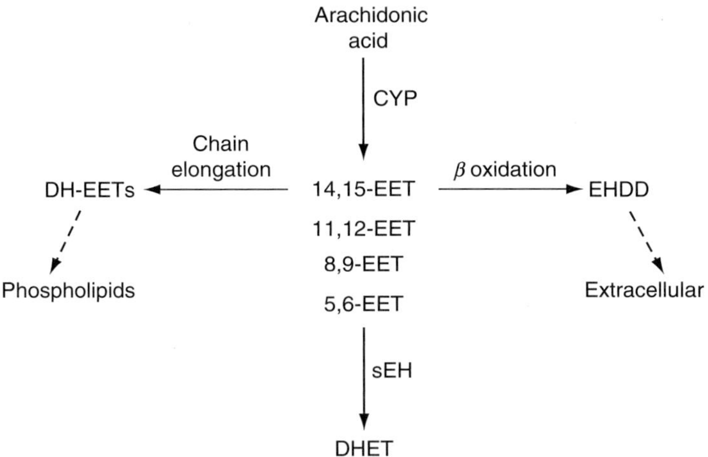

FIGURE 1 *Overview of EET biosynthesis and metabolism.* Arachidonic acid is metabolized by cytochrome P450 (CYP) epoxygenases to produce epoxyeicosatrienoic acids (EETs). There are four EET regioisomers which vary with the placement of the epoxide group. They include 5,6-, 8,9-, 11,12-, and 14,15-EET. EETs are esterified to cellular membrane phospholipids following chain elongation, metabolized by β-oxidation to shorter carbon chain molecules or metabolized by soluble epoxide hydrolase (sEH) to their corresponding vicinal diols, the dihydroxyeicosatrienoic acids (DHETs). DH-EET, dihomo EET; EHDD, epoxyheptadecadienoic acid.

(Wu et al., 1996; Zeldin, 2001; Zeldin et al., 1995). No EET production is detected in smooth muscle (Campbell et al., 2006).

The EETs are rapidly taken up by vascular cells and metabolized by soluble epoxide hydrolase (sEH) to form dihydroxyeicosatrienoic acids (DHETs) (Fig. 1; Spector & Norris, 2007; Spector et al., 2004). The DHETS are generally less bioactive than their corresponding EETs. Alternatively, EETs undergo secondary metabolism or β-oxidation, forming 16-carbon epoxy-derivatives that accumulate in the extracellular fluid, and they can be chain elongated to form 22-carbon 1α-1β-dihomo derivatives that are incorporated into phospholipids (Fig. 1; Fang et al., 2002). Understanding the chemical and biochemical mechanisms that contribute to EET function will play an important part in the future design of drugs that act through the EET pathway. This topic is discussed in more detail below.

III. EETs and Vascular Tone

One of the early descriptions of a biological action of the EETs was as vasodilators in the intestinal microcirculation (Proctor et al., 1987). This original observation eventually led to the identification of EETs as endothelium-dependent hyperpolarizing factors (EDHFs) (Figs. 2 and 3) (Archer et al., 2003; Campbell et al., 1996a; Coats et al., 2001; Gauthier et al., 2005; Huang et al.,

2005; Miura et al., 2001; Popp et al., 1998). There is a plethora of studies in humans and animals that support this observation and the topic is the focus of a number of recent reviews (Campbell & Fleming, 2010; Imig & Hammock, 2009; Revermann, 2010; Sudhahar et al., 2010). Key aspects are highlighted below.

A. Experimental Evidence

In response to chemical or physical stimuli such as ACH, thrombin, bradykinin, and fluid sheer stress, vascular endothelial cells regulate the tone of the underlying smooth muscle by producing and releasing various

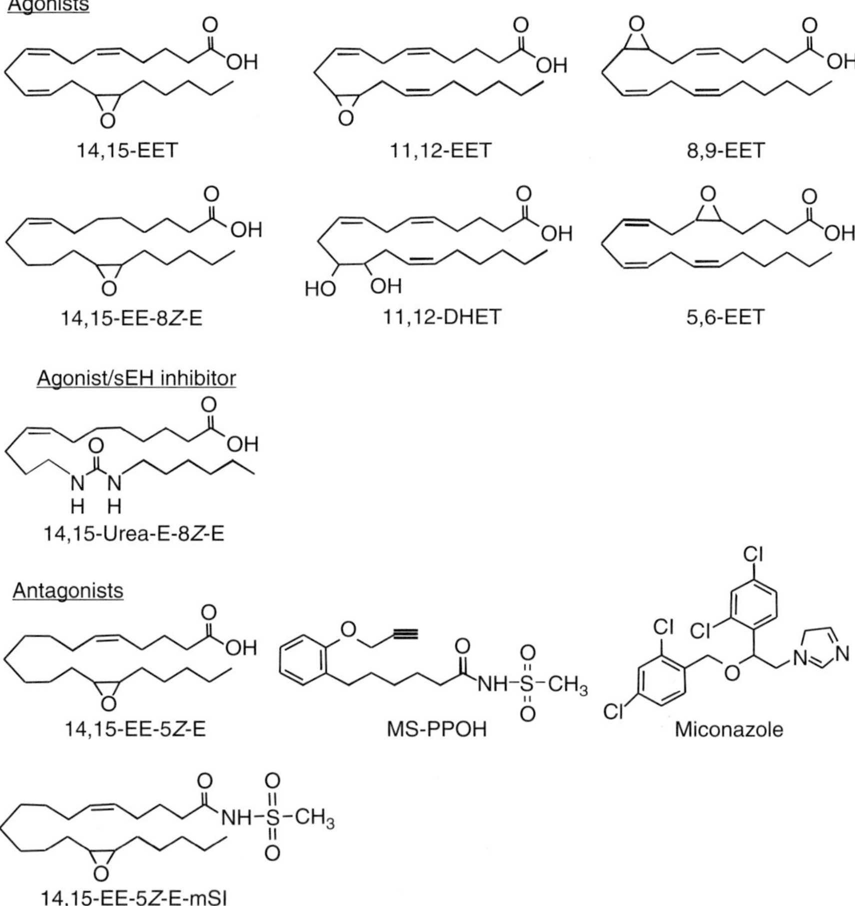

FIGURE 2 Chemical structures of epoxyeicosatrienoic acid (EET) agonists, an EET agonist/sEH inhibitor and EET antagonists. mSI, methylsulfonamide; EE-8Z-E, epoxyeicosa-8Z-enoic acid; EE-5Z-E, epoxyeicosa-5Z-enoic acid; E-8Z-E, eicosa-8Z-enoic acid; sEH, soluble epoxide hydrolase; MS-PPOH, N-(methylsulfonyl)-2-(2-propynyloxy)-benzenehexanamide.

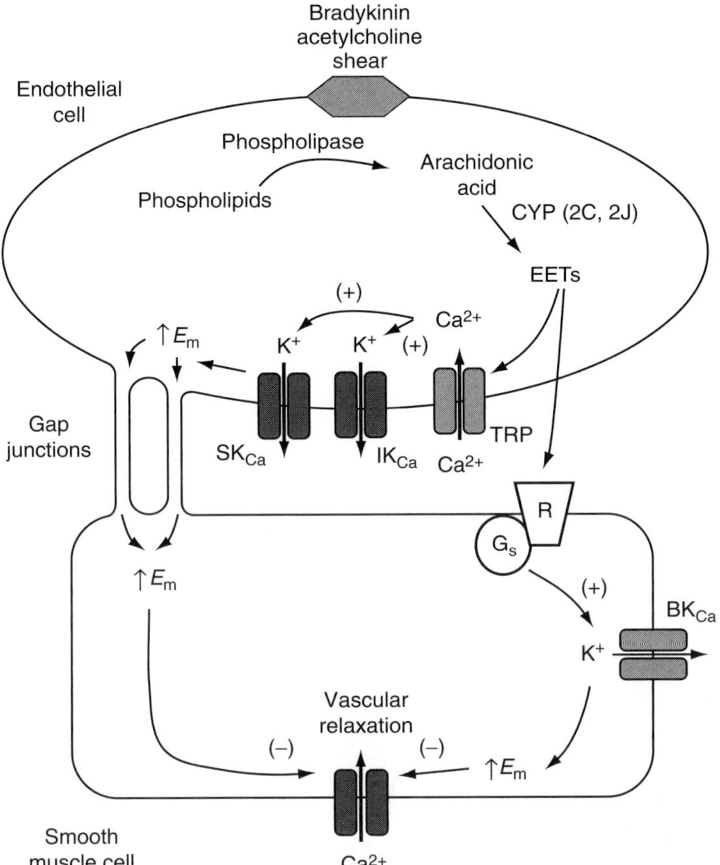

FIGURE 3 *Proposed mechanisms of epoxyeicosatrienoic acid (EET)-mediated hyperpolarization and relaxation.* Stimulation by bradykinin, acetylcholine, or shear activates phospholipase in endothelial cell membranes to elicit the release of arachidonic acid. Arachidonic acid is metabolized by cytochrome P450 (CYP) epoxygenases of the 2C and 2J families to EETs. EETs diffuse to the smooth muscle cell to induce membrane hyperpolarization via activation of large conductance, calcium-sensitive potassium (BK_{Ca}) channels. This causes K efflux, an increase in membrane potential (E_m) or hyperpolarization and relaxation. EETs also act intracellularly in endothelial cells to promote Ca efflux through transient receptor potential (TRP) channels. Calcium activates small conductance (SK) and intermediate conductance (IK) K_{Ca} to cause hyperpolarization. Endothelial hyperpolarization spreads to the smooth muscle cell via gap junctions. R, putative EET receptor; Gs, stimulatory guanine nucleotide-binding protein.

relaxing and contracting factors (Cohen & Vanhoutte, 1995; Feletou & Vanhoutte, 2006; Fleming & Busse, 2006; Furchgott & Vanhoutte, 1989; Furchgott & Zawadzki, 1980). The relaxing factors are the COX metabolite, prostacyclin (PGI_2), endothelium-derived relaxing factor identified as nitric oxide (NO), and EDHFs. EDHF was originally described as an endothelial factor that hyperpolarized the membrane of the underlying vascular smooth muscle. Hyperpolarization was mediated by the opening of potassium (K) channels because it was inhibited by increased extracellular K concentration and K channel blockers (Adeagbo & Triggle, 1993; Cohen & Vanhoutte, 1995). Inhibitors of NO synthase (NOS) or COX do not block hyperpolarizations. PGI_2, NO, and EDHF serve the same function to dilate the blood vessels in response to agonists, and to physical forces. These dilators antagonize the activity of vasoconstrictors and maintain organ blood flow. The activity of NO differs from EDHF along the vasculature. Endothelium-dependent dilation to NO is greatest in large arteries, whereas EDHF has its greatest effect in small arteries and arterioles (Nagao et al., 1992; Nishikawa et al., 1999). While a number of compounds have been proposed as mediators of EDHF activity, there is convincing evidence that EETs and EDHF have identical properties. Both are produced by the endothelium and open K channels, hyperpolarize, and relax smooth muscle. Similar to EDHF, EETs are more potent in relaxing small coronary arteries than large arteries (Campbell et al., 1996b; Oltman et al., 1998). While the primary role of the EETs is vasodilation in most vascular beds, EETs cause vasoconstriction in the pulmonary artery (Kerseru et al., 2008; Moreland et al., 2007; Zhu et al., 2000). The exact mechanism of the vasoconstriction has not been elucidated and remains a focus for future studies.

In a number of different vascular beds from a variety of species, endothelium-dependent vasodilator responses to arachidonic acid, ACH, or bradykinin that are not blocked by NOS and COX inhibitors are blocked by inhibitors of CYP, the enzyme responsible for EET synthesis. EETs were more completely characterized as EDHFs using bovine and porcine coronary arteries (Campbell et al., 1996a; Fisslthaler et al., 1999). The four regioisomeric EETs relaxed bovine coronary arteries in a concentration-related manner. Significant relaxations to the EETs occurred with nanomolar concentrations and were blocked by the K channel blockers TEA and charybdotoxin, and high K. When membrane potential (E_m) of coronary arterial vascular smooth muscle was measured, 11,12-EET caused membrane hyperpolarization. This hyperpolarization was also blocked by iberiotoxin. Similarly, others showed that in human coronary and internal mammary arteries, 11,12-EET relaxed and hyperpolarized the smooth muscle, and these effects were blocked by the K channel blocker iberiotoxin (Archer et al., 2003; Larsen et al., 2005). Additionally, arachidonic acid, bradykinin, and methacholine hyperpolarized the smooth muscle of endothelium-intact coronary arteries, and CYP inhibitors blocked the hyperpolarizations. Antisense oligonucleotides against CYP2C inhibited the relaxations and hyperpolarizations to bradykinin, whereas sense and scrambled oligonucleotides were without effect (Fisslthaler et al., 1999). This inhibition by the antisense oligonucleotides was accompanied by a reduction of the expression of CYP2C. Arachidonic acid and methacholine stimulated the release of EETs from perfused coronary arteries and coronary endothelial cells. Identical results were obtained with human internal mammary arteries (Archer et al., 2003). Related studies were performed with bradykinin in bovine and porcine coronary arteries and human internal mammary arteries (Archer et al., 2003; Campbell et al., 2001; Hecker et al., 1994; Weston et al., 2005). CYP inhibitors attenuated relaxations and hyperpolarization by bradykinin. Taken collectively, these studies indicate that the hyperpolarizations and relaxations to methacholine and bradykinin are mediated by the EETs. Furthermore, in arteries from experimental animals and humans, physical forces such as shear stress and increased flow caused endothelium-dependent relaxations that were reduced, but not blocked, by inhibitors of NOS and COX (Gauthier & Campbell, 2006; Huang et al., 2005; Miura et al., 2001; Popp et al., 1998). The non-NO, non-PG-dependent relaxations were blocked by CYP inhibitors and were associated with the release of EETs. While the non-NO-, non-PG-mediated relaxations to ACH were greater in vessels from females compared to males, few studies have measured responses in the presence of CYP inhibitors or EET antagonists (McCulloch & Randall, 1998; Scotland et al., 2005; Tagawa et al., 1997; White et al., 2000; Woodman & Boujaoude, 2004). Additional studies are necessary to establish the exact role of gender in EET-mediated regulation of vascular tone.

For EETs to be considered important in the regulation of vascular tone, EETs should function as EDHF in intact animals. In the microcirculation of anesthetized dogs, bradykinin and ACH caused non-NO, non-PG-mediated relaxations that were blocked by high extracellular K and iberiotoxin indicating a role of an EDHF in mediating the dilation (Nishikawa et al., 1999, 2000; Watanabe et al., 2005; Widmann et al., 1998). They were also blocked by the CYP inhibitors, miconazole and metyrapone. Similarly, following COX and NOS inhibition, arachidonic acid induced dilation of coronary microvessels. These dilations were blocked by K channel blockers and a CYP inhibitor suggesting that a CYP metabolite of arachidonic acid functions as EDHF in the canine coronary circulation. Using intravital microscopy of cremaster arterioles from conscious hamsters to measure changes in vessel diameter, high concentrations of ACH induced a dilation that was not altered by NOS and COX inhibitors (Watanabe et al., 2005). However, the relaxations were completely blocked by the elevation of extracellular K. Likewise, treatment with the CYP inhibitor, sulfaphenazole attenuated the non-NO, non-PG, ACH-induced relaxations. EET-mediated dilations were also demonstrated *in vivo* in canine coronary and kidney arterioles, hamster cheek pouch arterioles, rat cremaster arterioles, rat mesenteric and hind limb, and sciatic nerve circulations (Chu et al., 2000; Loeb et al., 1997; Matsuda et al., 2004; Nishikawa et al., 1999; Oltman et al., 1998, 2001; Parkington et al., 2002; Thomsen et al., 2000; Welsh and Segal, 2000).

The topic of EDHF in the regulation of vascular tone in humans was recently reviewed (Bellien et al., 2008a) with a careful evaluation of data implicating a specific role of EETs. To summarize, *ex vivo* studies in isolated arteries and a small number of *in vivo* studies in humans provide evidence that EETs may function as EDHF (Archer et al., 2003; Bellien et al., 2008b, 2010; Hatoum et al., 2005; Kemp and Cocks, 1997; Larsen et al., 2005, 2008; Lenasi, 2009; Lenasi and Strucl, 2008; Taddei et al., 2006; Virdis et al., 2010). In the isolated vessel studies, the existence of a CYP-related EDHF has been clearly reported for the human coronary arterioles, internal mammary artery, and radial artery. While *in vivo* data for EET-mediated vasodilation in humans are understandably limited, there is strong evidence to support an *in vivo* role of EETs as EDHF. Intra-arterial infusions of bradykinin induced increases in forearm blood flow in human subjects treated with the COX inhibitor, aspirin, and the NOS inhibitor, *N*(G)-monomethyl-L-arginine (LNA). These non-PG, non-NO increases in blood flow were blocked by potassium chloride infusion or TEA and inhibited by CYP inhibitor miconazole (Halcox et al., 2001; Honing et al., 2000). Forearm blood flow response to ACH or bradykinin has been compared in normotensive and essential hypertensive patients (Taddei et al., 2006). In normotensive subjects, vasodilation to ACH and bradykinin was blunted by NOS inhibition but not by the CYP 2C9 inhibitor sulfaphenazole. In contrast, in the hypertensive patients, the vasodilator responses to the agonists were reduced compared with the normotensive group. More interestingly, the relaxation responses in the hypertensive group were resistant to NO inhibition but were blunted by CYP inhibition. These findings suggest that EETs sustain endothelium-dependent vasodilation in hypertension. A similar compensatory role for EETs may exist in patients with congestive heart failure (Katz and Krum, 2001). Forearm blood flow responses to ACH were decreased by COX and NOS inhibition in the normal subjects but not in patients with congestive heart failure. While the effect of a CYP antagonist was not tested, the results again suggest the presence of a non-PG, non-NO relaxing factor that participates in the regulation of forearm blood flow in humans. More recently, EETs regulate arterial stiffness during changes in blood flow (Bellien et al., 2010). Heating the skin of the hand increases blood flow by promoting the release of vasorelaxing factors which decrease isometric tone and wall stiffness. The NOS inhibitor, LNA partially inhibited the increase in blood flow. The remaining blood flow increase was completely prevented with TEA and the CYP inhibitor fluconazole. Therefore, in human peripheral conduit arteries, the adaptation of smooth muscle tone and arterial stiffness during blood flow variations is regulated by the vascular endothelium through the release of both NO- and CYP-mediated EDHF.

In summary, EETs derived from endothelial CYP epoxygenases are diffusible vasodilator factors that act by hyperpolarizing vascular smooth muscle cells. Experiments performed both *in vitro* and *in vivo* in animals and humans support the contribution of EETs to endothelium-dependent relaxations.

B. Mechanism of Action

All four EET regioisomers, 14,15-, 11,12-, 8,9-, and 5,6-EET, equipotently relax bovine and canine coronary arteries (Campbell et al., 1996a; Rosolowsky and Campbell, 1993). This suggests that the position of the epoxy group is not critical for relaxation in these arteries. Alternatively, 5,6-EET is more active than the other EET regioisomers in relaxing rat-tail arteries, rabbit and pig cerebral arteries, and rat renal arteries (Carroll et al., 1987; Ellis et al., 1990; Leffler and Fedinec, 1997; Pomposiello et al., 2003). Therefore, regioisomer specificity of EET-induced relaxations varies with species and vascular bed. Structural modifications aimed specifically to characterize the molecular components critical for EET dilator activity and to identify analogs with EET-specific antagonist properties have been reported for 14,15-EET (Fig. 3; Chen et al., 2009; Falck et al., 2003a, 2009; Gauthier et al., 2002, 2003a, 2004; Yang et al., 2007, 2008). 14,15-EET analogs, with modifications in the epoxy and carboxyl groups, deletions of the double bonds, and variations in the carbon chain length, were synthesized and tested for their ability to cause relaxation. Specific structural features were identified for full agonist activity: an acidic group at carbon-1, a 20 carbon backbone, a Δ8 double bond, and a 14(*S*),15(*R*)-*cis* epoxide (Falck et al., 2003a). Thus, the basic full agonist was 14(*S*),15(*R*)-(*cis*)-epoxyeicosa-8*Z*-enoic acid (14,15-EE-8*Z*-E). Addition of a methylsulfonamide (mSI) group to carboxyl of 14,15-EET reduced its metabolism by β-oxidation and incorporation into membrane phospholipids (Gauthier et al., 2003a). More recently, Falck et al. have developed chimeric analogs which are capable of functioning as stable 14,15-EET surrogates and inhibitors of sEH (Falck et al., 2009). Having analogs that limit 14,15-EET auto-oxidation and metabolism to inactive DHETs by sEH will increase the potential therapeutic applications of these drugs.

Analogs with low agonist activity were tested for their ability to inhibit EET-induced relaxations (Falck et al., 2003a; Gauthier et al., 2003b). 14,15-Epoxyeicosa-5*Z*-enoic acid (14,15-EE-5*Z*-E) inhibited the relaxations to 14,15-, 11,12-, 8,9-, and 5.6-EET; however, it was most active in inhibiting 14,15-EET. The non-PG-mediated relaxations to arachidonic acid were also blocked by 14,15-EE-5*Z*-E (Fig. 2). In contrast, it did not alter the relaxations to the NO donor sodium nitroprusside, the PGI_2 analog iloprost, or the K channel openers bimakalim and NS1619. Thus, 14,15-EE-5*Z*-E is a selective EET antagonist that does not inhibit other endothelial factors or K channels (Gauthier et al., 2002). 14,15-EE-5*Z*-E-mSI inhibited relaxations to 14,15- and 5,6-EET but not to 11,12- or 8,9-EET (Gauthier et al., 2003a). 14,15-EE-5*Z*-E-mSI did not alter the relaxations to sodium nitroprusside, iloprost, bimakalim, or NS1619 and did not affect the metabolism of arachidonic acid. Thus, 14,15-EE-5*Z*-E-mSI is a selective antagonist of 14,15- and 5,6-EET. 14,15-EE-5*Z*-E-mSI inhibited the indomethacin-resistant relaxations to arachidonic acid; however, the inhibition was less than what occurred with 14,15-EE-5Z-E. The combination of 14,15-EE-5Z-E-mSi and the CYP inhibitor N-(methylsulfonyl)-2-(2-propynyloxy)-benzenehexamide (MS-PPOH) inhibited arachidonic acid-induced relaxations to the same extent as 14,15-EE-5Z-E alone. Thus, 11,12- and/or 8,9-EET must contribute to the relaxation response to arachidonic acid. The discovery of EET antagonists provided vital pharmacological tools to examine the role of EETs as EDHFs. Thus, in the presence of COX and NOS inhibitors, ACH- and bradykinin-mediated relaxations and hyperpolarization were inhibited by 14,15-EE-5Z-E (Archer et al., 2003; Gauthier et al., 2002, 2003a).

EETs activate smooth muscle large conductance, calcium-activated potassium (BK_{Ca}) channels (Campbell et al., 1996a; Li & Campbell, 1997; Li et al., 1997). This leads to K efflux, an increase in the E_m or hyperpolarization. The hyperpolarization inhibits activation of voltage-activated calcium channels reducing calcium entry and causing relaxation. EETs activate BK_{Ca} channels in nanomolar concentrations in cell-attached patches in which the cell cytosol is in contact with the channel. However, when this association is disrupted using inside-out patches, the EETs are without effect (Li et al., 1997). The addition of GTP, but not ATP, to the cytoplasmic side of the inside-out patch restores the ability of the EET to activate BK_{Ca} channels. This can be blocked by the guanine nucleotide-binding (G) protein inhibitor GDP-β-S. It is also blocked by the addition of an antibody against Gsα but not by anti-Giα or anti-Gβγ antibodies. Activation of Gsα by ADP ribosylation with cholera toxin also increases BK_{Ca} channel activity in cell-attached patches (Li et al., 1999). In inside-out patches, BK_{Ca} channel activity is increased by Gsα-GTP; however, Gsα-GDP and Gβγ do not alter channel activity (Kume et al., 1992; Scornik et al., 1993). Other evidence implicates Gs in EET action. 11,12-EET increases tissue plasminogen activator (tPA) activity and protein expression in endothelial cells (Node et al., 2001). This is associated with a 3.5-fold increase in GTP binding to Gsα but not Giα. These studies indicate that a G protein with the characteristics of Gsα mediates the EET activation of BK_{Ca} channels, and this occurs by a membrane-delimited mechanism (i.e., the action is confined to the membrane and does not require compounds from other parts of the cell; Li et al., 1997). In addition, in intact smooth muscle cells, EETs promote endogenous ADP ribosylation of Gsα to increase BK_{Ca} channel activation (Li et al., 1999). It is thought that EETs activate the BK_{Ca} channel by a Gsα-mediated, membrane-delimited mechanism, which is sustained by ADP ribosylation of Gsα.

Others have suggested that EET-mediated vascular smooth muscle cell relaxation occurs by endothelial-mediated hyperpolarization (Feletou & Vanhoutte, 2006; Fleming & Busse, 2006). Agonists such as ACH and bradykinin increase endothelial EET synthesis. EETs act in an autocrine manner to increase intracellular calcium and activate endothelial apamin-sensitive, small conductance (SK_{Ca}) and charybdotoxin-sensitive, intermediate conductance (IK_{Ca}) channels. This results in hyperpolarization of the endothelial cell.

Smooth muscle cell relaxation occurs because of the hyperpolarizing current spreading from the endothelium to smooth muscle through gap junctions. Fleming and coworkers provide evidence in CYP 2C9-overexpressing human umbilical vein endothelial cells (HUVECs) that EETs regulate intracellular calcium by inducing the translocation of "transient receptor potential" (TRP) channel proteins to caveolin-1-rich areas of the endothelial cell membrane (Fleming et al., 2007; Loot et al., 2008). NO- and PGI_2-independent flow-induced vasodilation of the murine carotid artery is inhibited by CYP inhibitor as well as to a TRP4 channel blocker (Loot et al., 2008). The exact molecular mechanisms by which EETs induce the membrane translocation of TRP channels remain to be elucidated, although protein kinase A (PKA) may be involved in this process.

The evidence that EET action is coupled to a G protein and EETs promote GTP binding to membranes suggests that EETs act through a G protein-coupled receptor (GPCR). This concept is supported by the following studies. First, as indicated above, 14(*S*),15(*R*)-*cis*-EE-8*Z*-E-enoic acid was the simplest structure with full agonist activity (Falck et al., 2003a). The requirement for a specific stereoisomer of the epoxide suggested a specific binding site for the EET. Second, using vascular smooth muscle cells, 14,15-EET was tethered to silica beads so that it could not enter the cell. Tethered-14,15-EET inhibited aromatase activity to a similar extent as untethered 14,15-EET (Snyder et al., 2002). Thus, 14,15-EET acted on the cell surface and not intracellularly. Finally, a high-affinity EET-binding site was described in intact cells and membrane preparations from guinea pig mononuclear cells and human U937 cells. By use of ^{3}H-14,15-EET as a radioligand, specific and saturable binding with a K_d of 5.7 nM was determined in guinea pig monocytes and a K_d of 13.84 nM in U937 cells (Wong et al., 1993, 1997, 2000). This binding site was further defined in the cell membranes by Yang et al. (2008) by use of 20-^{125}iodo-14,15-epoxyeicosa-8*Z*-enoic acid (20-^{125}I-14,15-EE-8*Z*-E). 20-^{125}I-14,15-EE-8*Z*-E bound U937 membranes in a specific, saturable, and reversible manner with high affinity. EET analogs, but not prostaglandins or lipoxygenase metabolites, displaced the 14,15-EET radioligands from their binding site. The binding was also inhibited by GTPγS suggesting G protein coupling. More recently, 20-^{125}I-14,15-EE-5*Z*-E was characterized as an antagonist. It showed specific, saturable, reversible binding to U937 membranes and bound with higher affinity than the agonist radioligand. 20-^{125}I-14,15-EE-5*Z*-E had a K_d of 1.11 nM, whereas 20-^{125}I-14,15-EE-8*Z*-E had a K_d of 11.8 nM (Yang et al., 2008). A series of CYP and sEH inhibitors were tested for their ability to displace 20-^{125}I-14,15-EE-5*Z*-E from its binding site. The CYP inhibitors miconazole and MS-PPOH inhibited binding of the ligand suggesting that some CYP inhibitors are also EET antagonists (Chen et al., 2009). Identifying the EET receptor remains a subject of active research and an area that has obvious future therapeutic implications since many clinically useful drugs target receptors in their mechanism of action.

C. Therapeutic Potential

One likely limitation to the vasodilator effect of EETs is their metabolism by sEH or β-oxidation to the less biologically active metabolites (Imig & Hammock, 2009). The development of metabolically stable EETs or compounds that block EET metabolism to enhance the action of endogenous EETs are current approaches to increase EET activity. For example, Falck and coworkers have developed urea analogs of the EETs with agonist activity that also inhibit sEH (Falck et al., 2009). These compounds have not been tested *in vivo*. Using isolated blood vessels, EET-mediated relaxation responses were enhanced in the presence of a sEH inhibitor (Hercule et al., 2009; Larsen et al., 2005; Olearczyk et al., 2009). This result provided motivation to explore a possible contribution of sEH to the development of hypertension and a potentially novel use of sEH inhibitors in the treatment of the cardiovascular diseases. sEH can be inhibited *in vitro* by a variety of urea, carbamate, and amide derivatives. Furthermore, novel inhibitors of sEH were developed such as 12-(3-adamantane-1-yl-ureido)-dodecanoic acid (AUDA) and AUDA butyl ester. The topic of sEH inhibitors is the subject of a recent review by Imig and Hammock (2009). The first experimental evidence that a sEH inhibitor increases EET concentrations and lowers blood pressure in an animal model of hypertension was reported by Yu et al. (2000). EET hydrolysis was increased in kidneys of the spontaneously hypertensive rat (SHR) model compared to Wistar Kyoto (WKY) controls. When the sEH inhibitor DCU was administered to 8-week-old SHRs daily for 4 days, systolic blood pressure decreased. Imig and colleagues extended this observation and were the first to demonstrate that sEH contributes significantly to the elevated blood pressure in angiotensin II (ANG II)-dependent hypertension (Imig et al., 2002). Kidney sEH protein levels were elevated in the rat model of ANG II hypertension compared to normotensive rats. Administration of the selective sEH inhibitor, *N*-cyclohexyl-*N*-dodecyl (NCND) urea for 4 days lowered arterial blood pressure by 30 mmHg. A different sEH inhibitor, *N*-adamantyl-*N*′-dodecylurea (ADU), lowered systolic blood pressure, normalized vascular endothelial function, and attenuated left-ventricular hypertrophy in the deoxycorticosterone acetate (DOCA) plus high salt model of hypertension in rats (Imig et al., 2005). In mice with a target disruption of the sEH gene, the increase in blood pressure to DOCA-salt was blunted compared to wild-type mice (Manhiani et al., 2009). Furthermore, treatment of the wild-type DOCA-salt mice with the sEH inhibitor had a blood pressure lowering effect. Finally, Arete Therapeutics Inc. has developed an orally administered sEH inhibitor, AR9281, which is currently in a Phase II clinical program for the treatment of prediabetic patients with impaired glucose tolerance, mild obesity, and mild to moderate hypertension. The Phase I clinical program for AR9281 demonstrated that the drug was safe and well tolerated in healthy volunteers. Results from the Phase II clinical trial will lay the groundwork for the future use of sEH inhibitors and EET agonists in the treatment of vascular disease.

IV. EETs and Inflammation

Endothelial cells release a number or pro- and anti-inflammatory mediators (Pober & Sessa, 2007). A strong link between inflammation and endothelial dysfunction has been established. Inflammation is a risk factor for cardiovascular diseases, like hypertension and atherosclerosis. EETs have anti-inflammatory properties implicating their potential therapeutic use in the treatment of inflammation (Fig. 4). The role of CYP epoxygenases, sEH, and cardiovascular inflammation has been recently reviewed (Deng et al., 2010). The following sections will summarize specific effects of EETs on cell adhesion and platelet activation providing updated evidence on the role of EETs in inflammation.

A. Cell Adhesion

Macrophage- and leukocyte-derived cytokines such as tumor necrosis factor α (TNF-α) and interleukin 1α (IL-1α) activate endothelial cells and promote the surface expression of cell adhesion molecules (CAMs) including vascular cell adhesion molecule 1 (VCAM-1), E-selectin, and intercellular adhesion molecule 1 (ICAM-1). These events are integral to the inflammatory response.

1. Experimental Evidence

Node et al. reported that EETs are potent inhibitors of CAM expression induced by TNF-α, IL-1α, and bacterial lipopolysaccharide (LPS; Node et al., 1999). Although EETs inhibited the expression of VCAM-1, E-selectin, and

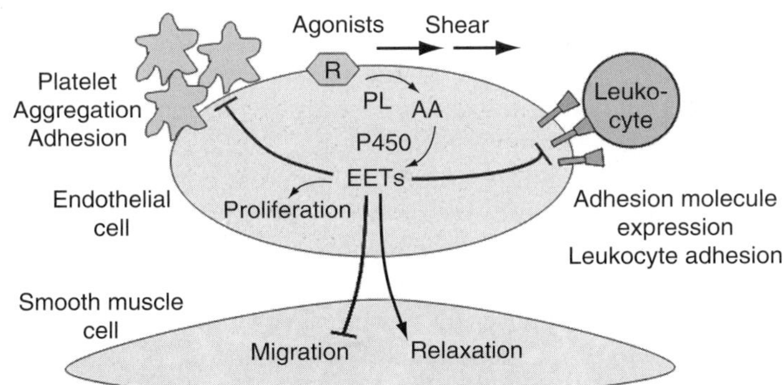

FIGURE 4 *A simplified schematic of vascular effects of epoxyeicosatrienoic acids (EETs).* Arrows denote stimulation and straight line without arrow denotes inhibition. Agonists or shear stress activate phospholipase (PL) in endothelial cell membranes. This releases arachidonic acid (AA) which is metabolized by cytochrome P450 (CYP) to EETs. EETs can diffuse to the vascular smooth muscle cell to cause relaxation and inhibit cell migration. EETs inhibit adhesion molecule expression on endothelial cells which decreases leukocyte adherence. EETs prevent platelet aggregation and adhesion. EETs promote angiogenesis by stimulating endothelial cell proliferation.

ICAM-1, the effect on VCAM-1 was the most pronounced. 11,12-EET was the most potent isomer causing 72% inhibition of TNF-α induced VCAM-1 expression. The IC_{50} for 11,12-EET-induced inhibition of VCAM-1 was 20 nM. 8,9-EET and 5,6-EET were less active, whereas 14,15-EET was without activity. Interestingly, 14,15-EET increased adherence of monocytes to endothelial cells suggesting a clear difference in activity between the EET regioisomers. The anti-inflammatory effect of EETs to decrease endothelial-leukocyte-adhesion has been confirmed in a number of subsequent cell and animal models (Falck et al., 2003b; Fleming et al., 2001b; Liu et al., 2005; Moshal et al., 2008; Pratt et al., 2002).

2. Mechanism of Action

The mechanism of action of EETs to inhibit monocyte and leukocyte adhesion is independent of membrane hyperpolarization. Inhibition of K_{Ca} channels with iberiotoxin or charybdotoxin blocked EET-induced vasodilation but did not block EET-induced inhibition of VCAM-1 expression (Node et al., 1999). Instead, EETs exert their anti-inflammatory effects in the vasculature by inhibiting cytokine-induced nuclear factor-κB (NF-κB). The proinflammatory transcription factor, NF-κB, is essential for the induction of numerous inflammatory mediators such as CAMs, COX-2, and inducible (i)NOS. NF-κB is normally bound to an inhibitory protein IκB and maintained as an inactive NF-κB–IκB complex in the cytoplasm. Cytokines like TNF-α activate IκB kinase (IKK), which phosphorylates Ser32 and Ser36 of IκB. Following polyubiquitination of the diphosphorylated IκB, the protein is degraded by the 26S proteasome. The free NF-κB subunits RelA (p65) and p50 are translocated to the nucleus where they bind to target genes that encode proinflammatory proteins and consequently regulate their transcription. Node et al. showed that 11,12-EET repressed VCAM-1 expression by inhibiting κB *cis*-acting elements in the promoter region of the VCAM-1 gene. In cells stimulated with TNF-α, the nuclear accumulation of Rel A was prevented by the coadministration of 11,12-EET. Stimulation of endothelial cells with TNF-α caused a rapid and almost complete disappearance of IκB-α that was prevented by cotreatment with 11,12-EET, but not 14,15-EET. Elevated concentrations of homocysteine contribute to inflammation and endothelial dysfunction by a mechanism that involves activation of NF-κB. This pathway induces matrix metalloproteinase (MMP)-9 expression and activity. MMPs participate in extracellular matrix degradation and may regulate CAM adhesion. Incubation of murine aortic endothelial cells with increasing concentration of homocysteine decreased CYP2J2 expression and activated MMP-9 (Moshal et al., 2008). Homocysteine induced MMP-9 activation by increasing NF-κB–DNA binding. CYP transfection or exogenous addition of 8,9-EET (1 μM) attenuated homocysteine-induced MMP-9 activation. 8,9-EET also increased IκB-α levels and attenuated the nuclear accumulation of Rel A. Exogenous 11,12-EET (up to 3 μM) had no effect on

MMP activation. Activation of the nuclear receptor peroxisome proliferator-activated receptor γ (PPARγ) in cultured endothelial cells suppresses the NF-κ B-mediated expression of inflammatory proteins such as VCAM-1 and ICAM-1. In bovine aortic endothelial cells, all four EET regioisomers blocked TNF-α mediated NF-κB activation and this was prevented in cells pretreated with GW9662, an antagonist of PPARγ (Liu et al., 2005). Competition and direct binding assays revealed that EETs bind to the ligand-binding domain of PPARγ with K_d in the micromolar range (Cowart et al., 2002). The relationship between PPARγ and NF-κB in EET-induced changes in CAM expression needs further study.

3. *Therapeutic Potential*

Inflammation is a common characteristic of numerous cardiovascular diseases such as hypertension and atherosclerosis. Just as sEH inhibitors increase vasodilatory effects of EETs, these inhibitors have potential in the treatment of certain inflammatory disorders by enhancing anti-inflammatory effects of EETs. As described above, a number of studies have looked at the effect of sEH inhibitors in animal models of hypertension (Imig & Hammock, 2009). One consequence of chronic hypertension is renal vascular and glomerular injury. The response to injury involves upregulation of inflammatory proteins. In rats with ANG II-induced hypertension, sEH inhibitors decreased collagen expression in glomeruli and tubular cells. Macrophage infiltration was also reduced (Imig et al., 2002). Similarly, in diabetic Goto–Kakizaki rats the sEH inhibitor, AUDA, blocked inflammation independently of lowering blood pressure (Olearczyk et al., 2009). Targeted disruption of the sEH gene prevented both renal inflammation and injury in DOCA-salt hypertensive mice (Manhiani et al., 2009). Reduction in renal inflammation and injury was also seen in wild-type DOCA-salt mice treated with a sEH inhibitor. Macrophage infiltration, renal NF-κB activation, and monocyte chemoattractant protein-1 excretion were all reduced in the sEH knockout mice. Finally, inhibition of sEH enhanced the anti-inflammatory effects of aspirin and 5-lipoxygenase activation protein inhibitor, MK886, in a LPS-challenged murine model of inflammation (Liu et al., 2010). The relative effectiveness of sEH inhibitors versus EET agonists versus stable EET agonists with sEH inhibitory activity is an area in need of study.

Atherosclerosis is a dynamic and progressive process with endothelial dysfunction and inflammation of the vascular wall. A recent study investigated the role of sEH inhibition in the hyperlipidemic ApoE$^{-/-}$ mouse following cuff placement around the femoral artery or ligation of left common carotid artery (Ulu et al., 2008). AUDA treatment reduced neointimal formation only in the femoral cuff model, which typically exhibits a more pronounced inflammatory phenotype than does the carotid artery ligation model. In the femoral cuff model, mRNA expressions for the cytokine, Gro-β, and for the proinflammatory enzyme, COX-2, were significantly lower in sEH$^{-/-}$/ApoE$^{-/-}$ mice compared with ApoE$^{-/-}$ mice. However, there were no differences in macrophage infiltration or the cytokines, TNF-α, Gro-α, and MCP-1 in the two groups.

B. Platelet Activation

Inflammatory disorders are often associated with platelet activation (Smyth et al., 2009). It is well established that endothelial NO and PGI_2 inhibit platelet aggregation and activation. There is also experimental evidence to support EET-mediated antiplatelet effects (Fig. 4).

1. Experimental Evidence

Platelet aggregation of washed human platelets induced by arachidonic acid was inhibited by EET isomers (1–10 μM) with no evidence of stereospecificity (Fitzpatrick et al., 1986). However, 11,12-EET (up to 10 μM) had no effect on platelet aggregation stimulated with collagen, ADP, or a thrombin-receptor activating peptide (VanRollins, 1995). Using an *in vivo* model of platelet aggregation in pial arterioles of mice, 14,15-EET, but not 8,9-EET, delayed platelet aggregation (Krotz et al., 2003, 2010). The dose of 14,15-EET (0.3 mg/kg) producing the effect had no effect on blood flow. The effect of EETs on adhesion of washed human platelets to confluent HUVEC was examined under both static and flow conditions (Krotz et al., 2010). 11,12-EET (1 μM) decreased endothelial cell adhesion of platelets and also decreased P-selectin expression. In an endothelial cell line that overexpressed CYP2C9, bradykinin increased the production of 11,12-, 8,9-, and 14,15-EET compared to controls. Media from the cell incubations with bradykinin inhibited platelet adhesion to HUVECs. Platelet adhesion and rolling was assessed in the dorsal skinfold chamber model in the microcirculation of hamsters (Krotz et al., 2010). The CYP2C9 inhibitor, sulfaphenazole, enhanced platelet adhesion to the endothelium. The addition of 11,12-EET (10 μM) reversed the effect of CYP inhibition. The view of EETs as important endogenous antiplatelet factors remains to be elucidated in future studies. Studies will need to clarify whether the endothelium *in vivo* releases EETs in amounts sufficient to control platelet activation.

2. Mechanism of Action

The exact mechanism whereby EETs inhibit platelet aggregation has not been established. In human platelets, inhibition of aggregation by EETs was not associated with a decrease in the production of the proaggregatory, thromboxane (TX) A_2 (Fitzpatrick et al., 1986). However, in mice, 14,15-EET, but not 8,9-EET, decreased serum TXB_2 concentrations (VanRollins, 1995). In human platelets, 11,12-EET caused a concentration-dependent increase in NOS activity and stimulated nitrite production (Zhang et al., 2008a). While NO inhibits platelet aggregation, there is no direct evidence that NO mediates the inhibition of aggregation by EETs. Platelets express K_{Ca} channels and EET-mediated inhibition of platelet adhesion likely targets these channels. Cultured HUVECs expressing CYP2C9 and stimulated by bradykinin released a factor that hyperpolarized human platelets (Krotz et al., 2003, 2010). The inhibitor of CYP2C9, sulfaphenazole, blocked this hyperpolarization. Hyperpolarization was also prevented by pretreatment of platelets with charybdotoxin or apamin, inhibitors of SK_{Ca} and IK_{Ca} channels. EET-mediated inhibition of platelet adhesion to endothelial cells was prevented by charybdotoxin but not apamin. Exogenous EETs (1 μM) were also tested for direct effects on platelet membrane hyperpolarization. Pronounced hyperpolarizations occurred with 11,12-EET and 8,9-EET. 14,15-EET was less active. These studies suggest that the EETs hyperpolarize the platelet membrane resulting in reduced adhesion.

3. *Therapeutic Potential*

There are no studies in which sEH inhibitors or EET analogs have been tested for antiplatelet effects in disease models. However, there is evidence that the sEH inhibitors increase the action of aspirin (Liu et al., 2010). In certain cardiovascular diseases, the COX product, TXA_2, increases platelet activation. Low-dose aspirin is an effective antiplatelet drug that limits adverse cardiovascular events by blocking TXA_2 synthesis and activity. In the LPS-induced mouse model of inflammation, the metabolism of arachidonic acid by COX is increased. A sEH inhibitor, *t*-AUCB, decreased TXA_2 but not PGE_2 in the plasma of LPS-treated mice. Aspirin treatment inhibited both TXA_2 and PGE_2. If *t*-AUCB was coadministered with aspirin, there was a greater reduction in PGE_2 and TXA_2 than with aspirin alone suggesting that EETs increase the activity of aspirin. While it still needs to be directly tested experimentally, it is possible that sEH inhibitors (or EET analogs) may be effective antiplatelet drugs on their own. Alternatively, since sEH inhibitors increase aspirin effects, these two drugs could be combined allowing the use of aspirin in doses that limit aspirin-related side effects.

V. EETs and Angiogenesis

Angiogenesis is a process involving a tightly regulated interaction between a number of different signaling molecules, which results in the sprouting of endothelial cells from existing blood vessels. While complex, angiogenesis is crucial for all tissue growth, expansion, and repair. Arachidonic acid metabolites contribute to angiogenesis by modulating endothelial cell proliferation or migration and capillary formation. EETs are regarded as proangiogenic compounds (Michaelis & Fleming, 2006; Fig. 4) and therapeutic manipulation of their effects may be important in diseases like ischemic heart disease, ischemic stroke, and atherosclerosis.

A. Evidence

The first report that EETs may be mitogens was performed in cultured rat glomerular mesengial cells (Harris et al., 1992). Following a 24-h exposure to exogenously administered 14,15- or 8,9-EET (1–10 μM), mesengial cells had a significant increase in ^{3}H-thymidine incorporation. Subsequent studies in various types of vascular endothelial cells confirmed EETs as proangiogenic mediators. Munzenmaier and Harder initially observed that CYP epoxygenase products were necessary for cerebral microvascular endothelial cells to form tubular structures *in vitro* (Munzenmaier & Harder, 2000). In porcine coronary artery endothelial cells, CYP2C8 overexpression increased 11,12-EET production and enhanced endothelial cell number (Michaelis et al., 2005b). Furthermore, in HUVECs, overexpression of CYP2C9 increased cell number and proliferation (Michaelis et al., 2005b). The specific CYP2C9 inhibitor, sulfaphenazole prevented the proliferative response. Using an *in vivo* model of angiogenesis, a single high concentration (150 μM) of 14,15-EET increased functional vasculature in a Matrigel plug (Medhora et al., 2003). More recently, the ability of all four EETs to regulate endothelial cell proliferation *in vitro* and angiogenesis *in vivo* was investigated (Pozzi et al., 2005; Yang et al., 2009). All four EETs induced significant increases in pulmonary murine endothelial cell proliferation with 5,6-EET eliciting the greatest effect. None of the DHETs stimulated endothelial cell proliferation. 5,6- and 8,9-EET, but not 11,12- or 14,15-EET, increased cell migration and capillary tube formation. To test the angiogenic activity of 5,6- and 8,9-EET *in vivo*, inert sponges were implanted subcutaneously in the back of adult mice and were injected every other day with either vehicle, 5,6-EET, or 8,9-EET (50 μM). After 14 days, sponges injected with 5,6- or 8,9-EET showed increased vessel density compared with sponges injected with vehicle only, demonstrating clearly *in vivo de novo* vascularization. Stimulation of endothelial cells with vascular endothelial growth factor (VEGF) induced the expression of CYP2C and the generation of 11,12-EET. Pretreatment with the CYP inhibitor, miconazole prevented the increase in 11,12-EET. VEGF-induced endothelial cell tube formation was prevented by the EET antagonist, 14,15-EE-Z-E. Hypoxia enhances cell proliferation and EET production. Exposure of CYP2C8- or 2C9-transfected HUVECs to hypoxia increased endothelial cell migration and tube formation (Michaelis et al., 2005a, 2005b). These effects were blocked by the EET antagonist, 14,15-EE-Z-E. Similar findings were obtained in porcine coronary artery endothelial cells. Bovine retinal endothelial cells expressed CYP2C protein under basal conditions (Michaelis et al., 2008). Hypoxia enhanced CYP2C protein expression and EET formation. Treatment with CYP2C antisense or the EET antagonist suppressed hypoxia-induced cell migration and *in vitro* tube formation.

In contrast to the studies in endothelial cells, exogenously administered EETs inhibit rat aortic smooth muscle cell migration in response to growth factors (Sun et al., 2002). Overexpression of CYP2J2 in these cells attenuated migration, and this effect was prevented by the CYP inhibitors, SKF525A and clotrimazole, but not by the K_{Ca} channel blocker, charybdotoxin. In these studies, EETs had no effect on proliferation as measured by ^{3}H-thymidine incorporation.

The ability of EETs to stimulate endothelial cell proliferation and decrease vascular smooth muscle cell migration may relate to tissue-specific effects of EETs. While these differing responses validate the underlying complexity of angiogenesis, it also supports the idea that EETs might be protective not only by inhibiting complications of smooth muscle cell migration in atherosclerosis but also by promoting endothelial cells proliferation and neovascularization in ischemic tissues.

B. Mechanism

Among the various mitogenic signaling pathways, the activation of extracellular-signal-regulated kinases (ERK), p38 mitogen-activated protein kinase (MAPK), and phosphoinositide 3-kinase/protein kinase B (PI3K/Akt) play important roles in endothelial cell function. Numerous studies have investigated signaling pathways involved in EET-mediated angiogenesis (Deng et al., 2010; Spector, 2009; Spector & Norris, 2007). EET-induced proliferation in a renal epithelial cell line was dependent on the activation of Src kinase and initiation of a tyrosine kinase phosphorylation cascade (Chen et al., 1998). An essential step in the signaling mechanism of EETs involved transactivation of epidermal growth factor-like (EGF) receptor. 14,15-EET induces EGF receptor activation and downstream signaling by induction of pro-heparin-binding (HB)-EGF processing through activation of a metalloproteinase that causes release of soluble HB-EGF. The released HB-EGF then binds to EGF receptor and activates its intrinsic receptor tyrosine kinase, leading to autophosphorylation. Transactivation of the EGF receptor following the cleavage of HB-EGF is a common event in cell activation and the mitogenic response to a variety of substances, including agonists acting through GPCRs. In human endothelial cells overexpressing CYP2C9, EETs also stimulated proliferation via a mechanism that involved activation of the EGF receptor (Michaelis et al., 2003). PI3K/Akt is a downstream target of the EGF receptor and all four EET regioisomers increased Akt phosphorylation and cell proliferation in murine endothelial cells (Pozzi et al., 2005). Incubation of human coronary artery endothelial cells with 11,12-EET activated ERK1/2 and the p38 MAPK (Fleming et al., 2001a). Activation could be observed in response to 11,12-EET at concentrations as low as 3 nM, whereas 14,15-EET had no effect. The K_{Ca} channel blockers charybdotoxin and apamin had no effect on the activation of ERK1/2 by 11,12-EET. The phosphorylation of Erk1/2 and p38 MAP kinase was also enhanced in actively proliferating endothelial cells that overexpressed CYP2C. In pulmonary murine microvascular endothelial cells, all three major pathways involved in endothelial cell-mediated mitogenic function, namely PI3K/Akt, p38 MAPK, and ERK1/2, were activated by the EETs (Pozzi et al., 2005). 8,9-EET was the most potent activator of p38 MAPK. Proper cell growth involves a series of cell-cycle regulatory proteins, cyclins that exert their function by binding to and activating a number of specific cyclin-dependent kinases (CDKs). 11,12-EET activation of the PI3-K/Akt elicited the subsequent phosphorylation and inhibition of forkhead transcription factors, FOXO1 and FOXO3b (Potente et al., 2003). This caused a downregulation of the CDK inhibitor, $p27^{Kip1}$, and increase in cyclin D1 expression. Overexpression of CYP2C9 in endothelial cells increased EET production that was associated with activation of MAPK phosphatase-1, decrease in c-Jun N-terminal kinase (JNK) activity, and increase in cyclin D1 expression. 11,12-EET (1 μM) also induced the expression of MAPK phosphatase-1(Potente et al., 2002).

EETs activate the cAMP/PKA pathway in endothelial cells (Imig et al., 1999). In relationship to angiogenesis, enhanced proliferation in HUVECs overexpressing the CYP2C9 enzyme was blocked by a PKA inhibitor (Michaelis et al., 2005a). These cells had increased concentrations of cAMP, COX-2 protein, and 11,12-EET. CYP2C9 overexpression stimulated endothelial tube formation, which was attenuated by the COX-2 inhibitor celecoxib. Thus, COX-2 may also contribute to CYP2C9-induced angiogenesis. In the rat aortic vascular smooth muscle cells, 11,12-EET inhibited migration and increased intracellular cAMP levels and PKA activity (Sun et al., 2002). Inhibitors of cAMP and PKA reversed the antimigratory effects of 11,12-EET.

Mechanisms to explain EET-mediated effects on angiogenesis are dependent on not only the cell type used but also the specific EET isomer. More comprehensive investigation of the signaling pathways using EET-specific mimetics is a necessary focus for future studies.

C. Therapeutic Potential

Recognizing EETs as signaling molecules that contribute to endothelial cell proliferation and vascular smooth muscle cell migration suggests that therapeutic manipulation of the EET pathway may aid in the treatment of diseases in which angiogenesis is involved. For example, angiogenesis is the key step for recovery after ischemia. Studies suggest that EETs are involved in the healing process of ischemic stroke (Iliff & Alkayed, 2009; Iliff et al., 2010; Simpkins et al., 2009; Zhang et al., 2007, 2008b). Cerebral ischemia was induced by a permanent middle cerebral artery (MCA) occlusion in the stroke-prone spontaneously hypertensive rats (SHRSP; Simpkins et al., 2009). Rats treated with the sEH inhibitor, AUDA, for 6 weeks had marked reduction in percentage of the hemisphere damaged by ischemia compared to untreated rats. MCA vessel wall thickness, wall to lumen ratio, and collagen deposition were reduced by AUDA in the SHRSP compared to WKY rats. Microvessel density in the untreated adult SHRSP was 33% lower than in the adult WKY rats. However, AUDA increased the microvessel density by 20% in the SHRSP providing further support that therapeutic manipulation of EETs with sEH inhibitors may provide cerebral protection by reducing the area at risk.

In contrast to beneficial effects of angiogenesis in ischemia is the view that smooth muscle cell migration and proliferation contribute to pathological processes in atherosclerosis. EETs inhibit vascular smooth muscle cell migration in cultured cells but it is not known if an increase in EETs with sEH inhibitors would offer protection in an *in vivo* atherosclerotic model by specifically attenuating vascular smooth muscle cell migration. In ANG II-treated apoE-deficient mice treated with a sEH inhibitor, there was a reduction in atherosclerotic lesion (Ulu et al., 2008). This indicated that while EETs may limit lesion development, the mechanism is independent of effects on vascular smooth muscle cell migration or proliferation. This idea was challenged using the carotid ligation technique in SHRSP (Simpkins et al., 2010). Unlike the ANG II ApoE-mice, treatment with the sEH inhibitor reduced vascular remodeling responses in the SHRSP model compared to WKY controls. A similar reduction in vascular remodeling occurred in mice lacking the sEH gene and subjected to carotid artery ligation. Neointimal area to medial ratio was reduced by the sEH inhibitor of wild-type mice. Interestingly, in a different model of vascular remodeling in which a wire is used to mechanically denude the femoral artery endothelium, sEH inhibitor treatment of rats, or in mice lacking the sEH gene, there was no reduction in neointimal hyperplasia indicating the importance of the endothelium as the source of EETs. While not yet proven, this contrasting result in the two different models suggests a role of the endothelium in EET-mediated protective effects. Having stable EET analogs with sEH inhibition versus sEH inhibitors or EET agonists to use for future *in vivo* studies will provide a better understanding of the mechanisms of vascular remodeling in atherosclerosis.

VI. Conclusion

A considerable and diverse body of evidence supports the role of EETs in vascular function. Several key points can be made. EETs are endothelial-derived CYP metabolites of arachidonic acid and function as autocrine and paracrine mediators to regulate vascular tone, cell adhesion, platelet activation, and angiogenesis. There are four regioisomers of EET: 5,6-, 8,9-, 11,12-, and 14,15-EET. EETs are hydrolyzed by sEH to corresponding DHETs, and this results in a decrease in EET activity. As vasodilators, EETs act on vascular smooth muscle to cause K activation, hyperpolarization, and relaxation through a G protein-coupled mechanism. Regioisomer specificity of EET-induced relaxations varies with species and vascular bed. The ability of EETs (primarily 11,12-EET) to inhibit monocyte- and leukocyte-adhesion to endothelial cells occurs through a signaling pathway that inhibits NF-κB activation. Platelet aggregation and platelet activation are attenuated by EETs by mechanisms that are still undefined. Endothelial cell proliferation is enhanced, whereas vascular smooth muscle migration is inhibited by EETs. Mechanisms for EET-mediated angiogenesis are diverse, complex, and dependent on cell type and the EET regioisomer involved. Comparisons of the threshold concentration of exogenously applied EETs that exerts a specific biological response are shown in Fig. 5. Some of the effects of EETs occur at low physiological concentrations (hyperpolarization and relaxation of vascular smooth muscle) while others require much greater concentrations (monocyte adhesion and smooth muscle cell migration).

The current data suggest that a GPCR for EETs exists. Identifying and characterizing the putative EET receptor(s) are an important focus of future studies. This will aid in clarification of mechanism of action of the EETs and novel drug design. Structural modifications of 14,15-EET provided important information regarding the molecular components critical for dilator activity and resulted in the identification of 14,15-EET-specific analogs with antagonist properties. Structural requirements for the activity of other EET regioisomers are needed. Inhibitors of sEH have been used in a number of different *in vitro* and *in vivo* systems as a way to increase EET concentrations and

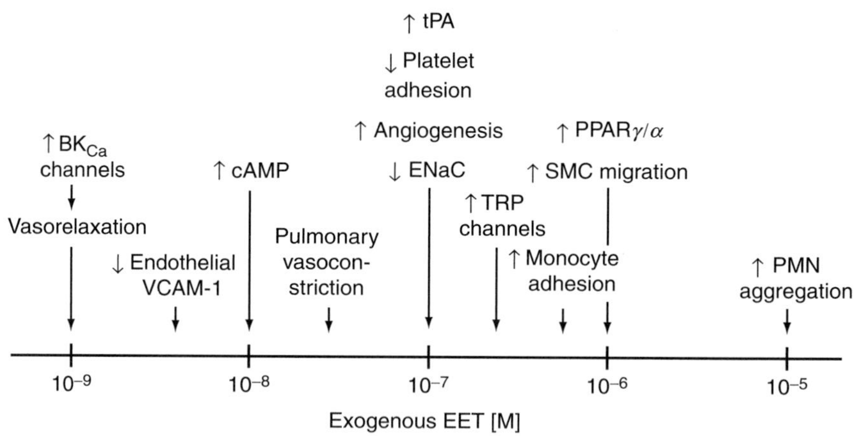

FIGURE 5 *Vascular effects of exogenous EETs.* The threshold concentration of EETs that produce the vascular effects is indicated on a concentration line. BK_{Ca}, large conductance, calcium-activated potassium; VCAM-1, vascular cell adhesion molecule 1; tPA, tissue plasminogen activator; ENaC, epithelial sodium channel; TRP, transient receptor potential; PMN, polymorphonuclear leukocyte.

prolong biological activity. The development of EET analogs that combine agonist activity with sEH inhibition is predicted to further advance the therapeutic applications of EETs in cardiovascular disease.

Acknowledgment

This study was supported by National Institutes of Health Grants HL-51055 (W. B. C.) and HL-093181 (S. L. P.).

Conflict of interest: The authors have no conflicts of interest to declare.

Abbreviations

ACH	acetylcholine
ADU	*N*-adamantyl-*N*′-dodecylurea
ANG II	angiotensin II
AUDA	12-(3-adamantane-1-yl-ureido)-dodecanoic acid
BK_{Ca}	large conductance, calcium-activated potassium
Ca	calcium
CAMs	cell adhesion molecules
CDKs	cyclin-dependent kinases
COX	cyclooxygenase
CYP	cytochrome P450
DHET	dihydroxyeicosatrienoic acids
DOCA	deoxycorticosterone acetate
EDHFs	endothelium-dependent hyperpolarizing factors
EDRF	endothelium-derived relaxing factor
EE-5Z-E	epoxyeicosa-5*Z*-enoic acid
EE-8Z-E	epoxyeicosa-8*Z*-enoic acid
EET	epoxyeicosatrienoic acid(s)
EGF	epidermal growth factor-like
E_m	membrane potential
ENaC	epithelial sodium channel
ERK	extracellular-signal-regulated kinases
G	guanine nucleotide-binding
GPCR	G protein-coupled receptor
HB	heparin binding
HUVECs	human umbilical vein endothelial cells
ICAM-1	intercellular adhesion molecule 1
IK_{Ca}	intermediate conductance, calcium-activated potassium
IL-1α	interleukin 1α
JNK	c-Jun N-terminal kinase
K	potassium

LNA	*N*(G)-monomethyl-L-arginine
LPS	lipopolysaccharide
MAPK	mitogen-activated protein kinase
MCA	middle cerebral artery
MMP	matrix metalloproteinase
mSI	methylsulfonamide
MS-PPOH	N-(methylsulfonyl)-2-(2-propynyloxy)-benzenehexamide
NCND	*N*-cyclohexyl-*N*-dodecyl
NF-κB	nuclear factor-κB
NO	nitric oxide
NOS	nitric oxide synthase
PGI_2	prostacyclin
PI3K/Akt	phosphoinositide 3-kinase/protein kinase B
PKA	protein kinase A
PMN	polymorphonuclear leukocyte
PPARγ	peroxisome proliferator-activated receptor
sEH	soluble epoxide hydrolase
SHR	spontaneously hypertensive rat
SHRSP	stroke-prone spontaneously hypertensive rats
SK_{Ca}	small conductance, calcium-activated potassium
TNF-α	tumor necrosis factor α
tPA	tissue plasminogen activator
TRP	transient receptor potential
TX	thromboxane
VCAM-1	vascular cell adhesion molecule 1
VEGF	vascular endothelial growth factor

References

Adeagbo, A. S. O., & Triggle, C. R. (1993). Varying extracellular [K]: A functional approach to separating EDHF- and EDNO-related mechanisms in perfused rat mesenteric arterial bed. *Journal of Cardiovascular Pharmacology*, *21*, 423–429.

Archer, S. L., Gragasin, F. S., Wu, X., Wang, S., McMurtry, S., Kim, D. H., et al. (2003). Endothelium-derived hyperpolarizing factor in human internal mammary artery is 11, 12-epoxyeicosatrienoic acid and causes relaxation by activating smooth muscle BKca channels. *Circulation*, *107*, 769–776.

Bellien, J., Favre, J., Iacob, M., Gao, J., Thuillez, C., Richard, V., et al. (2010). Arterial stiffness is regulated by nitric oxide and endothelium-derived hyperpolarizing factor during changes in blood flow in humans. *Hypertension*, *55*, 674–680.

Bellien, J., Thuillez, C., & Joannides, R. (2008a). Contribution of endothelium-derived hyperpolarizing factors to the regulation of vascular tone in humans. *Fundamental & Clinical Pharmacology*, *22*, 363–377.

Bellien, J., Thuillez, C., & Joannides, R. (2008b). Role of endothelium-derived hyperpolarizing factor in the regulation of radial artery basal diameter and endothelium-dependent dilatation in vivo. *Clinical and Experimental Pharmacology & Physiology*, *35*, 494–497.

Campbell, W. B., Falck, J. R., & Gauthier, K. (2001). Role of epoxyeicosatrienoic acids as endothelium-derived hyperpolarizing factor in bovine coronary arteries. *Medical Science Monitor: International Medical Journal of Experimental and Clinical Research*, *7*, 578–584.

Campbell, W. B., & Fleming, I. (2010). Epoxyeicosatrienoic acids and endothelium-dependent responses. *Pfluegers Archiv*, *459*, 881–895.

Campbell, W. B., Gebremedhin, D., Pratt, P. F., & Harder, D. R. (1996a). Identification of epoxyeicosatrienoic acids as endothelium-derived hyperpolarizing factors. *Circulation Research*, *78*, 415–423.

Campbell, W. B., Holmes, B. B., Falck, J. R., Capdevila, J. H., & Gauthier, K. M. (2006). Adenoviral expression of cytochrome P450 epoxygenase in coronary smooth muscle cells: Regulation of potassium channels by endogenous 14(S), 15(R)-EET. *The American Journal of Physiology*, *290*, H64–H71.

Campbell, W. B., Pratt, P. F., Gebremedhin, D., & Harder, D. R. (1996b). Epoxyeicosatrienoic acids are endothelium-derived hyperpolarizing and vasodilating factors in bovine coronary arteries. In P. M. Vanhoutte (Ed.), *Endothelium-Derived Hyperpolarizing Factor* (pp. 81–89). Amsterdam: Harwood Academic.

Capdevila, J., Chacos, N., Werringloer, J., Prough, R. A., & Estabrook, R. W. (1981). Liver microsomal cytochrome P450 and the oxidative metabolism of archidonic acid. *Proceedings of the National Academy of Sciences of the United States of America*, *78*, 5362–5366.

Carroll, M. A., Schwartzman, M., Capdevila, J., Falck, J. R., & McGiff, J. C. (1987). Vasoactivity of arachidonic acid epoxides. *European Journal of Pharmacology*, *138*, 281–283.

Chen, J.-K., Falck, J. R., Reddy, K. M., Capdevila, J., & Harris, R. C. (1998). Epoxyeicosatrienoic acids and their sulfonimide derivatives stimulate tyrosine phosphorylation and induce mitogenesis in renal epithelial cells. *The Journal of Biological Chemistry*, *273*, 29254–29261.

Chen, Y., Falck, J. R., Tuniki, V. R., & Campbell, W. B. (2009). 20-^{125}Iodo-14,15-epoxyeicosa-5Z-enoic acid: A high affinity radioligand used to characterize the epoxyeicosatrienoic acid antagonist binding site. *The Journal of Pharmacology and Experimental Therapeutics*, *331*, 1137–1145.

Chu, Z. M., Croft, K. D., Kingsbury, D. A., Falck, J. R., Reddy, K. M., & Beilin, L. J. (2000). Cytochrome P450 metabolites of arachidonic acid may be important mediators in angiotensin II-induced vasoconstriction in the rat mesentery in vivo. *Clin Science (London, England: 1979)*, *98*, 277–282.

Coats, P., Johnston, F., MacDonald, J., McMurray, J. J. V., & Hillier, C. (2001). Endothelium-derived hyperpolarizing factor: Identification and mechanism of action in human subcutaenous resistance arteries. *Circulation*, *103*, 1702–1708.

Cohen, R. A., & Vanhoutte, P. M. (1995). Endothelium-dependent hyperpolarization: Beyond nitric oxide and cyclic GMP. *Circulation*, *92*, 3337–3349.

Cowart, L. A., Wei, S., Hsu, M. H., Johnson, E. F., Krishna, U. M., Falck, J. R., et al. (2002). The CYP4A isoforms hydroxylate epoxyeicosatrienoic acids to form high affinity peroxisome proliferator-activated receptor ligands. *The Journal of Biological Chemistry*, *20*, 35105–35112.

Deng, Y., Theken, K. N., & Lee, C. R. (2010). Cytochrome P450 epoxygenases, soluble epoxide hydrolase, and the regulation of cardiovascular inflammation. *Journal of Molecular and Cellular Cardiology*, *48*, 331–341.

Ellis, E. F., Police, R. J., Yancy, L., McKinney, J. S., & Amruthesh, S. C. (1990). Dilation of cerebral arterioles by cytochrome P-450 metabolites of arachidonic acid. *The American Journal of Physiology*, *259*, H1171–H1177.

Falck, J. R., Kodela, R., Manne, R., Atcha, K. R., Puli, N., Dubasi, N., et al. (2009). 14,15-Epoxyeicosa-5,8,11-trienoic acid (14,15-EET) surrogates containing epoxide bioisosteres: Influence upon vascular relaxation and soluble epoxide hydrolase inhibition. *Journal of Medicinal Chemistry*, *52*, 5069–5075.

Falck, J. R., Krishna, U. M., Reddy, Y. K., Kumar, P. S., Reddy, K. M., Hittner, S. B., et al. (2003a). Comparison of the vasodilatory properties of 14,15-EET analogs: Structural requirements for dilation. *The American Journal of Physiology*, *284*, H337–H349.

Falck, J. R., Reddy, L. M., Reddy, Y. K., Bondlela, M., Krishna, U. M., Ji, Y., et al. (2003b). 11,12-Epoxyeicosatrienoic acid (11,12-EET): Structural determinants for inhibition of TNF-alpha-induced VCAM-1 expression. *Bioorganic & Medicinal Chemistry Letters*, *13*, 4011–4014.

Fang, X., Weintraub, N. L., Oltman, C. L., Stoll, L. L., Kaduce, T. L., Harmon, S., et al. (2002). Human coronary endothelial cells convert 14,15-EET to a biologically active chain-shortened epoxide. *American Journal of Physiology. Heart and Circulatory Physiology*, *283*, H2306–H2314.

Feletou, M., & Vanhoutte, P. M. (2006). Endothelium-derived hyperpolaizing factor. Where are we now? *Arteriosclerosis, Thrombosis, and Vascular Biology*, *26*, 1215–1225.

Fisslthaler, B., Popp, R., Kiss, L., Potente, M., Harder, D. R., Fleming, I., et al. (1999). Cytochrome P450 2C is an EDHF synthase in coronary arteries. *Nature*, *401*, 493–497.

Fisslthaler, B., Popp, R., Michaelis, U. R., Kiss, L., Fleming, I., & Busse, R. (2001). Cyclic stretch enhances the expression and activity of coronary endothelium-derived hyperpolarizing factor synthase. *Hypertension*, *38*, 1427–1432.

Fitzpatrick, F. A., Ennis, M. D., Baze, M. E., Wynalda, M. A., McGee, J. E., & Liggett, W. (1986). Inhibition of cyclooxygenase activity and platelet aggregation by epoxyeicosatrienoic acids. *The Journal of Biological Chemistry*, *261*, 15334–15338.

Fleming, I., & Busse, R. (2006). Endothelium-derived epoxyeicosatrienoic acids and vascular function. *Hypertension*, *47*, 629–633.

Fleming, I., Fisslthaler, B., Michaelis, U. R., Kiss, L., Popp, R., & Busse, R. (2001a). The coronary endothelium-derived hyperpolarizing factor (EDHF) stimulates multiple signalling pathways and proliferation of vascular cells. *Pfluegers Archiv*, *442*, 511–518.

Fleming, I., Michaelis, U. R., Bredenkotter, D., Fisslthaler, B., Dehghani, F., Brandes, R. P., et al. (2001b). Endothelium-derived hyperpolarizing factor synthase (cytochrome P450 2C9) is a functionally significant source of reactive oxygen species in coronary arteries. *Circulation Research*, *88*, 44–51.

Fleming, I., Rueben, A., Popp, R., Fisslthaler, B., Schrodt, S., Sander, A., et al. (2007). Epoxyeicosatrienoic acids regulate Trp-channel-dependent Ca signaling and hyperpolariztion in endothelial cells. *Arteriosclerosis, Thrombosis, and Vascular Biology*, *27*, 2612–2618.

Furchgott, R. F., & Vanhoutte, P. M. (1989). Endothelium-derived relaxing and contracting factors. *The FASEB Journal*, *3*, 2007–2018.

Furchgott, R. F., & Zawadzki, J. W. (1980). The obligatory role of endothelial cells in the relaxation of arterial smooth muscle by acetylcholine. *Nature*, *288*, 373–376.

Gauthier, K. M., & Campbell, W. B. (2006). Epoxyeicosatrienoic acids mediate flow-induced dilation of bovine small coronary arteries. *Hypertension*, *48*, e68.

Gauthier, K. M., Deeter, C., Krishna, U. M., Reddy, Y. K., Bondlela, M., Falck, J. R., et al. (2002). 14,15-Epoxyeicosa-5(Z)-enoic acid: A selective epoxyeicosatrienoic acid antagonist that inhibits endothelium-dependent hyperpolarization and relaxation in coronary arteries. *Circulation Research*, *90*, 1028–1036.

Gauthier, K. M., Edwards, E. M., Falck, J. R., Reddy, D. S., & Campbell, W. B. (2005). 14,15-Epoxyeicosatrienoic acid represents a transferable endothelium-dependent relaxing factor in bovine coronary arteries. *Hypertension*, *45*, 666–671.

Gauthier, K. M., Falck, J. R., Reddy, L. M., & Campbell, W. B. (2004). 14,15-EET analogs: Characterization of structural requirements for agonist and antagonist activity in bovine coronary arteries. *Pharmacological Research*, *49*, 515–524.

Gauthier, K. M., Jagadeesh, S. G., Falck, J. R., & Campbell, W. B. (2003a). 14,15-Epoxyeicosa-5(Z)-enoic-mSI: A 14, 15- and 5,6-EET antagonist in bovine coronary arteries. *Hypertension*, *42*, 555–561.

Gauthier, K. M., Pratt, P., Falck, J. R., & Campbell, W. B. (2003b). Inhibition of bradykinin-induced relaxations by an epoxeicosatrienoic acid antagonist: 14,15-epoxyeicosa-5Z-monoenoic acid. In P. M. Vanhoutte (Ed.), *EDHF 2002* (pp. 325–331). New York: Taylor and Francis.

Halcox, J. P. J., Narayanan, S., Crames-Joyce, L., Mincemoyer, R., & Quyyumi, A. A. (2001). Characterization of endothelium-derived hyperpolarizing factor in the human forearm microcirculation. *The American Journal of Physiology*, *280*, H2470–H2477.

Harris, R. C., Homma, T., Jacobson, H. R., & Capdevila, J. (1992). Epoxyeicosatrienoic acids activate Na/H exchange and are mitogenic in cultured rat glomerular mesangial cells. *Journal of Cellular Physiology*, *144*, 429–437.

Hatoum, O. A., Gauthier, K. M., Binion, D. G., Miura, H., Telford, G., Otterson, M. F., et al. (2005). Novel mechanism of vasodilation in inflammatory bowel disease. *Arteriosclerosis, Thrombosis, and Vascular Biology*, *25*, 2355–2361.

Hecker, M., Bara, A. T., Bauersachs, J., & Busse, R. (1994). Characterization of endothelium-derived hyperpolarizing factor as a cytochrome P450-derived arachidonic acid metabolite in mammals. *Journal de Physiologie*, *481*, 407–414.

Hercule, H. C., Schunck, W.-H., Gross, V., Seringer, J., Leung, F. P., Weldon, S. M., et al. (2009). Interaction between P450 eicosanoids and nitric oxide in the control of arterial tone in mice. *Arteriosclerosis, Thrombosis, and Vascular Biology*, *29*, 54–60.

Honing, M. L. H., Smits, P., Morrison, P. J., & Rabelink, T. J. (2000). Bradykinin-induced vasodilation of human forearm resistance vessels is primarily mediated by endothelium-dependent hyperpolarization. *Hypertension*, *35*, 1314–1318.

Huang, A., Sun, D., Jacobson, A., Carroll, M. A., Falck, J. R., & Kaley, G. (2005). Epoxyeicosatrienoic acids are released to mediate shear stress-dependent hyperpolarization of arteriolar smooth muscle. *Circulation Research*, *96*, 376–383.

Iliff, J. J., & Alkayed, N. J. (2009). Soluble epoxide hydrolase inhibition: Targeting multiple mechanisms of ischemic brain injury with a single agent. *Future Neurology*, *4*, 179–199.

Iliff, J. J., Jia, J., Nelson, J., Goyagi, T., Klaus, J., & Alkayed, N. J. (2010). Epoxyeicosanoid signaling in CNS function and disease. *Prostaglandins & Other Lipid Mediators*, *91*, 68–84.

Imig, J. D., & Hammock, B. D. (2009). Soluble epoxide hydrolase as a therapeutic target for cardiovascular diseases. *Nature Reviews Drug Discovery*, *8*, 794–805.

Imig, J. D., Inscho, E. W., Deichmann, P. C., Reddy, K. M., & Falck, J. R. (1999). Afferent arteriolar vasodilation to the sulfonimide analog of 11,12-epoxyeicosatrienoic acid involves protein kinase A. *Hypertension*, *33*, 408–413.

Imig, J. D., Zhao, X., Capdevila, J., Morisseau, C., & Hammock, B. D. (2002). Soluble epoxide hydrolase inhibition lowers blood pressure in angiotensin II hypertension. *Hypertension*, *39*, 690–694.

Imig, J. D., Zhao, X., Zaharis, C. Z., Olearczyk, J. L., Pollock, D. M., Newman, J. W., et al. (2005). An orally active epoxide hydrolase inhibitor lowers blood pressure and provides renal protection in salt-sensitive hypertension. *Hypertension*, *46*, 975–981.

Katz, S. D., & Krum, H. (2001). Acetylcholine-mediated vasodilation in the forearm circulation of patients with heart failure: Indirect evidence for the role of endothelium-derived hyperpolarizing factor. *The American Journal of Cardiology*, *87*, 1089–1092.

Kemp, B. K., & Cocks, T. M. (1997). Evidence that mechanisms dependent and independent of nitric oxide mediate endothelium-dependent relaxation to bradykinin in human small resistance-like coronary arteries. *British Journal of Pharmacology*, *120*, 757–762.

Kerseru, B., Barbosa-Sicard, E., Popp, R., Fisslthaler, B., Dietrich, A., Gudermann, T., et al. (2008). Epoxyeicosatrienoic acids and the soluble epoxide hydrolase are determinants of pulmonary artery pressure and the acute hypoxic pulmonary vasoconstriction response. *The FASEB Journal*, *22*, 4306–4315.

Krotz, F., Hellwig, N., Burkle, M. A., Lehrer, S., Riexinger, T., Mannell, H., et al. (2010). A sulfaphenazole-sensitive EDHF opposes platelet-endothelium interactions in vitro and in the hamster microcirculation in vivo. *Cardiovascular Research*, *85*, 542–550.

Krotz, F., Riexinger, T., Buerkle, M. A., Nithipatikom, K., Gloe, T., Sohn, H. Y., et al. (2003). Membrane potential-dependent inhibition of platelet adhesion to endothelial cells by epoxyeicosatrienoic acids. *Arteriosclerosis, Thrombosis, and Vascular Biology*, *24*, 595–600.

Kume, H., Graziano, M. P., & Kotlikoff, M. I. (1992). Stimulatory and inhibitory regulation of calcium-activated potassium channels by guanine nucleotide-binding proteins. *Proceedings of the National Academy of Sciences of the United States of America*, *89*, 11051–11055.

Larsen, B. T., Gutterman, D. D., Sato, A., Toyama, K., Campbell, W. B., Zeldin, D. C., et al. (2008). Hydrogen peroxide inhibits cytochrome p450 epoxygenases: Interaction between two endothelium-derived hyperpolarizing factors. *Circulation Research*, *102*, 59–67.

Larsen, B. T., Miura, H., Hatoum, O. A., Campbell, W. B., Hammock, B. D., Zeldin, D. C., et al. (2005). Epoxyeicosatrienoic and dihydroxyeicosatrienoic acids dilate human coronary arterioles via BKca channels: Implications for soluble epoxide hydrolase inhibition. *The American Journal of Physiology*, *290*, H491–H499.

Leffler, C. W., & Fedinec, A. L. (1997). Newborn piglet cerebral microvascular responses to epoxyeicosatrienoic acids. *The American Journal of Physiology*, *273*, H333–H338.

Lenasi, H. (2009). The role of nitric oxide- and prostacyclin-independent vasodilatation in the human cutaneous microcirculation: Effect of cytochrome P450 2C9 inhibition. *Clinical Physiology and Functional Imaging*, *29*, 263–270.

Lenasi, H., & Strucl, M. (2008). The effect of nitric oxide synthase and cyclooxygenase inhibition on cutaneous microvascular reactivity. *European Journal of Applied Physiology*, *103*, 719–726.

Li, P.-L., & Campbell, W. B. (1997). Epoxyeicosatrienoic acids activate potassium channels in coronary smooth muscle through guanine nucleotide binding protein. *Circulation Research*, *80*, 877–884.

Li, P., Chen, C.-L., Bortell, R., & Campbell, W. B. (1999). 11,12-Epoxyeicosatrienoic acid stimulates endogenous mono-ADP-ribosylation in bovine coronary arterial smooth muscle. *Circulation Research*, *85*, 349–356.

Li, P.-L., Zou, A.-P., & Campbell, W. B. (1997). Regulation of potassium channels in coronary arterial smooth muscle by endothelium-derived vasodilators. *Hypertension*, *29*, 262–267.

Liu, J. Y., Yang, J., Inceoglu, B., Qiu, H., Ulu, A., Hwang, S. H., et al. (2010). Inhibition of soluble epoxide hydrolase enhances the anti-inflammatory effects of aspirin and 5-lipoxygenase activation protein inhibitor in a murine model. *Biochemical Pharmacology*, *79*, 880–887.

Liu, Y., Zhang, Y., Schmelzer, K., Lee, T.-S., Fang, X., Zhu, Y., et al. (2005). The antiinflammatory effect of laminar flow: The role of PPARγ, epoxyeicosatrienoic acids and soluble epoxide hydrolase. *Proceedings of the National Academy of Sciences of the United States of America*, *102*, 16747–16752.

Loeb, A. L., Godeny, I., & Longnecker, D. E. (1997). Anesthetics alter relative contributions of NO and EDHF in rat cremaster muscle microcirculation. *The American Journal of Physiology*, *273*, H618–H627.

Loot, A. E., Popp, R., Fisslthaler, B., Vriens, J., Nilius, B., & Fleming, I. (2008). Role of cytochrome P450-dependent transient receptor potential V4 activation in flow-induced vasodilation. *Cardiovascular Research*, *80*, 445–452.

Manhiani, M., Quigley, J. E., Knight, S. F., Tasoobshirazi, S., Moore, T., Brands, M. W., et al. (2009). Soluble epoxide hydrolase gene deletion attenuates renal injury and inflammation with DOCA-salt hypertension. *American Journal of Physiology Renal Physiology*, *297*, F740–F748.

Matsuda, H., Hayashi, K., Wakino, S., Kubota, E., Honda, M., Tokuyama, H., et al. (2004). Role of endothelium-derived hyperpolarizing factor in ACE inhibitor-induced renal vasodilation in vivo. *Hypertension*, *43*, 603–609.

McCulloch, A. I., & Randall, M. D. (1998). Sex differences in the relative contributions of nitric oxide and EDHF to agonist-stimulated endothelium-dependent relaxations in the rat isolated mesenteric arterial bed. *British Journal of Pharmacology*, *123*, 1700–1706.

Medhora, M., Daniels, J., Mundey, K., Fisslthaler, B., Busse, R., Jacobs, E. R., et al. (2003). Epoxygenase-driven angiogenesis in human lung microvascular endothelial cells. *American Journal of Physiology. Heart and Circulatory Physiology*, *284*, H215–H224.

Michaelis, U. R., Falck, J. R., Schmidt, R., Busse, R., & Fleming, I. (2005a). Cytochrome P450 2C9-derived epoxyeicosatrienoic acids induce the expression of cyclooxygenase-2 in endothelial cells. *Arteriosclerosis, Thrombosis, and Vascular Biology*, *25*, 321–332.

Michaelis, U. R., Fisslthaler, B., Barbosa-Sicard, E., Falck, J. R., Fleming, I., & Busse, R. (2005b). Cytochrome P450 epoxygenases 2C8 and 2C9 are implicated in hypoxia-induced endothelial cell migration and angiogenesis. *Journal of Cell Science*, *118*, 5489–5498.

Michaelis, U. R., Fisslthaler, B., Medhora, M., Harder, D., Fleming, I., & Busse, R. (2003). Cytochrome P450 2C9-derived epoxyeicosatrienoic acids induce angiogenesis via cross-talk with the epidermal growth factor receptor (EGFR). *The FASEB Journal*, *17*, 770–772.

Michaelis, U. R., & Fleming, I. (2006). From endothelium-derived hyperpolarizing factor (EDHF) to angiogenesis: Epoxyeicosatrienoic acids (EETs) and cell signaling. *Pharmacology & Therapeutics*, *111*, 584–595.

Michaelis, U. R., Xia, N., Barbosa-Sicard, E., Falck, J. R., & Fleming, I. (2008). Role of cytochrome P450 2C epoxygenases in hypoxia-induced cell migration and angiogenesis in retinal endothelial cells. *Investigative Ophthalmology & Visual Science*, *49*, 1242–1247.

Miura, H., Wachtel, R. E., Liu, Y., Loberiza, F. R., Jr., Saito, T., Miura, M., et al. (2001). Flow-induced dilation of human coronary arterioles: Important role of Ca^{2+}-activated K^{+} channels. *Circulation*, *103*, 1992–1998.

Moreland, K. T., Procknow, J. D., Sprague, R. S., Iverson, J. L., Lonigro, A. J., & Stephenson, A. H. (2007). Cyclooxygenase (COX)-1 and COX-2 participate in 5,6-epoxyeicosatrienoic acid-induced contraction of rabbit intralobar pulmonary arteries. *The Journal of Pharmacology and Experimental Therapeutics*, *321*, 446–454.

Moshal, K. S., Zeldin, D. C., Sithu, S. D., Sen, U., Tyagi, N., Kumar, M., et al. (2008). Cytochrome P450 (CYP) 2J2 gene transfection attenuates MMP-9 via inhibition of NF-kappabeta in hyperhomocysteinemia. *Journal of Cellular Physiology*, *215*, 771–781.

Munzenmaier, D. H., & Harder, D. R. (2000). Cerebral microvascular endothelial cell tube formation: Role of astrocytic epoxyeicosatrienoic acid release. *The American Journal of Physiology*, *278*, H1163–H1167.

Nagao, T., Illiano, S., & Vanhoutte, P. M. (1992). Heterogenous distribution of endothelium-dependent relaxations resistant to N-nitro-L-arginine in rats. *The American Journal of Physiology*, *263*, H1090–H1094.

Nishikawa, Y., Stepp, D. W., & Chilian, W. M. (1999). In vivo location and mechanism of EDHF-mediated vasodilation in canine coronary microcirculation. *The American Journal of Physiology*, *277*, H1252–H1259.

Nishikawa, Y., Stepp, D. W., & Chilian, W. M. (2000). Nitric oxide exerts feedback inhibition on EDHF-induced coronary arteriolar dilation in vivo. *The American Journal of Physiology*, *279*, H459–H465.

Nithipatikom, K., Pratt, P. F., & Campbell, W. B. (2000). Determination of EETs using microbore liquid chromatography with fluorescence detection. *The American Journal of Physiology*, *279*, H857–H862.

Node, K., Huo, Y., Ruan, X., Yang, B., Spiecker, M., Ley, K., et al. (1999). Anti-inflammatory properties of cytochrome P450 epoxygenase-derived eicosanoids. *Science*, *285*, 1276–1279.

Node, K., Ruan, X.-L., Dai, J., Yang, S.-X., Graham, L., Zeldin, D. C., et al. (2001). Activation of Gas mediates induction of tissue-type plasminogen activator gene transcription by epoxyeicosatrienoic acids. *The Journal of Biological Chemistry*, *276*, 15983–15989.

Olearczyk, J. J., Quigley, J. E., Mitchell, B. C., Yamamoto, T., Kim, I. H., Newman, J. W., et al. (2009). Administration of a substituted adamantyl urea inhibitor of soluble epoxide hydrolase protects the kidney from damage in hypertensive Goto-Kakizaki rats. *Clinical Science (London, England: 1979)*, *116*, 61–70.

Oltman, C. L., Kane, N. L., Fudge, J. L., Weintraub, N. L., & Dellsperger, K. C. (2001). Endothelium-derived hyperpolarizing factor in coronary microcirculation: Responses to arachidonic acid. *The American Journal of Physiology*, *281*, H1553–H1560.

Oltman, C. L., Weintraub, N. L., VanRollins, M., & Dellsperger, K. C. (1998). Epoxyeicosatrienoic acids and dihydroxyeicosatrienoic acids are potent vasodilators in the canine coronary microcirculation. *Circulation Research*, *83*, 932–939.

Parkington, H. C., Chow, J. A., Evans, R. G., Coleman, H. A., & Tare, M. (2002). Role for endothelium-derived hyperpolarizing factor in vascular tone in rat mesenteric and hindlimb circulations in vivo. *Journal de Physiologie*, *542*, 929–937.

Pober, J. S., & Sessa, W. C. (2007). Evolving functions of endothelial cells in inflammation. *Nature Reviews. Immunology*, *7*, 803–815.

Pomposiello, S. I., Quilley, J., Carroll, M. A., Falck, J. R., & McGiff, J. C. (2003). 5, 6-epoxyeicosatrienoic acid mediates the enhanced renal vasodilation to arachidonic acid in the SHR. *Hypertension*, *42*, 548–554.

Popp, R., Fleming, I., & Busse, R. (1998). Pulsatile stretch in coronary arteries elicits release of endothelium-derived hyperpolarizing factor: A modulator of arterial compliance. *Circulation Research*, *82*, 696–703.

Potente, M., Fisslthaler, B., Busse, R., & Fleming, I. (2003). 11,12-Epoxyeicosatrienoic acid-induced inhibition of FOXO factors promotes endothelial proliferation by down-regulating p27Kip1. *The Journal of Biological Chemistry*, *278*, 29619–29625.

Potente, M., Michaelis, U. R., Fisslthaler, B., Busse, R., & Fleming, I. (2002). Cytochrome P450 2C9-induced endothelial cell proliferation involves induction of mitogen-activated protein (MAP) kinase phosphatase-1, inhibition of the c-Jun N-terminal kinase, and up-regulation of cyclin D1. *The Journal of Biological Chemistry*, *277*, 15671–15676.

Pozzi, A., Macias-Perez, I., Abair, T., Wei, S., Su, Y., Zent, R., et al. (2005). Characterization of 5,6- and 8,9-epoxyeicosatrienoic acids as potent in vivo angiogenic lipids. *The Journal of Biological Chemistry*, *280*, 27138–27146.

Pratt, P. F., Rosolowsky, M., & Campbell, W. B. (2002). Effect of epoxyeicosatrienoic acids on polymorphonuclear leukocyte function. *Life Sciences*, *70*, 2521–2533.

Proctor, K. G., Falck, J. R., & Capdevila, J. (1987). Intestinal vasodilation by epoxyeicosatrienoic acids: Arachidonic acid metabolites produced by a cytochrome P450 monooxygenase. *Circulation Research*, *60*, 50–59.

Revermann, M. (2010). Pharmacological inhibition of the soluble epoxide hydrolase-from mouse to man. *Current Opinion in Pharmacology*, *10*, 173–178.

Rosolowsky, M., & Campbell, W. B. (1993). Role of PGI2 and EETs in the relaxation of bovine coronary arteries to arachidonic acid. *The American Journal of Physiology*, *264*, H327–H335.

Rosolowsky, M., & Campbell, W. B. (1996). Synthesis of hydroxyeicosatetraenoic acids (HETEs) and epoxyeicosatrienoic acids (EETs) by cultured bovine coronary artery endothelial cells. *Biochimica et Biophysica Acta*, *1299*, 267–277.

Scornik, F. S., Codina, J., Birnbaumer, L., & Toro, L. (1993). Modulation of coronary smooth muscle Kca channels by Gsa independent of phosphorylation by protein kinase A. *The American Journal of Physiology*, *265*, H1460–H1465.

Scotland, R. S., Madhani, M., Chauhan, S. D., Moncada, S., Andresen, J., Nilsson, H., et al. (2005). Investigation of vascular responses in endothelial nitric oxide synthase/cyclooxygenase-1 double knockout mice: Key role for endothelium-derived hyperpolarizing factor in the regulation of blood pressure in vivo. *Circulation*, *111*, 796–803.

Simpkins, A. N., Rudic, R. D., Roy, S., Tsai, H. J., Hammock, B. D., & Imig, J. D. (2010). Soluble epoxide hydrolase inhibition modulates vascular remodeling. *American Journal of Physiology. Heart and Circulatory Physiology*, *298*, H795–H806.

Simpkins, A. N., Rudic, R. D., Schreihofer, D. A., Roy, S., Manhiani, M., Tsai, H. J., et al. (2009). Soluble epoxide inhibition is protective against cerebral ischemia via vascular and neural protection. *The American Journal of Pathology*, *174*, 2086–2095.

Smyth, S. S., McEver, R. P., Weyrich, A. S., Morrell, C. N., Hoffman, M. R., Arepally, G. M., et al. (2009). Platelet functions beyond hemostasis. *Journal of Thrombosis and Haemostasis*, *7*, 1759–1766.

Snyder, G. D., Krishna, U. M., Falck, J. R., & Spector, A. A. (2002). Evidence for a membrane site of action for 14,15-EET on expression of aromatase in vascular smooth muscle. *The American Journal of Physiology*, *283*, H1936–H1942.

Spector, A. A. (2009). Arachidonic acid cytochrome P450 epoxygenase pathway. *Journal of Lipid Research*, *50*(Suppl), S52–S56.

Spector, A. A., Fang, X., Snyder, G. D., & Weintraub, N. L. (2004). Epoxyeicosatrienoic acids (EETs): Metabolism and biochemical function. *Progress in Lipid Research*, *43*, 55–90.

Spector, A. A., & Norris, A. W. (2007). Action of epoxyeicosatrienoic acids on cellular function. *The American Journal of Physiology*, *292*, C996–C1012.

Sudhahar, V., Shaw, S., & Imig, J. D. (2010). Epoxyeicosatrienoic acid analogs and vascular function. *Current Medicinal Chemistry*, *17*, 1181–1190.

Sun, J., Sui, X.-X., Bradbury, A., Zeldin, D. C., Conte, M. S., & Liao, J. K. (2002). Inhibition of vascular smooth muscle cell migration by cytochrome P450 epoxygenase-derived eicosanoids. *Circulation Research*, *90*, 1020–1027.

Taddei, S., Varsari, D., Cipriano, A., Ghiadoni, L., Glaetta, F., Franzoni, F., et al. (2006). Identification of a cytochrome P450 2C9-derived endothelium-derived hyperpolarizing factor in human essential hypertensive patients. *Journal of the American College of Cardiology*, *48*, 508–515.

Tagawa, H., Shimokawa, H., Tagawa, T., Kuroiwa-Matsumoto, M., Hirooka, Y., & Takeshita, A. (1997). Short-term estrogen augments both nitric oxide-mediated and non-nitric oxide-mediated endothelium-dependent forearm vasodilation in postmenopausal women. *Journal of Cardiovascular Pharmacology*, *30*, 481–488.

Thomsen, K., Rubin, I., & Lauritzen, M. (2000). In vivo mechanisms of acetylcholine-induced vasodilation in rat sciatic nerve. *American Journal of Physiology. Heart and Circulatory Physiology*, *279*, H1044–H1054.

Ulu, A., Davis, B. B., Tsai, H. J., Kim, I. H., Morisseau, C., Inceoglu, B., et al. (2008). Soluble epoxide hydrolase inhibitors reduce the development of atherosclerosis in apolipoprotein e-knockout mouse model. *Journal of Cardiovascular Pharmacology*, *52*, 314–323.

VanRollins, M. (1995). Epoxygenase metabolites of docosahexaenoic and eicosapentaenoic acids inhibit platelet aggregation at concentrations below those affecting thromboxane synthesis. *The Journal of Pharmacology and Experimental Therapeutics*, *274*, 798–804.

Virdis, A., Cetani, F., Giannarelli, C., Banti, C., Ghiadoni, L., Ambrogini, E., et al. (2010). The sulfaphenazole-sensitive pathway acts as a compensatory mechanism for impaired nitric oxide availability in patients with primary hyperparathyroidism. Effect of surgical treatment. *Journal of Clinical Endocrinology & Metabolism*, *95*, 920–927.

Watanabe, S., Yashiro, Y., Mizuno, R., & Ohhashi, T. (2005). Involvement of NO and EDHF in flow-induced vasodilation in isolated hamster cremasteric arterioles. *Journal of Vascular Research*, *42*, 137–147.

Welsh, D. G., & Segal, S. S. (2000). Role of EDHF in conduction of vasodilation along hamster cheek pouch arterioles in vivo. *American Journal of Physiology. Heart and Circulatory Physiology*, *278*, H1832–H1839.

Weston, A. H., Feletou, M., Vanhoutte, P. M., Falck, J. R., Campbell, W. B., & Edwards, G. (2005). Bradykinin-induced, endothelium-dependent responses in porcine coronary arteries: Involvement of potassium channel activation and epoxyeicosatrienoic acids. *British Journal of Pharmacology*, *145*, 775–784.

White, R. M., Rivera, C. O., & Davison, C. A. (2000). Nitric oxide-dependent and-independent mechanisms account for gender differences in vasodilation to acetylcholine. *The Journal of Pharmacology and Experimental Therapeutics*, *292*, 375–380.

Widmann, M. D., Weintraub, N. L., Fudge, J. L., Brooks, L. A., & Dellsperger, K. C. (1998). Cytochrome P-450 pathway in acetylcholine-induced canine coronary microvascular vasodilation in vivo. *The American Journal of Physiology*, *274*, H283–H289.

Wong, P. Y.-K., Lai, P.-S., & Falck, J. R. (2000). Mechanism and signal transduction of 14(R), 15 (S)-epoxyeicosatrienoic acid (14,15-EET) binding in guinea pig monocytes. *Prostaglandins & Other Lipid Mediators*, *62*, 321–333.

Wong, P. Y.-K., Lai, P.-S., Shen, S.-Y., Belosludtsev, Y. Y., & Falck, J. R. (1997). Post-receptor signal transduction and regulation of 14(R), 15(S)-epoxyeicosatrienoic acid (14,15-EET) binding in U-937 cells. *Journal of Lipid Mediators and Cell Signalling*, *16*, 155–169.

Wong, P. Y., Lin, K. T., Yan, Y. T., Ahern, D., Iles, J., Shen, S. Y., et al. (1993). 14(R),15(S)-Epoxyeicosatrienoic acid receptor in guinea pig mononuclear cell membranes. *Journal of Lipid Mediators and Cell Signalling*, *6*, 199–208.

Woodman, O. L., & Boujaoude, M. (2004). Chronic treatment of male rats with daidzein and 17 beta-oestradiol induces the contribution of EDHF to endothelium-dependent relaxation. *British Journal of Pharmacology*, *141*, 322–328.

Wu, S., Moomaw, C. R., Tomer, K. B., Falck, J. R., & Zeldin, D. C. (1996). Molecular cloning and expression of CYP2J2, a human cytochrome P450 arachidonic acid epoxygenase highly expressed in heart. *The Journal of Biological Chemistry*, *271*, 3460–3468.

Yang, W., Holmes, B. B., Gopal, V. R., Kishore, R. V. K., Sangras, B., Yi, X.-Y., et al. (2007). Characterization of 14, 15-epoxyeicosatrienoyl-sulfonamides as 14,15-epoxyeicosatrienoic acid agonists: Use for studies of metabolism and ligand binding. *The Journal of Pharmacology and Experimental Therapeutics*, *321*, 1023–1031.

Yang, W., Tuniki, V. R., Anjaiah, S., Falck, J. R., Hillard, C. J., & Campbell, W. B. (2008). Characterization of epoxyeicosatrienoic acid binding site in U937 membranes using a novel radiolabeled agonist, 20–125I-14,15-epoxyeicosa-8(Z)-enoic acid. *The Journal of Pharmacology and Experimental Therapeutics*, *324*, 1019–1027.

Yang, S., Wei, S., Pozzi, A., & Capdevila, J. H. (2009). The arachidonic acid epoxygenase is a component of the signaling mechanisms responsible for VEGF-stimulated angiogenesis. *Archives of Biochemistry and Biophysics*, *489*, 82–91.

Yu, Z., Xu, F., Huse, L. M., Morisseau, C., Draper, A. J., Newman, J. W., et al. (2000). Soluble epoxide hydrolase regulates hydrolysis of vasoactive epoxyeicosatrienoic acids. *Circulation Research*, *87*, 992–998.

Zeldin, D. C. (2001). Epoxygenase pathways of arachidonic acid metabolism. *The Journal of Biological Chemistry*, *276*, 36059–36062.

Zeldin, D. C., Dubois, R. N., Falck, J. R., & Capdevila, J. H. (1995). Molecular cloning, expression and characterization of an endogenous human cytochrome P450 arachidonic acid epoxygenase isoform. *Archives of Biochemistry and Biophysics*, *322*, 7686.

Zhang, L., Cui, Y., Geng, B., Zeng, X., & Tang, C. (2008a). 11,12-Epoxyeicosatrienoic acid activates the L-arginine/nitric oxide pathway in human platelets. *Molecular and Cellular Biochemistry*, *308*, 51–56.

Zhang, W., Koerner, I. P., Noppens, R., Grafe, M., Tsai, H. J., Morisseau, C., et al. (2007). Soluble epoxide hydrolase: A novel therapeutic target in stroke. *Journal of Cerebral Blood Flow and Metabolism*, *27*, 1931–1940.

Zhang, W., Otsuka, T., Sugo, N., Ardeshiri, A., Alhadid, Y. K., Iliff, J. J., et al. (2008b). Soluble epoxide hydrolase gene deletion is protective against experimental cerebral ischemia. *Stroke*, *39*, 2073–2078.

Zhu, D., Bousamra, M., 2nd, Zeldin, D. C., Falck, J. R., Townsley, M., Harder, D. R., et al. (2000). Epoxyeicosatrienoic acids constrict isolated pressurized rabbit pulmonary arteries. *American Journal of Physiology. Lung Cellular and Molecular Physiology*, *278*, L335–L343.

Siu Ling Wong, Wing Tak Wong, Xiao Yu Tian, Chi Wai Lau, and Yu Huang

Institute of Vascular Medicine, Li Ka Shing Institute of Health Sciences, and School of Biomedical Sciences, Chinese University of Hong Kong, Hong Kong SAR, China

Prostaglandins in Action: Indispensable Roles of Cyclooxygenase-1 and -2 in Endothelium-Dependent Contractions

Abstract

Endothelium regulates local vascular tone by means of releasing relaxing and contracting factors, of which the latter have been found to be elevated in vascular pathogenesis of hypertension, diabetes, hypercholesterolemia, and

Advances in Pharmacology, Volume 60

1054-3589/10 $35.00
10.1016/S1054-3589(10)60007-5

aging. Endothelium-derived contracting factors (EDCFs) are mainly metabolites of arachidonic acid generated by cyclooxygenase (COX), as vasodilatations in patients with hypertension, metabolic diseases, or advancing age are improved by acute treatment with COX inhibitor indomethacin. COX is presented in two isoforms, COX-1 and COX-2, with the former regarded as constitutive and the latter mainly expressed upon induction. Experiments with animal models of vascular dysfunctions, however, reveal that both isoforms have similar capacity to participate in endothelium-dependent contractions, with augmented expression and activity. COX-derived prostaglandin (PG) H_2, $PGF_{2\alpha}$, PGE_2, prostacyclin (PGI_2), and thromboxane A_2 (TxA_2) are the proposed EDCFs that mediate endothelium-dependent contractions via the activation of thromboxane–prostanoid (TP) receptor in various vascular beds from different species. Although COX inhibition seems to be a possible strategy in combating COX-associated vascular complications, the incidence of adverse cardiovascular effects of Vioxx has greatly antagonized this concept. Further review of COX inhibitors is required, especially toward the selectivity of coxibs and whether it directly inhibits prostacyclin synthase activity. Meanwhile, TP receptor antagonism may emerge as a therapeutic alternative to reverse prostanoid-mediated vascular dysregulations.

I. Introduction

Normal vascular tone is attained by a fine-tuning balance between relaxing and contracting factors, and endothelium is actively involved in their production and release. Endothelium-derived relaxing factors (EDRFs) are well defined to be nitric oxide (NO), endothelium-derived hyperpolarizing factor, and prostacyclin (PGI_2), with NO being the major vasodilator in conduit arteries. As for endothelium-derived contracting factors (EDCFs), their chemical identities are less clear because of their heterogeneity in different arteries among species. A large body of converging evidence at least suggests that the majority of EDCFs are cyclooxygenase (COX)-derived arachidonic acid metabolites, although arachidonic acid can also be converted to other vasoactive factors via lipoxygenase and cytochrome *P*450 monooxygenase as reviewed by Bogatcheva et al. (2005). In pathological states, the elevated COX activity results in an increased production of vasoconstrictors. The balance between EDRFs and EDCFs is disturbed which favors the action of the latter, leading to persistent vasoconstriction that contributes to enhanced endothelium-dependent contractions and blood pressure elevation (Fig. 1), of which the former is discussed in detail in this chapter.

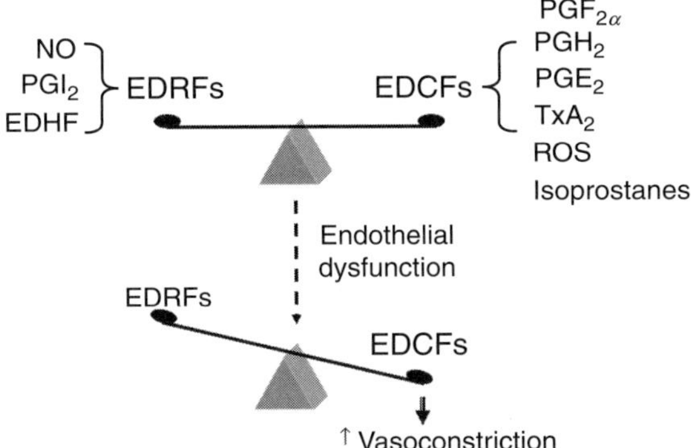

FIGURE 1 Balance of endothelium-derived relaxing factors (EDRFs) and contracting factors (EDCFs) in the regulation of vascular tone. When the production of EDRFs diminishes and/or EDCFs are overgenerated, the imbalance favors vasoconstrictions. NO, nitric oxide; PGI_2, prostacyclin; EDHF, endothelium-derived hyperpolarizing factor; $PGF_{2\alpha}$, prostaglandin $F_{2\alpha}$; PGH_2, prostaglandin H_2; PGE_2, prostaglandin E_2; TxA_2, thromboxane A_2; ROS, reactive oxygen species.

II. COX Isozymes

A. COX-Mediated Arachidonic Acid Metabolism

COX exists mainly in two isoforms, termed COX-1 and COX-2, localized in the endoplasmic reticulum or nuclear envelope (Smith et al., 2000). COX-1 is expressed constitutively in many cell types, including endothelial cells and vascular smooth muscle cells (Hla & Neilson, 1992), while COX-2 is generally regarded as an inducible enzyme that responds to stimuli such as shear stress (Topper et al., 1996), tumor necrosis factor-α, and lipopolysaccharide (Williams et al., 1999). COX first converts arachidonic acid to prostaglandin (PG) G_2 and H_2; the latter is then catalyzed enzymatically by respective synthases or isomerases into conventional PGs, namely PGD_2, PGE_2, $PGF_{2\alpha}$, PGI_2, and thromboxane A_2 (TxA_2; Fig. 2). These prostaglandins bind to their respective G protein-coupled receptors of D-prostanoid (DP), E-prostanoid (EP), F-prostanoid (FP), prostacyclin (IP) and thromboxane–prostanoid (TP). COX-1 and COX-2 share similar capacity to generate PGH_2 preceding its chemical conversion into the five prostaglandins (Smith et al., 1996). Although COX-2 is a predominant mediator for the systemic formation of PGI_2, recent studies show that local vascular PGI_2 production is derived mainly from endothelial COX-1 but not COX-2, which is not expressed in human arteries (Flavahan, 2007). The relative expression of COX isoforms may determine which isozyme is involved in prostaglandin production. In contrast to human arteries, COX-2 is constitutively coexpressed with COX-1 in the hamster aorta (Wong et al., 2009) and rat aorta, albeit the amount of COX-2 transcripts is significantly less than that of COX-1 in the latter preparation (Tang & Vanhoutte, 2008).

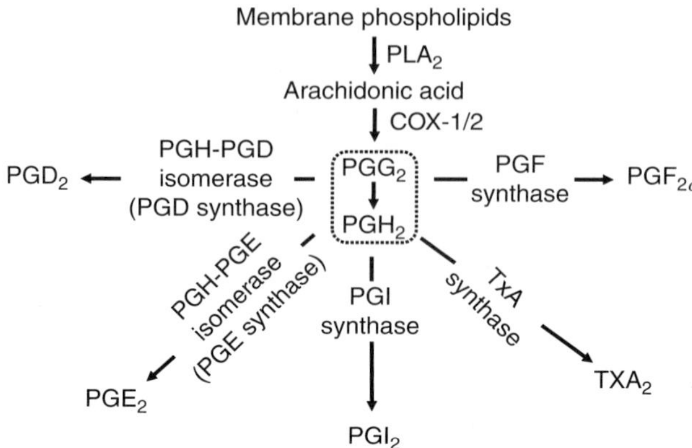

FIGURE 2 Arachidonic acid metabolism via cyclooxygenase (COX). Membrane phospholipids are converted to arachidonic acid by phospholipase A_2 (PLA_2). COX-1 or COX-2 then utilizes arachidonic acid as the substrate to generate prostaglandin (PG) G_2 and PGH_2, which is further transformed into various PGs and thromboxane A_2 (TxA_2) under the action of respective synthases.

The type(s) of prostaglandin generated also depends on the expression and activity of prostaglandin isozymes/synthases. For example, endothelial cells of spontaneously hypertensive rats (SHR) show an elevated level of prostacyclin synthase (PGIS) (Tang & Vanhoutte, 2008), thus PGH_2 is more readily converted to PGI_2 under the action of the overexpressed PGIS.

B. COX-Independent Production of Isoprostanes

Utilizing the same substrate as COX, isoprostanes are formed from arachidonic acid but via a COX-independent free radical-catalyzed mechanism in humans (Davì et al., 1997, 1999). Morrow et al. (1990) was the first to demonstrate the production of isoprotanes in humans through peroxidation of arachidonic acid. The level of isoprostanes has been taken as a reliable clinical biomarker for cardiovascular risk. Isoprostanes also actively modulate the vascular tone. For example, 15-F_{2t}-IsoP evokes TP receptor-mediated vasocontractions *in vitro* and *in vivo*, and such effect is potentiated in arteries without endothelium (Cracowski et al., 2001). In addition, isoprostanes can stimulate the release of TxA_2 and endothelin-1 from the endothelial cells (Daray et al., 2006; Fukunaga et al., 1995; Yura et al., 1999), which might in turn function as EDCFs.

C. Crosstalk Between EDCFs and NO

As the release of endothelium-derived relaxing and contracting factors is likely to occur simultaneously, is the net effect between vasodilatation and vasoconstriction a result of their mutual inhibition of production or chemical inactivation? This hypothesis is not experimentally supported at least in hamster arteries (unpublished data). The action of EDCFs and NO appears to be independent of each other, although the COX-2 expression can be modulated by NO as reviewed by Pérez-Sala and Lamas (2001). Endothelium-dependent contractions are best observed in the presence of NO synthase inhibitor N^G-nitro-L-arginine methyl ester (L-NAME; Auch-Schwelk et al., 1992) which prevents NO production. However, treatment with ODQ or NS2028, downstream inhibitors of soluble guanylate cyclase, also unmasked endothelium-dependent contractions (Yang et al., 2004), indicating that NO *per se* may not inhibit the release of EDCFs. It seems that the vasodilatory effect of NO overrides EDCF-induced vasoconstrictions not only under physiological situations but also in pathological states. This explains why inhibition of NO signaling permits endothelium-dependent contractions to appear. Actually, our unpublished results show that the release of EDCF is unaltered regardless of the presence or absence of L-NAME or NO donors such as sodium nitroprusside, further suggesting that there is unlikely to be a mutual interference between the release of EDCFs and NO in the endothelial cells.

III. COX-1 and COX-2 in Patho/Physiological States

Hypertension, diabetes, hypercholesterolemia, and aging are the four most prominent pathological states which lead to increased COX activity. The roles of COX-1 and COX-2 are indispensable in terms of which isoform is involved in which pathological condition due to the heterogeneity of vascular beds and species. The involvement of COX isoforms and the associated prostaglandins reported from key studies in relation to endothelium-dependent contractions are summarized in Table I.

A. Hypertension

As first demonstrated in the human brachial artery in hypertensive patients, nonselective COX inhibitor indomethacin improves vasodilatations (Taddei et al., 1997a), suggesting that COX is involved in the pathogenesis of hypertension. SHR is commonly used to elucidate the mechanism underlying endothelium-dependent contractions in hypertension. The aortae of SHR exhibit impaired acetylcholine-induced endothelium-dependent relaxations and exaggerated endothelium-dependent contractions compared with those from age-matched normotensive Wistar-Kyoto rats (WKY; Lüscher & Vanhoutte, 1986). Indomethacin normalizes the attenuated endothelium-dependent relaxations in SHR aortae and mesenteric arteries, suggesting that the reduced relaxations to acetylcholine are probably not caused by a reduced production of EDRFs, rather due to an enhanced level of the

TABLE I Indispensable Contributions of COX Isoforms and Prostanoids in Different Pathological States

Pathology	*Vascular bed/model of study*	*COX isoform*	*Prostanoids*	*References*
Hypertension	Aortae of SHR	COX-1	TxA_2, PGI_2	Gluais et al. (2005, 2006), Ge et al. (1995), Taddei and Vanhoutte (1993)
	Cultured endothelial cells from arteries of SHR	COX-2	$PGF_{2\alpha}$, 8-isoprostane	Alvarez et al. (2005), Garcia-Cohen et al. (2000)
Diabetes	Rabbit aortae exposed to high glucose	Unspecified	TxA_2, $PGF_{2\alpha}$, 15-HETE	Tesfamariam et al. (1990, 1995)
	Aortae of alloxan-induced diabetic rabbits	Unspecified	TxA_2, PGH_2, 15-HETE	Tesfamariam et al. (1989, 1995)
	Mesenteric arteries of alloxin-induced diabetic female Wistar rats	COX-2	$PGF_{2\alpha}$	Akamine et al. (2006)
	Mesenteric arteries of type 2 diabetic OLETF rats	COX-1/COX-2	TxA_2, PGE_2	Matsumoto et al. (2007)
	Aortae of streptozotocin-induced diabetic rats	Unspecified	TxA_2	Peredo et al. (2006)
	Femoral arteries of streptozotocin-induced diabetic rats	COX-1	TxA_2, PGE_2	Shi et al. (2007)
	Renal arteries from diabetic patients	COX-2	$PGF_{2\alpha}$	Wong et al. (2009)
Hypercholesterolemia	Nonobese and nondiabetic hypercholesterolemic rabbits	Likely COX-1	Not assayed	Jerez et al. (2008)
Aging	Femoral artery of rats	COX-1/COX-2	Not assayed	Shi et al. (2008)
	Aortae of hamsters	COX-2	$PGF_{2\alpha}$	Wong et al. (2009)

COX, cyclooxygenase; 15-HETE, 15-hydroxyeicosatetraenoic acid; OLETF, Otsuka Long-Evans Tokushima Fatty; PG, prostaglandin; SHR, spontaneously hypertensive rat; TxA_2, thromboxane A_2.

simultaneously released EDCFs which counteract the effects of relaxing factors on vascular smooth muscle cells (Lüscher & Vanhoutte, 1986; Lüscher et al., 1990). Not only does acetylcholine stimulate greater indomethacin-sensitive endothelium-dependent contractions in SHR aortae, but also the contractions are stimulated by serotonin (Auch-Schwelk & Vanhoutte, 1991), endothelin (Taddei & Vanhoutte, 1993), adenosine triphosphate (Mombouli & Vanhoutte, 1993), and calcium ionophore (Yang et al., 2004), indicating that augmented endothelium-dependent contraction is unlikely to be related to an increased sensitivity of muscarinic receptors to acetylcholine in arteries from hypertensive animals. More serious investigation into which COX isoform is responsible for the contractions began with the use of more selective COX inhibitors targeting either isoforms. Endothelium-dependent contractions in SHR arteries are preferentially inhibited by COX-1 inhibitor (Ge et al., 1995).

Oxygen radicals enhance endothelium-independent contractions which are sensitive to COX-1 inhibition and TP receptor antagonism in SHR aortae, as the contractions are prevented by superoxide dismutase mimetic (Yang et al., 2002), implying a possible link between reactive oxygen species (ROS) and the COX-1 activity and that ROS might be the upstream stimulator of COX-1, resulting in the release of constrictive prostaglandins. Since oxygen radical-induced vasoconstrictions are attenuated by the COX-1 inhibitor, it is unlikely that ROS or hydroxyl radicals *per se* act as an EDCF. Indeed, oxygen-derived free radicals do not trigger a release of prostaglandins (Auch-Schwelk et al., 1990). However, it is yet to be confirmed whether the release of ROS-catalyzed but COX-independent isoprostanes are involved in endothelium-dependent contractions. It is explainable if ROS trigger the production of COX-1-derived ROS, which in turn catalyze the formation of isoprostanes that cause vasoconstrictions via interaction with the TP receptor. Of note, Tang et al. (2007) demonstrated that in SHR endothelial cells indomethacin inhibits the acetylcholine-induced COX-mediated production of ROS.

The release of contracting factor(s) can be agonist-specific even in the same vascular bed. While acetylcholine stimulates the release of PGI_2 in SHR aortae (Gluais et al., 2005), endothelin increases the level of TxA_2 (Taddei & Vanhoutte, 1993) and calcium ionophore elevates the amount of both prostanoids (Gluais et al., 2006).

Although the aforementioned endothelium-dependent contractions in SHR aortae are mediated primarily by COX-1, evidence exists that SHR endothelial cells can synthesize and liberate COX-2-derived prostaglandins. For example, endothelial COX-2-derived $PGF_{2\alpha}$ and 8-isoprostane account for the augmented α-adrenoceptor-mediated contractions in SHR arteries (Alvarez et al., 2005), and vasoconstrictions in response to *tert*-butyl hydroperoxide, an oxidative stress from lipid peroxidation, are COX-2-mediated (Garcia-Cohen et al., 2000). In deoxycorticosterone acetate salt-induced hypertensive rats, COX-2 protein level is elevated and intraperitoneal administration of a selective COX-2 inhibitor NS-398 attenuates hypertension (Adeagbo et al., 2005).

B. Diabetes

The flow-mediated vasodilatation in the brachial artery is impaired in subjects with visceral obesity and diabetes (Hashimoto et al., 1998; Ihlemann et al., 2002), and this observation attracts attention as to whether diabetic vasculopathies are attributed to an enhanced release of contracting factors. A role of COX and the associated generation of vasoconstricting prostanoids, TxA_2 and $PGF_{2\alpha}$, under hyperglycemic conditions are suggested by the *in vitro* exposure of rabbit aortae to high glucose (Tesfamariam et al., 1990). Early studies showed a significant reduction in endothelium-dependent relaxations accompanied by augmented contractions in the aortae of alloxan-induced diabetic rabbits, and nonselective COX inhibition or TP receptor antagonism restores the impaired relaxations and abolishes the contractions. TxA_2, or its precursor PGH_2 (Tesfamariam et al., 1989), and 15-hydroxyeicosatetraenoic acid (15-HETE) are all proposed as the contracting factors in this preparation while the release of 15-HETE is also elevated in high glucose-treated aortae of normal rabbits (Tesfamariam et al., 1995).

Diabetic vasculopathies are not limited to aortae. Severely attenuated endothelium-dependent relaxations and enhanced contractions are also observed in mesenteric arteries of type 2 diabetic Otsuka Long-Evans Tokushima Fatty (OLETF) rats compared with the age-matched control Long-Evans Tokushima Otsuka (LETO) rats. Acetylcholine-stimulated production of TxA_2 and PGE_2 is increased, accompanied by an elevated expression of COX-1 and COX-2, while the activity of endothelial NO synthase and protein expression of extracellular superoxide dismutase are reduced in OLETF rats (Matsumoto et al., 2007). Chronic oral administration of eicosapentaenoic acid to the OLETF rats reverses the imbalance between vasoconstrictions and relaxations, possibly through restoring the NO production and suppressing the COX-2 activity via inhibiting extracellular signal-regulated kinase and nuclear factor kappa B (NF-κB) activation (Matsumoto et al., 2009b). Chronic treatment with metformin (an oral antidiabetic drug), pyrrolidine dithiocarbamate (a thiol antioxidant), or ozagrel (thromboxane synthase inhibitor) improves the NO- or EDHF-mediated relaxations and reduces the EDCF-mediated contractions by reducing the production of TxA_2, PGE_2, superoxide anion, and normalizing the NF-κB activity (Matsumoto et al., 2008, 2009a, 2009c). Mesenteric arteries from diabetic db/db mice at age over 12 weeks exhibit a greater transmural pressure-induced myogenic tone, which is sensitive to COX inhibition and TP receptor antagonism, indicating a positive attribution from vasoconstricting prostaglandins (Lagaud et al., 2001).

Streptozotocin induces diabetes in rats by destroying β-cells in the pancreas. Peredo et al. (2006) examined the profile of the released prostaglandins upon the induction of diabetic condition with streptozotocin and found that the prostanoid release in rat aortae remains unaltered in the first month of streptozotocin treatment. As the diabetic condition progresses, PGI_2 production starts to decline with an increased release of vasoconstricting metabolites in both aortae and mesenteric arteries, indicating that long-term diabetic condition can lead to unfavorable modification in prostanoid production favoring the vasoconstrictors. In the femoral arteries of streptozotocin-treated rats, the augmented calcium ionophore (A23187)-induced contraction is inhibited by indomethacin, TP receptor antagonist S18886 (terutroban), and thromboxane synthase inhibitor dazoxiben, implying a role of TxA_2; from 4 weeks onward after streptozotocin injection, COX-1 expression is increased and EP-1 receptor antagonists are required together with terutroban to prevent endothelium-dependent contractions (Shi et al., 2007), implying that PGE_2 may be another prostanoid generated by the dysfunctional endothelium. While the EDCF identity of PGE_2 is yet to be confirmed in the streptozotocin-induced diabetic rats, Rutkai et al. (2009) has recently shown that a 4-day oral administration of EP-1 receptor antagonist AH6809 markedly lowers systolic blood pressure in db/db mice and reduces the augmented pressure- and angiotensin II-induced tone of pressurized gracilis muscle arterioles from untreated db/db mice. Exogenous PGE_2 or selective EP-1 receptor agonist 17-phenyl-trinor-PGE_2 triggers greater contractions in the arterioles and EP-1 expression is higher in the aortae of db/db mice (Rutkai et al., 2009), which is indicative of a contributory role of PGE_2 in diabetic vascular dysfunction. In alloxan-induced diabetic Wistar rats, relaxation to acetylcholine in the perfused mesenteric arteriolar bed is attenuated accompanied by an increased generation of superoxide anions. Both harmful effects are inhibited by COX-2 inhibition with diclofenac (Akamine et al., 2006), suggesting that COX is not only involved in the formation of constrictive prostaglandins, $PGF_{2\alpha}$ in this case, but also actively exerting oxidative stress in the vasculatures.

Renal pathophysiology is common in diabetic patients (Kamgar et al., 2006; Mogensen & Schmitz, 1988). Renal COX-2 expression and activity are upregulated in streptozotocin-induced type 1 diabetic rats or Zucker type 2 diabetic fatty rats, while the latter also exhibits a reduced renal COX-1 expression and an increased urinary excretion of PGE_2 and TxB_2 (Komers et al., 2001, 2005). Preliminary studies on human renal arteries showed that acetylcholine evokes contractions in arteries from diabetic patients but not from nondiabetic subjects and that the acetylcholine-stimulated release of $PGF_{2\alpha}$ in these arteries is prevented by acute treatment with celecoxib, a specific and clinically in-use COX-2 inhibitor (Wong et al., 2009), thus supporting a critical role of COX-2 in diabetic renovascular pathologies.

C. Hypercholesterolemia

Hypercholesterolemic patients show a reduced NO-dependent flow-mediated vasodilatation (Mullen et al., 2001) and impairment in forearm vasodilatation in response to acetylcholine (Noon et al., 1998), while the latter can be partially restored with aspirin, suggesting that hypercholesterolemia may lead to an elevated generation of contracting prostaglandins that counteract the effects of vasodilators. *In vivo* production of aspirin-insensitive 8-epi-prostaglandin $F_{2\alpha}$ is significantly augmented in patients with hypercholesterolemia, which is suppressed by vitamin E supplementation (Davì et al., 1997). Taken together, hypercholesterolemia enhances COX-2 activity and/or poses oxidative stress; both situations result in the increased generation of vasoconstrictors.

Indomethacin but not free radical scavenger tempol reverses the impaired relaxations to acetylcholine in nonobese and nondiabetic hypercholesterolemic rabbits (Jerez et al., 2008). Although COX-2 inhibitor NS 398 does not improve instead attenuates the relaxations, the exact role of COX-1 needs to be further established. Jerez et al. pointed out that ROS is not involved in vasoconstrictor production in this animal model. The generation of vasoactive factors is not confined to the vascular wall in hypercholesterolemia. Supernatant containing products from polymorphonuclear leukocytes of rabbits fed with high cholesterol diet causes endothelium-dependent contractions and impairs acetylcholine-induced relaxations in the aortae of control rabbits (Hart et al., 1995), indicating a systemic activation of inflammatory responses. Since atherosclerosis is closely related to hypercholesterolemia, Shimokawa and Vanhoutte (1989) adopted a model of atherosclerosis in which endothelium of the porcine left anterior descending coronary artery was denuded by a balloon and then the pigs were fed high cholesterol diet for 10 weeks. They found that serotonin-induced contractions are greater and the endothelium-dependent component is sensitive to COX inhibition.

In contrast to the aforementioned observations of an increase in vasoconstrictions in hypercholesterolemia, there are frequent reports of a reduction in contractions from aortae of cholesterol-fed animal models (Cohen et al., 1988; Dam et al., 1997; Foudi et al., 2009; Pfister & Campbell, 1996; Van Diest et al., 1996). The reduction of contraction is neither due to vascular remodeling, as acute treatment of inhibitors can abrogate the attenuation, nor due to an elevated production of NO, since NO synthase inhibition is without effect (Pfister & Campbell, 1996). The role of arachidonic acid-metabolizing enzymes in these cases, however, remains obscure perhaps owing to different cholesterol treatment schemes. While Dam et al. proposed that the attenuation of contractions may result from a loss of endothelium-derived lipoxygenase constrictive products that mediate angiotensin II-induced contractions in rabbit aortae, Pfister and Campbell hypothesized a shunt of vasoactive factor production from COX-derived constrictive prostanoids to vasodilatory lipoxygenase or cytochrome *P*450 metabolites under hypercholesterolemic conditions. Foudi et al.'s results indicate a positive role for COX-2-derived PGI_2 in decreasing vasoconstrictions.

While COX-2 is considered harmful in hypercholesterolemia, discrete evidence does indicate that COX-2 may confer vascular protection when it is upregulated by specific treatments. Impaired acetylcholine-induced relaxations in the aortae of hypercholesterolemic rabbits are rescued by an 8-week treatment with 17β-estradiol, and this restoration is prevented by nimesulide, another COX-2 inhibitor (Ghanam et al., 2000), suggesting that COX-2-derived vasodilators may help to improve relaxations. This prompts a proposal that COX-2 has a dual role in the vascular system—whether it is beneficial or harmful actually depends on what the triggering agonists are.

D. Health and Youth

Pronounced endothelium-dependent contractions are most observable in animal models of diseases. If a delicate balance of vasoconstrictors and vasodilators exists in the normal vasculature, blockade of the major vasodilatory pathway should allow the appearance of endothelium-dependent contractions; so are endothelium-dependent contractions evident in healthy young animals? A recent study from Wong et al. (2009) demonstrated the acetylcholine-induced endothelium-dependent contractions in the aortae from healthy young hamsters. These contractions are abolished by COX-2 inhibitors but not by COX-1 inhibitors, and COX-2 expression is clearly detectable in quiescent aortic rings, localized mainly in the endothelium. COX-2-derived $PGF_{2\alpha}$ is identified as the most probable EDCF (Wong et al., 2009). Against a common belief that COX-1 is constitutive whereas COX-2 is inducible, Wong et al.'s study indicates a crucial and housekeeping role of endothelial COX-2 in attaining normal vascular tone (Fig. 3).

E. Aging

The brachial artery NO-dependent vasodilatation is usually taken as an index in evaluating vascular function in humans. Even among normotensive subjects, the dilatation decreases with advancing age. Infusion of indomethacin remarkably potentiates the vasodilatation (Taddei et al., 1997b), indicating an increased release of COX-derived products which counteract the dilatory effect of NO during aging. Hypertension facilitates the early onset of aging in the vascular wall, such that in patients with essential hypertension, impaired vasodilatations that are responsive to COX inhibition occur at a comparatively younger age (Taddei et al., 1997b). Over-production of COX-derived vasoconstrictors contributes prominently to the development of vascular dysfunctions during aging. Premature aging is also well documented

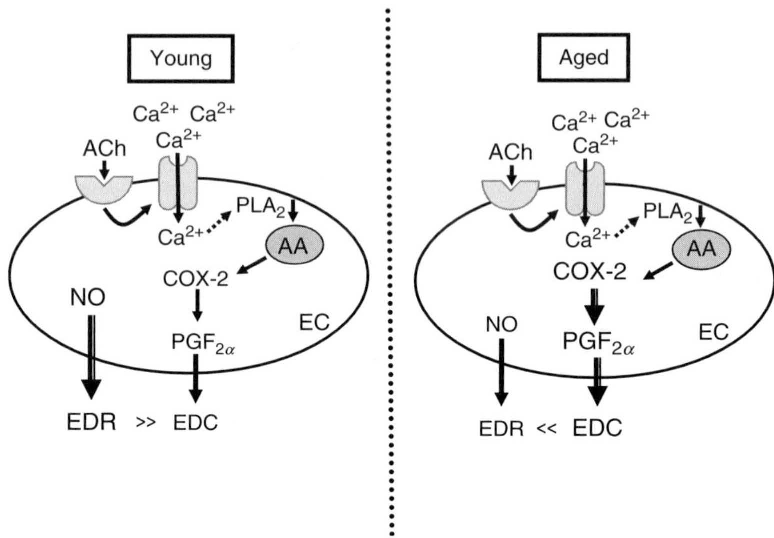

FIGURE 3 In the aorta from young hamsters, acetylcholine (ACh) triggers an influx of extracellular calcium (Ca^{2+}) in the endothelial cells (EC), and COX-2 is activated to produce prostaglandin (PG) $F_{2\alpha}$. During aging, COX-2 expression is augmented which leads to increased release of $PGF_{2\alpha}$, resulting in greater endothelium-dependent contraction (EDC). Since nitric oxide (NO) production is diminished, EDC overrides endothelium-dependent relaxation (EDR), such that EDC is observable without pretreatment of NO synthase inhibitor, which is required in the aortic rings from the young hamsters. PLA_2, phospholipase A_2; AA, arachidonic acid.

in animal experiments involving SHR (Abeywardena et al., 2002; Arribas et al., 1994; Fujii et al., 1993; Ibarra et al., 2006, Küng & Lüscher, 1995).

Since the vasodilatory effect of NO diminishes with age owing to a reduction of its bioavailability, endothelium-dependent contractions become more pronounced and readily observable in aged animals even in the absence of L-NAME (Wong et al., 2009; Fig. 3). In the aortae of aged rats, particularly SHR, the endothelium exerts less inhibition on the contractile responses toward 5-HT, resulting in greater contractions. While indomethacin does not modify 5-HT-induced contractions in arteries from younger WKY and in endothelium-denuded arteries from young SHR and aged WKY, it prevents the contractions in arteries with intact endothelium from young SHR and aged WKY. Surprisingly, 5-HT-induced contractions in endothelium-denuded arterial rings from aged SHR are inhibited by indomethacin, indicating that senescent vascular smooth muscles may be another source of COX-derived vasoconstricting factors (Ibarra et al., 2006).

Endothelium-dependent contractions to calcium ionophore are significantly greater in the femoral arteries from aged rats when compared with their younger counterparts. While these contractions are abolished by indomethacin, they can be partially inhibited by specific inhibitors to COX-1 and

COX-2, suggesting a joint activation of both COX isoforms during aging. Of note, the protein level of both COX-1 and COX-2 is augmented, whereas the latter is actually undetectable in arteries from younger rats (Shi et al., 2008), indicating an emerging role of COX-2 in aging-related vascular dysfunction, which may be responsible for the enhanced endothelium-dependent contractions in aged animals. Genomic studies on the endothelial cells show an increase in the mRNA levels of COX-1, COX-2, thromboxane synthase, PGF synthase, hematopoietic-type PGD synthase, and membrane PGE synthase-2 in aged rats, indirectly supporting the exaggerated importance of arachidonic acid metabolism through COX during aging. The study from Wong et al. (2009) further points to the enhanced contribution of COX-2 in endothelium-dependent contractions in hamster aortae, accompanied by an upregulation of COX-2 expression and augmented release of and vascular sensitivity to $PGF_{2\alpha}$, while the COX-1 expression remains unaltered.

F. Endothelial Regeneration

Endothelial cells may undergo apoptosis in response to pathological insults and senescence. Regeneration takes place to reendothelialize the injured area in an attempt to compensate for the lost of endothelial function. The function of the regenerated endothelial cells, however, does not fully recover to the native one as revealed in several animal studies. Relaxations to serotonin are attenuated with the appearance of greater serotonin-induced endothelium-dependent contractions in the reendothelialized porcine coronary artery after denudation (Shimokawa et al., 1987), possibly due to depressed endothelium-dependent hyperpolarization in response to the agonist (Thollon et al., 1999). These *in vitro* results are supported by *in vivo* demonstrations of hypercontractions to intracoronary administration of serotonin or aggregating platelets, which are not observed in arteries with native endothelium (Shimokawa & Vanhoutte, 1991). Unlike native endothelial cells, vasoconstrictors are released preferentially in response to serotonin in the regenerated endothelium and the protection conferred by the endothelium against aggregating platelets diminishes. In fact, the expression of COX-1 is elevated while that of endothelial NO synthase decreases in the regenerated endothelial cells (Lee et al., 2007). Impaired endothelium-dependent pertussis toxin-sensitive relaxations are also reported in these reendothelialized arteries, possibly in relation to a reduced amount or functionality of Gi proteins (Borg-Capra et al., 1997; Shibano et al., 1994). Chronic treatment with a combined 5-hydroxytryptamine-2 receptor antagonist and calcium channel inhibitor LU49938 inhibits vasoconstrictions triggered by intracoronary injection of serotonin and eccentric myointimal thickening in the reendothelialized coronary arteries, and restores the endothelium-dependent pertussis toxin-sensitive G protein-mediated responses

(Park et al., 1995). Taken together, regenerated endothelial cells following vascular injury cannot fully replace the native endothelial function.

G. Common Requirements of Endothelium-Dependent Contractions: Calcium and TP Receptor

Although different prostanoids may take part in mediating endothelium-dependent contractions under various patho/physiological conditions, extracellular calcium ion is the common requirement for generation and release of contracting prostanoids and subsequent vasoconstrictions. COX-1 and COX-2 are not calcium-dependent enzymes, but cytosolic phospholipase A_2 (PLA_2) requires calcium ions to convert membrane phospholipids to arachidonic acid as the substrate for COX. This explains why receptor-independent calcium ionophore can also trigger prostanoid synthesis and endothelium-dependent contractions (Shi et al., 2007, 2008; Tang et al., 2007). Wong et al. (2009) demonstrate clearly that the presence of extracellular calcium ions is essential to endothelium-dependent contractions in hamster aortae, as acetylcholine fails to elicit contractions in the absence of extracellular calcium, while pronounced contractions occur after calcium ions are reintroduced. Indeed, acetylcholine-stimulated calcium influx into the endothelial cells is prevented by a nonselective cation channel blocker. A recent study from Wong et al. (2010) further substantiates that calcium influx is critical to endothelium-dependent contractions and store operated calcium channel (SOCC) activated by calcium-independent PLA_2 may be also involved.

In addition, TP receptor appears to be the common target for the released prostanoids as endothelium-dependent contractions are sensitive to TP receptor antagonism, even though each prostanoid has its own natural receptor. This may be attributed to a nonselective affinity of the TP receptor toward the structurally similar prostanoids. Thus PGI_2, which conventionally acts on its IP receptor to produce a vasodilatory effect, activated the TP receptor resulting in contractions in the aortae of SHR (Gluais et al., 2005). Likewise, endogenous $PGF_{2\alpha}$ happens to stimulate the TP receptor instead of its own FP receptor to trigger contractions (Wong et al., 2009; Fig. 4).

IV. Controversies over COX-2 Inhibitors and Cardiovascular Events

Acute infusion of indomethacin augments vasodilatation in hypertensive and aged patients, and substantial amount of experimental data on animal studies indicates indispensable roles of COX-1 and COX-2 in improving endothelium-dependent relaxations and inhibiting the exaggerated vasoconstrictions. The concept that chronic COX inhibition may correct the vascular

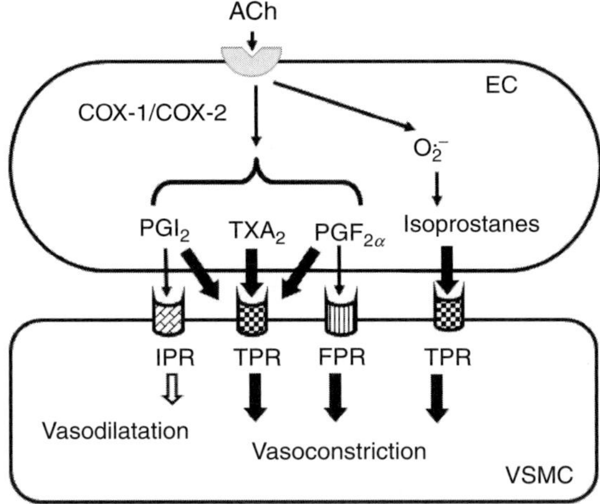

FIGURE 4 Nonselectivity of TP receptor (TPR) to prostanoids. Except its natural agonist thromboxane A_2 (TxA_2), prostacyclin (PGI_2), prostaglandin $F_{2\alpha}$ ($PGF_{2\alpha}$), and isoprostanes can also activate TPR. ACh, acetylcholine; COX, cyclooxygenase; $O_2^{\bullet-}$, superoxide anion; IPR, prostacyclin receptor; FPR, FP receptor; EC, endothelial cell; VSMC, vascular smooth muscle cell.

imbalance of endothelium-derived vasoactive factors, however, is antagonized by the outcomes of several large randomized clinical trials particularly with the use of specific COX-2 inhibitors, rofecoxib and diclofenac (McGettigan & Henry, 2006). Instead of conveying vascular benefit, COX-2 inhibition causes adverse cardiovascular events including the increased incidence of thrombosis, myocardial infarction, and stroke, leading to the withdrawal of Vioxx from the market in 2004 (Marnett, 2009). Explanation of such detrimental effects of COX-2 inhibition is attributed to a general belief that endothelium-derived COX-2-mediated production of PGI_2 is suppressed while the level of COX-1-mediated TxA_2 in platelets is unaltered, hence favoring platelet activation and aggregation.

Intriguingly, clinical trials do not demonstrate an increased cardiovascular risk in patients treated with another selective COX-2 inhibitor, celecoxib (Silverstein et al., 2000; White et al., 2007). Retrospective studies from Cho et al. (2003) and meta-analysis from Aw et al. (2005) actually show that patients receiving celecoxib therapy have a lower risk of hypertension development and a slightly decreased systolic blood pressure relative to those treated with rofecoxib. Patients taking celecoxib have a significantly reduced risk for nonfatal myocardial infarction as compared with those not taking any NSAIDs or using rofecoxib (Kimmel et al., 2005). In patients with intermittent claudication associated with peripheral arterial disease, 1-week celecoxib treatment enhances flow-mediated dilatation in the brachial artery and reduces levels of inflammatory biomarkers such as high-sensitivity C-reactive protein (Flórez et al., 2009). These clinical findings are supported and perhaps, accounted by the therapeutic effects of celecoxib against vascular dysfunctions in animal models (Abdelrahman & Al Suleimani. 2008; Cheng et al., 2002; Hermann et al., 2003). It is clear that inhibition of COX-2 activity may not necessarily correlate with the adverse cardiovascular effects; rather, COX-2 inhibitors may be of potential to treat vascular complications in diabetes and hypertension. In fact, it is possible that the side effects of rofecoxib are related to its direct inhibition on PGIS (Griffoni et al., 2007). Using the production of PGI_2 (measured in form of its metabolite 6-keto $PGF_{1\alpha}$) as an indicator of COX-2 activity with PGH_2 supplied exogenously to bypass the action of COX-2, PGIS activity in human umbilical vein endothelial cells is examined. While nonselective NSAIDs, acetylsalicylic acid and naproxen, and selective COX-2 inhibitor celecoxib do not inhibit the PGIS activity even at a concentration as high as 0.1 mM, rofecoxib reduces the PGIS activity by $\sim$30% at 0.1 nM, a concentration that does not inhibit COX-2. Rofecoxib inhibition of the PGIS activity is over $\sim$60% at 10 μM. For a patient taking a single dose of rofecoxib (25 mg), the plasma level of rofecoxib ranges between 50 nM and 1 μM, it is thus likely that the PGIS activity may have been suppressed by rofecoxib even within the therapeutic dosage. Inevitably, the use of coxibs should be reviewed extensively with confirmation on whether the inhibitors directly suppress the PGI_2 production via PGIS.

V. Conclusion

Endothelium plays a pivotal role in the regulation of vascular tone, and the imbalance between the release of relaxing and contracting factors in diseased states or aging favors hyperconstrictions of blood vessels. While inhibition on the enhanced COX-mediated endothelium-dependent contractions may ameliorate vascular dysfunction in diabetes and hypertension and celecoxib appears promising in improving cardiovascular function, extensive research and trials are required to prove what COX inhibitors are not risk prone in the general population.

In the past decade, a lot of focus has been directed to the role of COX in endothelial dysfunction, but seldom has investigations looked deeply into how prostanoid synthases or isomerases contribute to the production of corresponding prostanoids. This is possibly due to a lack of commercially available specific inhibitors against these synthases or isomerases. With the development and advances in the technique of using small interfering RNA to knock down specific proteins, actions of the synthases and isomerases and hence their therapeutic potentials can hopefully be unveiled. Actually, it is the action of these downstream synthases and isomerases that eventually determine what kind of prostaglandins will be produced and liberated upon specific triggers.

Meanwhile, TP receptor antagonism with terutroban could provide an alternative to correct hyperconstrictions resulting from the augmented release of prostanoids, as the TP receptor is a common converging target for contracting factors. As aforementioned, the nonselectivity of the TP receptor allows the binding of various COX-derived prostanoids and ROS-catalyzed isoprostanes. Therefore, there is actually an advantage of antagonizing the TP receptor over inhibiting specific COX isozymes in reducing prostanoid-induced hyperconstrictions as isoprostanes are insensitive to COX inhibition but inhibited by TP receptor antagonism. Liu et al. (2009, 2010) demonstrated that vascular dysfunction due to the activation of the TP receptor is reversed by terutroban. Indeed, constrictive prostanoids also contribute to smooth muscle proliferation, vascular inflammation, and progression of atherosclerosis (Cyrus et al., 2010; Ishizuka et al., 1998; Zucker et al., 1998), thus terutroban also exhibits antithrombotic, antiatherosclerotic, and anti-inflammatory properties as reviewed by Chamorro (2009). TP receptor antagonism could become a therapeutic option to reverse prostanoid-mediated vascular dysfunction (Schrör, 2009), and development of the third generation NSAIDs, dual cyclooxygenase 2 inhibitor (COXIB)/TP antagonists, may represent a new strategy in reducing cardiovascular risks while retaining the gastrointestinal benefits of COXIBs (Rovati et al., 2010).

Acknowledgments

This study was supported by Hong Kong Research Grant Council (4653/08M), CUHK Focused Investment Scheme, and CUHK Li Ka Shing Institute of Health Sciences.

Conflict of interest: The authors have no conflicts of interest to declare.

References

Abdelrahman, A. M., & Al Suleimani, Y. M. (2008). Four-week administration of nimesulide, a cyclooxygenase-2 inhibitor, improves endothelial dysfunction in the hindlimb vasculature of streptozotocin-induced diabetic rats. *Archives of Pharmacal Research*, *31*, 1584–1589.

Abeywardena, M. Y., Jablonskis, L. T., & Head, R. J. (2002). Age- and hypertension-induced changes in abnormal contractions in rat aorta. *Journal of Cardiovascular Pharmacology*, *40*, 930–937.

Adeagbo, A. S., Zhang, X., Patel, D., et al. (2005). Cyclo-oxygenase-2, endothelium and aortic reactivity during deoxycorticosterone acetate salt-induced hypertension. *Journal of Hypertension*, *23*, 1025–1036.

Akamine, E. H., Urakawa, T. A., de Oliveira, M. A., et al. (2006). Decreased endothelium-dependent vasodilation in diabetic female rats: Role of prostanoids. *Journal of Vascular Research*, *43*, 401–410.

Alvarez, Y., Briones, A. M., Balfagón, G., et al. (2005). Hypertension increases the participation of vasoconstrictor prostanoids from cyclooxygenase-2 in phenylephrine responses. *Journal of Hypertension*, *23*, 767–777.

Arribas, S., Marín, J., Ponte, A., et al. (1994). Norepinephrine-induced relaxations in rat aorta mediated by endothelial beta adrenoceptors. Impairment by ageing and hypertension. *Journal of Pharmacology and Experimental Therapeutics*, *270*, 520–527.

Auch-Schwelk, W., Katusic, Z. S., & Vanhoutte, P. M. (1990). Thromboxane A_2 receptor antagonists inhibit endothelium-dependent contractions. *Hypertension*, *15*, 699–703.

Auch-Schwelk, W., Katusić, Z. S., & Vanhoutte, P. M. (1992). Nitric oxide inactivates endothelium-derived contracting factor in the rat aorta. *Hypertension*, *19*, 442–445.

Auch-Schwelk, W., & Vanhoutte, P. M. (1991). Endothelium-derived contracting factor released by serotonin in the aorta of the spontaneously hypertensive rat. *American Journal of Hypertension*, *4*, 769–772.

Aw, T. J., Haas, S. J., Liew, D., et al. (2005). Meta-analysis of cyclooxygenase-2 inhibitors and their effects on blood pressure. *Archives of Internal Medicine*, *165*, 490–496.

Bogatcheva, N. V., Sergeeva, M. G., Dudek, S. M., et al. (2005). Arachidonic acid cascade in endothelial pathobiology. *Microvascular Research*, *69*, 107–127.

Borg-Capra, C., Fournet-Bourguignon, M. P., Janiak, P., et al. (1997). Morphological heterogeneity with normal expression but altered function of G proteins in porcine cultured regenerated coronary endothelial cells. *British Journal of Pharmacology*, *122*, 999–1008.

Chamorro, A. (2009). TP receptor antagonism: A new concept in atherothrombosis and stroke prevention. *Cerebrovascular Diseases*, *27*(Suppl 3), 20–27.

Cheng, H. F., Wang, C. J., Moeckel, G. W., et al. (2002). Cyclooxygenase-2 inhibitor blocks expression of mediators of renal injury in a model of diabetes and hypertension. *Kidney International*, *62*, 929–939.

Cho, J., Cooke, C. E., & Proveaux, W. (2003). A retrospective review of the effect of COX-2 inhibitors on blood pressure change. *American Journal of Therapeutics*, *10*, 311–317.

Cohen, R. A., Zitnay, K. M., Haudenschild, C. C., et al. (1988). Loss of selective endothelial cell vasoactive functions caused by hypercholesterolemia in pig coronary arteries. *Circulation Research*, *63*, 903–910.

Cracowski, J. L., Devillier, P., Durand, T., et al. (2001). Vascular biology of the isoprostanes. *Journal of Vascular Research*, *38*, 93–103.

Cyrus, T., Ding, T., & Praticò, D. (2010). Expression of thromboxane synthase, prostacyclin synthase and thromboxane receptor in atherosclerotic lesions: Correlation with plaque composition. *Atherosclerosis*, *208*, 376–381.

Dam, J. P., Vleeming, W., Riezebos, J., et al. (1997). Effects of hypercholesterolemia on the contractions to angiotensin II in the isolated aorta and iliac artery of the rabbit: Role of arachidonic acid metabolites. *Journal of Cardiovascular Pharmacology*, *30*, 118–123.

Daray, F. M., Colombo, J. R., Kibrik, J. R., et al. (2006). Involvement of endothelial thromboxane A_2 in the vasoconstrictor response induced by 15-E2t-isoprostane in isolated human umbilical vein. *Naunyn-Schmiedeberg's Archives of Pharmacology*, *373*, 367–375.

Davì, G., Alessandrini, P., Mezzetti, A., et al. (1997). In vivo formation of 8-Epi-prostaglandin $F_{2\alpha}$ is increased in hypercholesterolemia. *Arteriosclerosis, Thrombosis, and Vascular Biology*, *17*, 3230–3235.

Davì, G., Ciabattoni, G., Consoli, A., et al. (1999). In vivo formation of 8-iso-prostaglandin $F_{2\alpha}$ and platelet activation in diabetes mellitus: Effects of improved metabolic control and vitamin E supplementation. *Circulation*, *99*, 224–229.

Flavahan, N. A. (2007). Balancing prostanoid activity in the human vascular system. *Trends in Pharmacological Sciences*, *28*, 106–110.

Flórez, A., de Haro, J., Martínez, E., et al. (2009). Selective cyclooxygenase-2 inhibition reduces endothelial dysfunction and improves inflammatory status in patients with intermittent claudication. *Revista Española de Cardiología*, *62*, 851–857.

Foudi, N., Norel, X., Rienzo, M., et al. (2009). Altered reactivity to norepinephrine through COX-2 induction by vascular injury in hypercholesterolemic rabbits. *American Journal of Physiology. Heart and Circulatory Physiology*, *297*, H1882–H1888.

Fujii, K., Ohmori, S., Tominaga, M., et al. (1993). Age-related changes in endothelium-dependent hyperpolarization in the rat mesenteric artery. *The American Journal of Physiology*, *265*, H509–H516.

Fukunaga, M., Yura, T., & Badr, K. F. (1995). Stimulatory effect of 8-Epi-$PGF_{2\alpha}$, an F2-isoprostane, on endothelin-1 release. *Journal of Cardiovascular Pharmacology*, *26*(Suppl 3), S51–S52.

Garcia-Cohen, E. C., Marin, J., Diez-Picazo, L. D., et al. (2000). Oxidative stress induced by tert-butyl hydroperoxide causes vasoconstriction in the aorta from hypertensive and aged rats: Role of cyclooxygenase-2 isoform. *The Journal of Pharmacology and Experimental Therapeutics*, *293*, 75–81.

Ge, T., Hughes, H., Junquero, D. C., et al. (1995). Endothelium-dependent contractions are associated with both augmented expression of prostaglandin H synthase-1 and hypersensitivity to prostaglandin H_2 in the SHR aorta. *Circulation Research*, *76*, 1003–1010.

Ghanam, K., Lavagna, C., Burgaud, J. L., et al. (2000). Involvement of cyclooxygenase 2 in the protective effect of 17β-estradiol on hypercholesterolemic rabbit aorta. *Biochemical and Biophysical Research Communications*, *275*, 696–703.

Gluais, P., Lonchampt, M., Morrow, J. D., et al. (2005). Acetylcholine-induced endothelium-dependent contractions in the SHR aorta: The Janus face of prostacyclin. *British Journal of Pharmacology*, *146*, 834–845.

Gluais, P., Paysant, J., Badier-Commander, C., et al. (2006). In SHR aorta, calcium ionophore A-23187 releases prostacyclin and thromboxane A_2 as endothelium-derived contracting factors. *American Journal of Physiology. Heart and Circulatory Physiology*, *291*, H2255–H2264.

Griffoni, C., Spisni, E., Strillacci, A., et al. (2007). Selective inhibition of prostacyclin synthase activity by rofecoxib. *Journal of Cellular and Molecular Medicine*, *11*, 327–338.

Hart, J. L., Sobey, C. G., & Woodman, O. L. (1995). Cholesterol feeding enhances vasoconstrictor effects of products from rabbit polymorphonuclear leukocytes. *The American Journal of Physiology*, *269*, H1–H6.

Hashimoto, M., Akishita, M., Eto, M., et al. (1998). The impairment of flow-mediated vasodilatation in obese men with visceral fat accumulation. *International Journal of Obesity and Related Metabolic Disorders*, *22*, 477–484.

Hermann, M., Camici, G., Fratton, A., et al. (2003). Differential effects of selective cyclooxygenase-2 inhibitors on endothelial function in salt-induced hypertension. *Circulation*, *108*, 2308–2311.

Hla, T., & Neilson, K. (1992). Human cyclooxygenase-2 cDNA. *Proceedings of the National Academy of Sciences of the United States of America*, *89*, 7384–7388.

Ibarra, M., López-Guerrero, J. J., Mejía-Zepeda, R., et al. (2006). Endothelium-dependent inhibition of the contractile response is decreased in aorta from aged and spontaneously hypertensive rats. *Archives of Medical Research*, *37*, 334–341.

Ihlemann, N., Stokholm, K. H., & Eskildsen, P. C. (2002). Impaired vascular reactivity is present despite normal levels of von Willebrand factor in patients with uncomplicated type 2 diabetes. *Diabetic Medicine*, *19*, 476–481.

Ishizuka, T., Kawakami, M., Hidaka, T., et al. (1998). Stimulation with thromboxane A_2 (TXA_2) receptor agonist enhances ICAM-1, VCAM-1 or ELAM-1 expression by human vascular endothelial cells. *Clinical and Experimental Immunology*, *112*, 464–470.

Jerez, S., Sierra, L., Coviello, A., et al. (2008). Endothelial dysfunction and improvement of the angiotensin II-reactivity in hypercholesterolemic rabbits: Role of cyclooxygenase metabolites. *European Journal of Pharmacology*, *580*, 182–189.

Kamgar, M., Nobakhthaghighi, N., Shamshirsaz, A. A., et al. (2006). Impaired fibrinolytic activity in type II diabetes: Correlation with urinary albumin excretion and progression of renal disease. *Kidney International*, *69*, 1899–1903.

Kimmel, S. E., Berlin, J. A., Reilly, M., et al. (2005). Patients exposed to rofecoxib and celecoxib have different odds of nonfatal myocardial infarction. *Annals of Internal Medicine*, *142*, 157–164.

Komers, R., Lindsley, J. N., Oyama, T. T., et al. (2001). Immunohistochemical and functional correlations of renal cyclooxygenase-2 in experimental diabetes. *Journal of Clinical Investigation*, *107*, 889–898.

Komers, R., Zdychová, J., Cahová, M., et al. (2005). Renal cyclooxygenase-2 in obese Zucker (fatty) rats. *Kidney International*, *67*, 2151–2158.

Küng, C. F., & Lüscher, T. F. (1995). Different mechanisms of endothelial dysfunction with aging and hypertension in rat aorta. *Hypertension*, *25*, 194–200.

Lagaud, G. J., Masih-Khan, E., Kai, S., et al. (2001). Influence of type II diabetes on arterial tone and endothelial function in murine mesenteric resistance arteries. *Journal of Vascular Research*, *38*, 578–589.

Lee, M. Y., Tse, H. F., Siu, C. W., et al. (2007). Genomic changes in regenerated porcine coronary arterial endothelial cells. *Arteriosclerosis, Thrombosis, and Vascular Biology*, *27*, 2443–2449.

Liu, C. Q., Leung, F. P., Wong, S. L., et al. (2009). Thromboxane prostanoid receptor activation impairs endothelial nitric oxide-dependent vasorelaxations: The role of Rho kinase. *Biochemical Pharmacology*, *78*, 374–381.

Liu, C. Q., Wong, S. L., Leung, F. P., et al. (2010). Prostanoid TP receptor-mediated impairment of cyclic AMP-dependent vasorelaxation is reversed by phosphodiesterase inhibitors. *European Journal of Pharmacology*, *632*, 45–51.

Lüscher, T. F., Aarhus, L. L., & Vanhoutte, P. M. (1990). Indomethacin improves the impaired endothelium-dependent relaxations in small mesenteric arteries of the spontaneously hypertensive rat. *American Journal of Hypertension*, *3*, 55–58.

Lüscher, T. F., & Vanhoutte, P. M. (1986). Endothelium-dependent contractions to acetylcholine in the aorta of the spontaneously hypertensive rat. *Hypertension*, *8*, 344–348.

Marnett, L. J. (2009). The COXIB experience: A look in the rearview mirror. *Annual Review of Pharmacology and Toxicology*, *49*, 65–90.

Matsumoto, T., Ishida, K., Kobayashi, T., et al. (2009a). Pyrrolidine dithiocarbamate reduces vascular prostanoid-induced responses in aged type 2 diabetic rat model. *Journal of Pharmacological Sciences*, *110*, 326–333.

Matsumoto, T., Kakami, M., Noguchi, E., et al. (2007). Imbalance between endothelium-derived relaxing and contracting factors in mesenteric arteries from aged OLETF rats, a model of type 2 diabetes. *American Journal of Physiology. Heart and Circulatory Physiology*, *293*, H1480–H1490.

Matsumoto, T., Nakayama, N., Ishida, K., et al. (2009b). Eicosapentaenoic acid improves imbalance between vasodilator and vasoconstrictor actions of endothelium-derived factors in mesenteric arteries from rats at chronic stage of type 2 diabetes. *The Journal of Pharmacology and Experimental Therapeutics*, *329*, 324–334.

Matsumoto, T., Noguchi, E., Ishida, K., et al. (2008). Metformin normalizes endothelial function by suppressing vasoconstrictor prostanoids in mesenteric arteries from OLETF rats, a model of type 2 diabetes. *American Journal of Physiology. Heart and Circulatory Physiology*, *295*, H1165–H1176.

Matsumoto, T., Takaoka, E., Ishida, K., et al. (2009c). Abnormalities of endothelium-dependent responses in mesenteric arteries from Otsuka Long-Evans Tokushima Fatty (OLETF) rats are improved by chronic treatment with thromboxane A_2 synthase inhibitor. *Atherosclerosis*, *205*, 87–95.

McGettigan, P., & Henry, D. (2006). Cardiovascular risk and inhibition of cyclooxygenase: A systematic review of the observational studies of selective and nonselective inhibitors of cyclooxygenase 2. *JAMA*, *296*, 1633–1644.

Mogensen, C. E., & Schmitz, O. (1988). The diabetic kidney: From hyperfiltration and microalbuminuria to end-stage renal failure. *The Medical Clinics of North America*, *72*, 1465–1492.

Mombouli, J. V., & Vanhoutte, P. M. (1993). Purinergic endothelium-dependent and -independent contractions in rat aorta. *Hypertension*, *22*, 577–583.

Morrow, J. D., Hill, K. E., Burk, R. F., et al. (1990). A series of prostaglandin F2-like compounds are produced in vivo in humans by a non-cyclooxygenase, free radical-catalyzed mechanism. *Proceedings of the National Academy of Sciences of the United States of America*, *87*, 9383–9387.

Mullen, M. J., Kharbanda, R. K., Cross, J., et al. (2001). Heterogenous nature of flow-mediated dilatation in human conduit arteries in vivo: Relevance to endothelial dysfunction in hypercholesterolemia. *Circulation Research*, *88*, 145–151.

Noon, J. P., Walker, B. R., Hand, M. F., et al. (1998). Impairment of forearm vasodilatation to acetylcholine in hypercholesterolemia is reversed by aspirin. *Cardiovascular Research*, *38*, 480–484.

Park, S. J., Lee, J. J., Kirchengast, M., et al. (1995). The combined 5-HT2 receptor and calcium channel inhibitor LU49938 restores endothelium dependent responses in the regenerated endothelium of the porcine coronary artery. *Cardiovascular Research*, *29*, 95–101.

Peredo, H. A., Rodríguez, R., Susemihl, M. C., et al. (2006). Long-term streptozotocin-induced diabetes alters prostanoid production in rat aorta and mesenteric bed. *Autonomic & Autacoid Pharmacology*, *26*, 355–360.

Pérez-Sala, D., & Lamas, S. (2001). Regulation of cyclooxygenase-2 expression by nitric oxide in cells. *Antioxidants and Redox Signaling*, *3*, 231–248.

Pfister, S. L., & Campbell, W. B. (1996). Contribution of arachidonic acid metabolites to reduced norepinephrine-induced contractions in hypercholesterolemic rabbit aortas. *Journal of Cardiovascular Pharmacology*, *28*, 784–791.

Rovati, G. E., Sala, A., Capra, V., et al. (2010). Dual COXIB/TP antagonists: A possible new twist in NSAID pharmacology and cardiovascular risk. *Trends in Pharmacological Sciences*, *31*, 102–107.

Rutkai, I., Feher, A., Erdei, N., et al. (2009). Activation of prostaglandin E_2 EP1 receptor increases arteriolar tone and blood pressure in mice with type 2 diabetes. *Cardiovascular Research*, *83*, 148–154.

Schrör, K. (2009). Cyclooxygenase-2-derived prostaglandin $F_{2\alpha}$: An endothelium-derived contractile factor acting independently of other endothelium-derived contractile factors via vascular thromboxane receptors. *Circulation Research*, *104*, 141–143.

Shi, Y., Feletou, M., Ku, D. D., et al. (2007). The calcium ionophore A23187 induces endothelium-dependent contractions in femoral arteries from rats with streptozotocin-induced diabetes. *British Journal of Pharmacology*, *150*, 624–632.

Shi, Y., Man, R. Y., & Vanhoutte, P. M. (2008). Two isoforms of cyclooxygenase contribute to augmented endothelium-dependent contractions in femoral arteries of 1-year-old rats. *Acta Pharmacologica Sinica*, *29*, 185–192.

Shibano, T., Codina, J., Birnbaumer, L., et al. (1994). Pertussis toxin-sensitive G proteins in regenerated endothelial cells of porcine coronary artery. *The American Journal of Physiology*, *267*, H979–H981.

Shimokawa, H., Aarhus, L. L., & Vanhoutte, P. M. (1987). Porcine coronary arteries with regenerated endothelium have a reduced endothelium-dependent responsiveness to aggregating platelets and serotonin. *Circulation Research*, *61*, 256–270.

Shimokawa, H., & Vanhoutte, P. M. (1989). Impaired endothelium-dependent relaxation to aggregating platelets and related vasoactive substances in porcine coronary arteries in hypercholesterolemia and atherosclerosis. *Circulation Research*, *64*, 900–914.

Shimokawa, H., & Vanhoutte, P. M. (1991). Angiographic demonstration of hyperconstriction induced by serotonin and aggregating platelets in porcine coronary arteries with regenerated endothelium. *Journal of the American College of Cardiology*, *17*, 1197–1202.

Silverstein, F. E., Faich, G., Goldstein, J. L., et al. (2000). Gastrointestinal toxicity with celecoxib vs nonsteroidal anti-inflammatory drugs for osteoarthritis and rheumatoid arthritis: The CLASS study: A randomized controlled trial. Celecoxib Long-term Arthritis Safety Study. *JAMA*, *284*, 1247–1255.

Smith, W. L., DeWitt, D. L., & Garavito, R. M. (2000). Cyclooxygenases: Structural, cellular, and molecular biology. *Annual Review of Biochemistry*, *69*, 145–182.

Smith, W., Garavito, R., & DeWitt, D. (1996). Prostaglandin endoperoxide H synthases (cyclooxygenases)-1 and -2. *The Journal of Biological Chemistry*, *271*, 33157–33160.

Taddei, S., & Vanhoutte, P. M. (1993). Endothelium-dependent contractions to endothelin in the rat aorta are mediated by thromboxane A_2. *Journal of Cardiovascular Pharmacology*, *22*(Suppl. 8), S328–S331.

Taddei, S., Virdis, A., Ghiadoni, L., et al. (1997a). Cyclooxygenase inhibition restores nitric oxide activity in essential hypertension. *Hypertension*, *29*, 274–279.

Taddei, S., Virdis, A., Mattei, P., et al. (1997b). Hypertension causes premature aging of endothelial function in humans. *Hypertension*, *29*, 736–743.

Tang, E. H., Leung, F. P., Huang, Y., et al. (2007). Calcium and reactive oxygen species increase in endothelial cells in response to releasers of endothelium-derived contracting factor. *British Journal of Pharmacology*, *151*, 15–23.

Tang, E. H., & Vanhoutte, P. M. (2008). Gene expression changes of prostanoid synthases in endothelial cells and prostanoid receptors in vascular smooth muscle cells caused by aging and hypertension. *Physiological Genomics*, *32*, 409–418.

Tesfamariam, B., Brown, M. L., & Cohen, R. A. (1995). 15-Hydroxyeicosatetraenoic acid and diabetic endothelial dysfunction in rabbit aorta. *Journal of Cardiovascular Pharmacology*, *25*, 748–755.

Tesfamariam, B., Brown, M. L., Deykin, D., et al. (1990). Elevated glucose promotes generation of endothelium-derived vasoconstrictor prostanoids in rabbit aorta. *Journal of Clinical Investigation*, *85*, 929–932.

Tesfamariam, B., Jakubowski, J. A., & Cohen, R. A. (1989). Contraction of diabetic rabbit aorta caused by endothelium-derived PGH_2-TxA_2. *The American Journal of Physiology*, *257*, H1327–H1333.

Thollon, C., Bidouard, J. P., Cambarrat, C., et al. (1999). Alteration of endothelium-dependent hyperpolarizations in porcine coronary arteries with regenerated endothelium. *Circulation Research*, *84*, 371–377.

Topper, J. N., Cai, J., Falb, D., et al. (1996). Identification of vascular endothelial genes differentially responsive to fluid mechanical stimuli: Cyclooxygenase-2, manganese superoxide dismutase, and endothelial cell nitric oxide synthase are selectively up-regulated by steady laminar shear stress. *Proceedings of the National Academy of Sciences of the United States of America*, *93*, 10417–10422.

Van Diest, M. J., Herman, A. G., & Verbeuren, T. J. (1996). Influence of hypercholesterolaemia on the reactivity of isolated rabbit arteries to 15-lipoxygenase metabolites of arachidonic acid: Comparison with platelet-derived agents and vasodilators. *Prostaglandins Leukotrienes and Essential Fatty Acids*, *54*, 135–145.

White, W. B., West, C. R., Borer, J. S., et al. (2007). Risk of cardiovascular events in patients receiving celecoxib: A meta-analysis of randomized clinical trials. *The American Journal of Cardiology*, *99*, 91–98.

Williams, C. S., Mann, M., & DuBois, R. N. (1999). The role of cyclooxygenases in inflammation, cancer and development. *Oncogene*, *18*, 7908–7916.

Wong, S. L., Leung, F. P., Lau, C. W., et al. (2009). Cyclooxygenase-2-derived prostaglandin $F_{2\alpha}$ mediates endothelium-dependent contractions in the aortae of hamsters with increased impact during aging. *Circulation Research*, *104*, 228–235.

Wong, M. S., Man, R. Y., & Vanhoutte, P. M. (2010). Calcium-independent phospholipase A_2 plays a key role in the endothelium-dependent contractions to acetylcholine in the aorta of the spontaneously hypertensive rat. *American Journal of Physiology. Heart and Circulatory Physiology*, *298*, H1260–H1266.

Yang, D., Félétou, M., Boulanger, C. M., et al. (2002). Oxygen-derived free radicals mediate endothelium-dependent contractions to acetylcholine in aortas from spontaneously hypertensive rats. *British Journal of Pharmacology*, *136*, 104–110.

Yang, D., Gluais, P., Zhang, J. N., et al. (2004). Nitric oxide and inactivation of the endothelium-dependent contracting factor released by acetylcholine in spontaneously hypertensive rat. *Journal of Cardiovascular Pharmacology*, *43*, 815–820.

Yura, T., Fukunaga, M., Khan, R., et al. (1999). Free-radical-generated F2-isoprostane stimulates cell proliferation and endothelin-1 expression on endothelial cells. *Kidney International*, *56*, 471–478.

Zucker, T. P., Bönisch, D., Muck, S., et al. (1998). Thrombin-induced mitogenesis in coronary artery smooth muscle cells is potentiated by thromboxane A_2 and involves upregulation of thromboxane receptor mRNA. *Circulation*, *97*, 589–595.

Michel Félétou*, Richard A. Cohen†, Paul M. Vanhoutte‡,§, and Tony J. Verbeuren*

*Department of Angiology, Institut de Recherches Servier, Suresnes, France

†Vascular Biology Unit, Whitaker Cardiovascular Institute, Boston University School of Medicine, Boston, Massachusetts, USA

‡Department Pharmacology and Pharmacy, Li Ka Shing Faculty Medicine, University of Hong Kong, Hong Kong, China

§Department BIN Fusion Technology, Chonbuk National University, Jeonju, Korea

TP Receptors and Oxidative Stress: Hand in Hand from Endothelial Dysfunction to Atherosclerosis

Abstract

Thromboxane A_2 and the activation of TP receptors that it causes play an important role in platelet aggregation and therefore in thrombosis. However, TP receptors are also involved in the pathologies of the vascular wall including impaired endothelium-dependent vasodilation, increased oxidant generation, and increased expression of adhesion molecules. The beneficial

Advances in Pharmacology, Volume 60

1054-3589/10 $35.00
10.1016/S1054-3589(10)60006-3

effects of TP antagonists on the vascular wall attenuate these features of vascular disease. They are not shared by aspirin. In fact, TP antagonists are active in patients treated with aspirin, indicating that their potential beneficial effects are mediated by mechanisms different from the antithrombotic actions of aspirin. Our studies have demonstrated the vascular benefits of TP antagonists in experimental animals, particularly in models of diabetes mellitus, in which elevated levels of eicosanoids play a role not only in vascular pathologies but also in those of the kidney and other tissues. They suggest that TP blockade protects against fundamental and widespread tissular dysfunction associated with metabolic disease including hyperlipidemia and hyperglycemia. TP receptor antagonists represent a promising avenue for the prevention of vascular disease in part because of these pleiotropic actions that extend beyond their antithrombotic properties.

I. Introduction

The excess burden of atherosclerotic cardiovascular disease in the Western world can be attributed to the increased incidence of risk factors such as hyperlipidemia, diabetes, and hypertension associated with obesity and metabolic syndrome. Smoking, sedentary life, and inappropriate diet are important, but avoidable, accelerating risk factors. Aging is a progressive "risk" factor that becomes increasingly more prominent with the ever advancing life span, and a family history of cardiovascular disease is a predictor that constitutes an unavoidable burden for the individual patient. The importance of prevention of cardiovascular disease by risk factor control (particularly hyperlipidemia, diabetes, smoking, and hypertension) cannot be overstressed. In addition, pharmacological therapies have now been proved to add significant protection against cardiovascular disease. This is in particular the case for HMG CoA reductase inhibitors (statins) and inhibitors of the renin–angiotensin system (especially angiotensin converting enzyme inhibitors [ACEI]). The antiatherosclerotic effects of statins exceed their ability to lower cholesterol, and those of ACEI go beyond their antihypertensive effects. In each of these two cases, a large body of experimental evidence supports pleiotropic effects, exerted at the level of the blood vessel wall, resulting in inhibition of inflammation and growth factor signaling, that are key to the progression of vascular disease. However, since neither of these two classes of drugs has important antithrombotic effects, in large numbers of patients at risk of cardiovascular disease, aspirin is administered to prevent thrombotic and embolic complications of atherosclerosis.

TP antagonists were developed as antiplatelet agents principally to prevent recurrent embolic stroke. Animal studies showed that TP antagonists have potent antiplatelet actions that are attributed to their ability to specifically target the platelet TP receptors, which are stimulated by platelet-derived thromboxane A_2. In addition, animal and cell studies have revealed other advantageous effects of those TP antagonists that may add significantly to the treatment of patients with arteriosclerotic cardiovascular disease (Rosenfeld et al., 2001; Verbeuren, 2006; Verbeuren et al., 1995). The benefits of blocking TP receptors may arise as a result of activation of eicosanoid production that accompanies the widespread vascular and organ inflammation and the increased oxidant stress associated with vascular disease (Fig. 1). As a result of inflammation, activation of phospholipases releases arachidonic acid, which serves as substrate for the production of eicosanoids. As reviewed below, the inflammation also increases the production of oxidants that shift the production/effects of the eicosanoids generated from vasodilation and antithrombosis to vasoconstriction, inflammation, and prothrombosis.

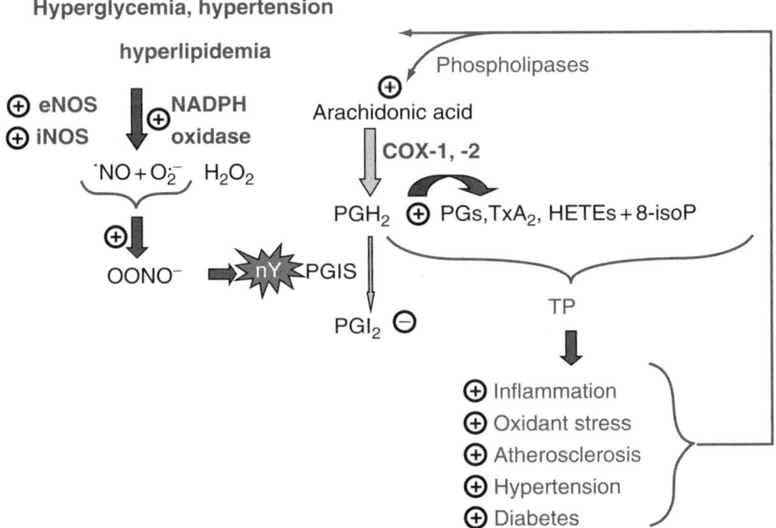

FIGURE 1 Role of TP receptors in cardiovascular disease. Cardiovascular risk factors, including hyperglycemia, hypertension, and hyperlipidemia, increase (plus sign) the expression and activity of endothelial and inducible nitric oxide synthase (eNOS, iNOS, respectively) and NADPH oxidase. As a result, the levels of nitric oxide ($^{\bullet}NO$), superoxide ($O_2^{-\bullet}$), and hydrogen peroxide (H_2O_2) increase. Peroxynitrite ($OONO^-$) the reaction product of $^{\bullet}NO$ and $O_2^{-\bullet}$ also increases. $OONO^-$ tyrosine nitrates (nY) PGI_2 synthase (PGIS) decreasing the production of PGI_2. As a result, arachidonic acid derived endoperoxide (PGH_2) is shunted to other prostaglandins (PGs), thromboxane A_2 (TxA_2), hydroxyeicosatetraenoic acids (HETEs), and 8-isoprostanes (isoP), all of which stimulate TP receptors. TP activation can increase inflammation and oxidants, accentuating the pathology of atherosclerosis, hypertension, and diabetes. These pathologies contribute to the production of oxidants as well as to the activation of phospholipases that are responsible for an increased generation of arachidonic acid products.

II. Arachidonic Acid Derivatives and Oxidative Stress

Arachidonic acid is released from membrane phospholipids by phospholipases and metabolized by cyclooxygenases (COXs), lipoxygenases, and cytochrome P450 monooxygenases. Two different COXs (COX-1 and COX-2) have been cloned and characterized. In most tissues, COX-1 is expressed constitutively, while COX-2 is often induced at sites of inflammation. However, COX-2 is also expressed constitutively in several organs and cell types, including endothelial cells. In the vascular wall, both endothelial and vascular smooth muscle cells express COXs; however, in healthy blood vessels, endothelial cells contain much more of the enzyme than the surrounding smooth muscle cells. Various biologically active eicosanoids are formed from short-lived but biologically active endoperoxides (PGH_2), through the action of various prostaglandin (PG) synthases PGD, PGE, PGF, PGI, and thromboxane synthases. PGs interact with G-protein-coupled receptors, classified in five subtypes DP, EP, FP, IP, and TP receptors in function of their preferential affinity toward the five primary prostanglandins PGD_2, E_2, $F_{2\alpha}$, I_2 (prostacyclin), and thromboxane A_2, respectively (Tsuboi et al., 2002).

In most blood vessels, prostacyclin is the principal metabolite of arachidonic acid, the endothelium being the major site of its synthesis. By stimulating its preferential IP receptor, PGI_2 is a potent inhibitor of platelet adhesion to the endothelial cell surface and of platelet aggregation, and generally acts as an endothelium-derived vasodilator and inhibitor of vascular smooth muscle migration and proliferation (Fetalvero et al., 2007; Moncada & Vane, 1979). The genetic deletion of IP receptors is associated with increased injury-induced restenosis (Cheng et al., 2002), thrombotic events (Murata et al., 1997), atherosclerosis (Egan et al., 2004; Kobayashi et al., 2004), and reperfusion injury (Xiao et al., 2001).

In the cardiovascular system, thromboxane A_2 is predominantly derived from platelet COX-1, but can also be produced by other cell types including endothelial cells. The stimulation of TP receptors elicits not only platelet aggregation and smooth muscle contraction but also the expression of adhesion molecules and the adhesion and infiltration of monocytes/macrophages (Nakahata, 2008). Thromboxane A_2 is by far the preferential physiological ligand of TP receptors but PGH_2 and other PGs as well as isoprostanes and hydroxyeicosatetraenoic acids (HETEs), although at higher concentrations, can activate this receptor with a various range of potency (Félétou et al., 2010b). By contrast, epoxyeicosatrienoic acids), which act as endothelium-derived hyperpolarizing factors in some vascular beds (Félétou & Vanhoutte, 2006b) and their dihydro-derivatives have been identified as selective endogenous antagonists of the TP receptors (Behm et al., 2009).

Mice deficient in TP receptors are normotensive but have abnormal vascular responses to thromboxane A_2 and show a tendency to bleeding (Thomas et al., 1998). The deletion of TP receptors decreases vascular proliferation and platelet activation in response to intimal lesions (Cheng et al., 2002), delays atherogenesis in apo$E^{-/-}$ mice (Kobayashi et al., 2004), and prevents angiotensin II- and L-NAME-induced hypertension and the associated cardiac hypertrophy (Francois et al., 2004, 2008). TP knockout mice are also protected against various LPS-induced responses such as the increase in iNOS expression (Yamada et al., 2003), acute renal failure (Boffa et al., 2004), and inflammatory tachycardia (Takayama et al., 2005).

Reactive oxygen species, such as superoxide anion ($O_2^{-\bullet}$) and hydrogen peroxide (H_2O_2), are derived from multiple sources within inflammatory leukocytes and vascular tissues including NADPH oxidase, uncoupled endothelial and inducible endothelial nitric oxide ($^\bullet$NO) synthase (eNOS, iNOS), xanthine oxidase, COXs, lipoxygenases, cytochrome P450 monooxygenases, and excess substrate utilization by mitochondria. Additionally, $^\bullet$NO react with $O_2^{-\bullet}$ to form the extremely potent oxidant, peroxynitrite ($ONOO^-$). Reactive oxygen species can inhibit endothelium-dependent vasodilator pathways (i.e., the NO and the endothelium-derived hyperpolarizing factor [EDHF] pathways) and shift the balance in response to eicosanoids from vasodilation and antithrombosis toward vasoconstriction and prothrombosis. Superoxide anions reduce the bioavailability of NO, inhibit its main target, soluble guanylyl cyclase, and inactivate calcium activated potassium channels. Peroxynitrites inhibit guanylyl cyclase, superoxide dismutases, and decrease the EDHF component of flow-mediated vasodilation (Félétou & Vanhoutte, 2006a). PGI_2 synthase is among the most sensitive targets of peroxynitrites and is inactivated by concentrations as low as 50 nM (Schmidt et al., 2003; Zou et al., 2002a, 2002b). When PGI_2 synthase is inactivated, the excess PGH_2 is shunted toward other metabolic pathways leading to a variety of products that can activate TP receptors and are, in general, deleterious to vascular function.

Reactive oxygen species enhance the stability and increase the density of functional TP receptors at the cell membrane (Valentin et al., 2004; Wilson et al., 2009) and, in endothelial cells, the activation of TP receptors inhibits NO production (Liu et al., 2009). The generation of deleterious eicosanoids, the posttranscriptional stabilization of TP receptors, and the decreased production of NO are reactive oxygen species-dependent feed-forward loops further altering the unbalance between relaxing/antithrombosis and contracting/prothrombosis pathways. Taken in conjunction, this experimental evidence indicates that TP receptors are likely to play a pivotal role in cardiovascular diseases (Félétou et al., 2010a).

III. Vascular Function

The understanding of vascular regulation was revolutionized by the discovery of Robert Furchgott, who recognized that the normal arterial endothelium released the vasodilator, nitric oxide ($^\bullet$NO) when stimulated with agents that include acetylcholine, bradykinin, or the calcium ionophore, A23187 (Furchgott & Zawadzki, 1980; Martin et al., 1985). It was soon recognized that diseased arteries had diminished endothelium-dependent vasodilator responses while, at least in the early stages of the diseases, retaining the ability of their smooth muscle cells to relax normally to •NO donors such as nitroglycerin and sodium nitroprusside. In many respects, Furchgott's discovery brought the focus of pathophysiology of vascular disease to the endothelium and generated the concept of endothelial dysfunction.

A. Hypertension

In the aorta of spontaneously hypertensive rats (SHR), when compared to that of normotensive Wistar-Kyoto rats (WKY), it was first shown that the impaired endothelium-dependent relaxations were restored by the presence of COX inhibitors (Lüscher & Vanhoutte, 1986). The endothelial dysfunction was associated with the generation of a diffusible endothelium-derived contracting factor(s) (EDCFs) that opposes the relaxing effect of nitric oxide with no or little alteration in its production. In healthy blood vessels, the release of EDCF is tempered by the presence of NO (Tang et al., 2005) and EDHF (Michel et al., 2008).

In the SHR aorta, the sequence of events leading to endothelium-dependent contractions requires an exacerbated increase in endothelial intracellular calcium concentration, the phospholipase A2-dependent mobilization of arachidonic acid, the activation of COX-1, and the resulting production of reactive oxygen species along with that of eicosanoids. The latter diffuses toward the vascular smooth muscle cells and directly activates the TP receptors while the former enhances the stability and increases the density of this receptor, as it was indicated above. Furthermore, reactive oxygen species can stimulate COX-1 in the smooth muscle (with subsequent stimulation of TP receptors by the produced prostanoids) and/or are involved in a positive feedback loop on the endothelial cells by further activating COX (Félétou et al., 2010a, 2010b; Harlan & Callahan, 1984).

Inhibition of thromboxane A_2 synthesis does not affect the endothelium-dependent contractions to acetylcholine but partially inhibits those in response to the calcium ionophore, A23187, to ADP and to endothelin-1, indicating that thromboxane A_2 is only one of the EDCFs that can be released from SHR aortic endothelial cells. The other EDCFs released by acetylcholine have been identified as PGH_2 and prostacyclin. The contribution of the latter is due importantly to the abundance of prostacyclin synthase in the endothelial cells, compared to the other specific synthases, and thus to the overwhelming production and release of prostacyclin (Tang & Vanhoutte, 2008). The contribution of prostacyclin to EDCF-mediated responses may seem paradoxical, as one would expect the prostanoid to rather contribute to endothelium-dependent relaxations. However, a characteristic of the SHR (but also of aged normotensive rats) is that their vascular smooth muscle cells (but not their platelets) have lost the responsiveness to IP receptor activation, and that prostacyclin, being a weak activator of the TP receptors, produces contraction (Gluais et al., 2005). At present, in the SHR, the origin of the nonfunctionality of the IP receptor is unknown. In human, polymorphism of the IP receptor has been associated with venous thrombosis and cardiovascular diseases (Patrignani et al., 2008; Stitham et al., 2007) but is unlikely to explain the dysfunction in the SHR aorta, since it is vascular smooth muscle selective. However, extracellular, transmembrane and cytosol-located cysteines play an important role in the trafficking, addressing, and structural integrity of the IP receptor and therefore in its function (Miggin et al., 2003; Stitham et al., 2006). Considering the susceptibility of cysteines to redox stress, this mechanism could be at the origin of the IP receptor dysfunction in hypertension.

In the SHR aorta, the endothelial cells also produce PGE_2 and $PGF_{2\alpha}$. These PGs can be produced actively from PGH_2 by their specific synthases or even spontaneously, in a nonenzymatic way. In the rat aorta, from either WKY or SHR, the mRNA expression of all the PG receptors can be detected, although at very low levels (Tang & Vanhoutte, 2008). However, the contractions in response to PGE_2 and $PGF_{2\alpha}$ involve the TP receptor and not their preferential EP and FP receptors, respectively (Gluais et al., 2005), suggesting that, in the rat aorta, either these proteins are not properly expressed or that these receptors are not functional. Nevertheless, in the SHR aorta the involvement of PGE_2 as an EDCF has been ruled out (Tang et al., 2008), although in the femoral artery of a rat model of diabetes, this PG activates EP receptors and contributes to endothelium-dependent contractions (Shi et al., 2007). Similarly, although in genetically modified mice $PGF_{2\alpha}$ and the FP receptor have been associated to hypertension and atherosclerosis (Yu et al., 2009), and in the hamster aorta, $PGF_{2\alpha}$, via the activation of the TP receptors, is the predominant EDCF (Wong et al., 2009), in the SHR aorta, the contribution of $PGF_{2\alpha}$ to EDCF-mediated responses, by activation of either FP or TP receptors, is at best marginal. In the SHR aorta, the involvement of isoprostanes, at least 8-iso-$PGF_{2\alpha}$, is unlikely since its production cannot be detected and since the contractile responses elicited by this compound (as those produced by PGE_2 and $PGF_{2\alpha}$) do not mimic the endothelium-dependent contractions in term of amplitude and kinetic (Gluais et al., 2005). Nevertheless, the involvement of isoprostane(s) cannot be excluded in the vascular dysfunction observed in diabetes and/or atherosclerosis, which are associated with elevated levels of oxidative stress.

The contribution of EDCF to endothelial dysfunction was first observed in the SHR but has since been reported in numerous other models of hypertension. In SHR endothelial cells, the mRNA and protein expression of COX-1 are enhanced when compared to that of WKY, and in both strains, they are augmented by aging. However, COX-2-derived contractile prostanoids can also be produced in arteries of both WKY and SHR as well as in various other models of hypertension and/or aging. The identity of the eicosanoids associated with EDCF-mediated responses depends on the model and the stimulating agent, but the activation of the TP receptors is ultimately involved in the endothelial dysfunction (Félétou et al., 2010b). In most rat models of hypertension (SHR, Dahl-salt sensitive, DOCA-salt, and renovascular hypertensive rats), the production of prostacyclin from the aortic wall is enhanced compared to that of normotensive controls (Ishimitsu et al., 1991), but this PG will be involved only in EDCF-mediated responses when the IP receptor is defective.

In hypertensive patients, but not in healthy subjects, the reduced vasodilation in response to acetylcholine is improved by the administration of indomethacin, a nonspecific COX inhibitor (Taddei et al., 2000). Interestingly, indomethacin and vitamin C also restore the inhibitory effect of a NO-synthase inhibitor on acetylcholine-induced vasodilation, indicating that, as in the SHR, the activation of COX generates not only EDCF but also reactive oxygen species that reduce the bioavailability of NO. In normotensive subjects, aging mainly affects the formation of NO and EDCF production only appears in old age. However, the presence of hypertension seems to cause an earlier onset of alteration in the L-arginine–NO pathway and also earlier formation of vasoconstrictor prostanoids, suggesting that the endothelial dysfunction observed in essential hypertension could be a mere acceleration of the changes seen in aging (Virdis et al., 2010). In these hypertensive patients, the selective inhibition of COX-1 partially restores the impaired acetylcholine-induced increase in forearm blood flow while the selective inhibition of COX-2, which does not produce any adverse effects in the forearm of healthy subjects (Verma et al., 2001), further reduces the response to the muscarinic agonist. These results indicate that COX-1-derived contractile PGs contribute to the endothelial dysfunction and that the reduced production of vasodilator PGs (prostacyclin) secondary to COX-2 inhibition is of minor importance in subjects with normal endothelial function, but becomes relatively more important in hypertensive patients with endothelial dysfunction, where they could play a beneficial compensatory role (Bulut et al., 2003). Similarly, a diminished prostacyclin receptor signaling, as observed in patients with a dysfunctional IP receptor mutation, results in accelerated atherothrombosis (Arehart et al., 2008).

In hypertensive patients, when compared to normotensive subjects, prostacyclin plasma levels are generally decreased (Frolich, 1990; Kuklinska et al., 2009). Since the endothelial generation of prostacyclin is only a fraction of the total synthesis of this PG, the plasma levels may not adequately represent the endothelial production. Alternatively, in hypertensive patients, the duration of the disease far exceeds what is generally observed in rodent models of hypertension and a prolonged exposure to oxidative stress may lead to tyrosine nitration of the PGI_2 synthase and the redistribution of PGH_2 metabolism toward other PGs, such as PGE_2 and $PGF_{2\alpha}$, that also activate TP receptors (Bachschmid et al., 2003; Zou et al., 1999).

In line with the phenotype observed in TP receptor knockout mice, TP receptor antagonists given *in vivo* evoke no or only minor changes in arterial blood pressure, but they limit the endothelial dysfunction associated not only with hypertension (Gelosa et al., 2010; Tesfamariam & Ogletree, 1995) but also, as described in the following sections, with diabetes and atherosclerosis.

B. Diabetes

It is now generally accepted that hyperglycemia-induced reactive species generation contributes to the pathogenesis of the cardiovascular diseases associated with diabetes (Fatehi-Hassanabad et al., 2010).

Arteries from diabetic rabbits (Tesfamariam & Cohen, 1992a, 1992b; Tesfamariam et al., 1989, 1990, 1991, 1995) and diabetic atherosclerotic mice also demonstrated abnormal acetylcholine-induced relaxations, and in mice these alterations were prevented by oral treatment with the TP antagonist, S18886 (Fig. 2, Zuccollo et al., 2005). The fact that the TP antagonist added *in vitro* could immediately prevent the abnormal relaxations in arteries from untreated diabetic mice, strongly suggested that the release of a vasoconstrictor eicosanoid is responsible (Zuccollo et al., 2005). As in arteries from hypertensive animals, it became clear early on that the prostanoid that countered the effects of $^{\bullet}NO$ in arteries from diabetic animals was not thromboxane A_2, because thromboxane synthase inhibitors did not prevent the abnormality. Instead, the vasoconstrictor activity could be ascribed to the product of COX, prostaglandin endoperoxide (PGH_2; Cohen et al., 1988; Pagano et al., 1991; Tesfamariam & Cohen, 1992a, 1992b) or other eicosanoids, such as 12- and 15-HETE (Tesfamariam et al., 1991), whose production increases as a result of shifting eicosanoid production away from PGI_2 synthase. As mentioned above, the cause of this shift in PGH_2 levels has been attributed to the increased production of peroxynitrite and the resulting inactivation of PGI_2 synthase in diseased arteries. For instance, in the arterial wall of atherosclerotic carotid arteries taken from type II diabetic patients, an enhanced tyrosine nitration of prostacyclin synthase is observed and is associated with excessive vascular inflammation (He et al., 2010). Depending on the type of vascular pathology, mitochondria, NADPH oxidase (Wang et al., 1999, 2001, 2002), or eNOS (Zou et al., 2002a, 2002b) can produce increased amounts of $O_2^{-\bullet}$. High levels of oxidants also increase the formation of nonenzymatic oxidation products of arachidonic acid, the isoprostanes, which are potent activators of TP receptors. In addition, we found that exposure of human endothelial cells to inflammatory cytokines or high glucose decreases the expression of eNOS, and that the decrease can be

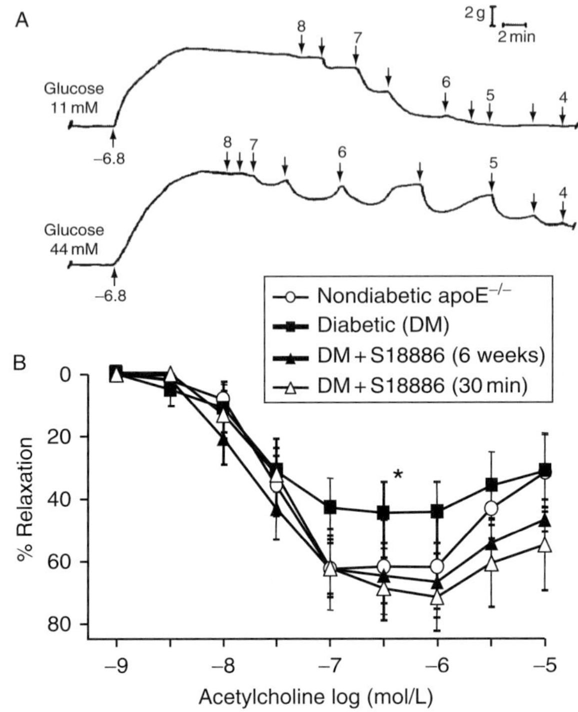

FIGURE 2 High glucose and diabetes impair endothelium-dependent vasodilatation. (A) Normal rabbit aortic ring (top) is contracted with phenylephrine and then fully relaxed by increasing concentrations of acetylcholine which releases •NO from the endothelium. After the ring below was exposed to 800 mg/dL glucose (44 mmol/L) for 6 h, phenylephrine caused a similar contraction, but acetylcholine caused less relaxation and each concentration caused a contraction due to the release of vasoconstrictor eicosanoids (from Tesfamariam et al., 1990). (B) Rings of aorta of apolipoprotein E deficient (apoE$^{-/-}$) mice made diabetic for 6 weeks with streptozotocin were contracted and relaxed by acetylcholine, following a similar protocol. Aortae of diabetic apolipoprotein E deficient mice relaxed significantly less than those of nondiabetic mice (∗). Treatment of the diabetic mice with S18886 during the 6 weeks of diabetes or incubation of the ring of aorta from an untreated diabetic mouse with S18886 *in vitro* improved the vasodilator response to acetylcholine so that there was no longer a significant difference with that of nondiabetic mice (from Zuccollo et al., 2005).

prevented by S18886. Therefore, it is likely that multiple mechanisms contribute to the improvement in vascular function associated with TP receptor blockade.

Although these studies were conducted in experimental animals, it is likely that vasoconstrictor eicosanoids contribute to vascular dysfunction in human patients. This is no better demonstrated by the fact that impaired acetylcholine-induced vasodilation in patients with coronary artery disease are immediately improved by TP blockade with S18886 (Belhassen et al., 2003). The fact that the patients in this study were already treated with aspirin suggests that COX-2 activity, rather than COX-1, may be the main source of the vasoconstrictor prostanoids involved in diminishing vasodilation in the patients with coronary artery disease. Indeed, in patients with severe coronary artery disease, COX-2 inhibition improved flow-mediated dilatation (Chenevard et al., 2003). Nevertheless, for the reasons mentioned above, it is also possible that HETE's or other eicosanoids, such as isoprostanes, are involved.

IV. Vascular Inflammation

Activation of TP receptors may be directly implicated in the chronic inflammatory response (Cayatte et al., 2000; Zuccollo et al., 2005) which contributes to the advancement of atherosclerotic vascular disease. TP agonists such as U46619 are potent stimulators of the expression of vascular cell adhesion molecule-1 (VCAM-1), a principal mediator of leukocyte adhesion to the endothelium (Cayatte et al., 2000; Zuccollo et al., 2005). Nitric oxide, oxidants, and eicosanoids also modulate the inflammatory response of the endothelium to cytokines and metabolic factors such as elevated glucose and fatty acids. As an integral part of the inflammatory response, iNOS is induced which is responsible for the production of both $^{\bullet}NO$ and $O_2^{-\bullet}$, NADPH oxidase is activated which produces more $O_2^{-\bullet}$ and H_2O_2, and phospholipases are stimulated which liberate more arachidonic acid. This being the case, it is thus not surprising that TP antagonists decrease the inflammatory response in endothelial cells (Ishizuka et al., 1996; Zuccollo et al., 2005). For example, human endothelial cells cultured in high glucose media, that simulate the proinflammatory diabetic milieu, show increased surface expression of VCAM-1, and TP blockade inhibits this increased expression (Fig. 3; Zuccollo et al., 2005). Our studies thus stress the importance of the endothelial TP receptor for the regulation of adhesion molecules that are essential in mediating inflammation.

Stimulating TP receptors also increases VCAM-1 expression in smooth muscle cells (Bayat et al., 2008). Although leukocyte adhesion to endothelium is the primary event in inflammation, vascular smooth muscle inflammation is presumably important in enhancing the deleterious influx and retention of leukocytes in the vascular wall. TP receptor stimulation enhanced VCAM-1 expression in smooth muscle cells, not by stimulating NFκB, the prototypical inflammatory transcription factor, but rather by stimulating JUN kinase and the transcription factor, AP1. The importance of this mechanism is seen in the aorta of diabetic hyperlipidemic apolipoprotein E deficient mice in which atherosclerosis is dramatically accelerated compared with nondiabetic mice, and which express VCAM-1 throughout the aortic wall (Fig. 3). Treating the mice with S18886 prevented the VCAM-1 expression indicating that TP receptors are stimulated by endogenous eicosanoids throughout the vasculature and that this

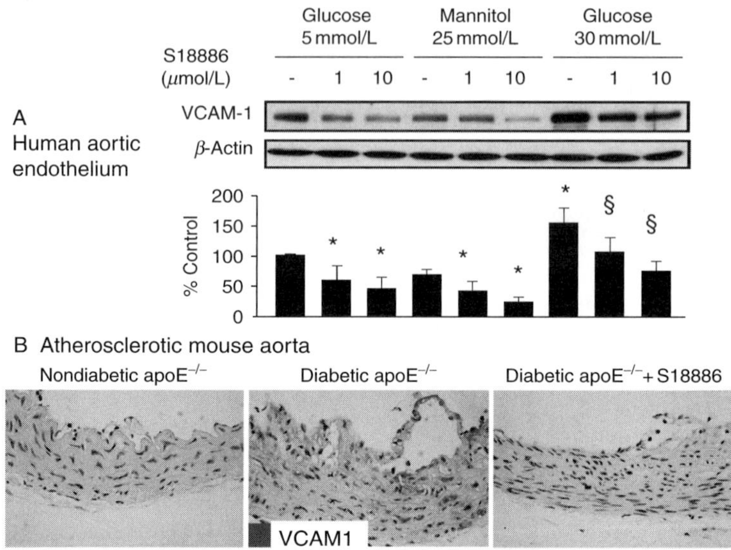

FIGURE 3 S18886 decreases VCAM-1 expression. (A) Human aortic endothelial cells in culture were incubated for 3 days in normal (5 mmol/L) or elevated (30 mmol/L or 600 mg/dL) glucose. High glucose significantly increased VCAM-1 expression assessed by immunoblotting, whereas a similar osmotic concentration of mannitol had no effect. S18886 (1–10 μmol/L) decreased the expression and normalized the expression of VCAM-1 in cells exposed to high glucose (from Zuccollo et al., 2005). (B) VCAM-1 expression demonstrated by immunohistochemistry is increased throughout the vascular wall of diabetic apolipoprotein E deficient mice, but treatment of such mice with S18886 prevents the increase. * values significantly different from control; § values significantly different from high glucose value (from Zuccollo et al., 2005).

contributes to the pathology and explains the antiatherogenic actions of TP blockade (Bayat et al., 2008).

V. Atherosclerosis

Because of the role of TP receptors in regulating the vasomotor and inflammatory events in blood vessels in general, and the endothelium in particular, it may be suspected that changes in •NO and adhesion molecules observed in *in vitro* studies of cultured cells and isolated arteries might also be found in longer term *in vivo* studies. Indeed, treating hyperlipidemic apolipoprotein E deficient mice with S18886 led to a 25% decrease in early atherosclerotic lesion development in the root of the aorta without affecting the hypercholesterolemia in these animals (Fig. 4; Cayatte et al., 2000). Because of the antiplatelet actions of the TP antagonist, mice were also treated with aspirin. The high dose of aspirin used inhibited thromboxane

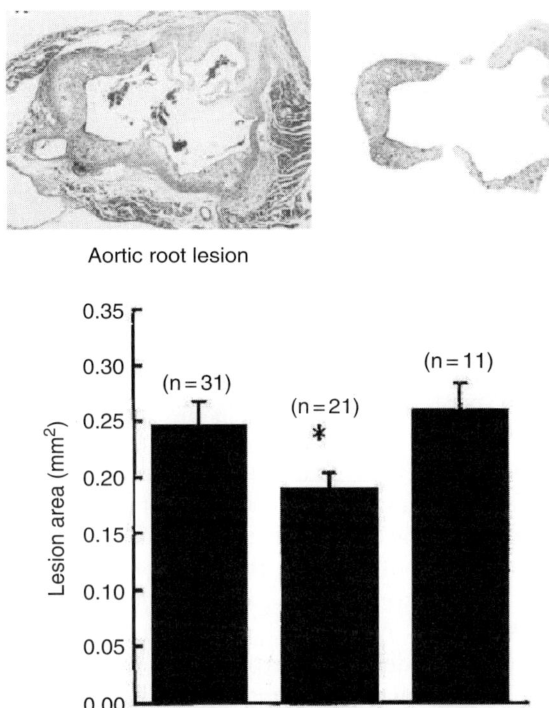

FIGURE 4 S18886 decreases atherosclerosis in the apolipoprotein E deficient mouse. Apolipoprotein E deficient mice were fed normal mouse chow and treated with either S18886 or aspirin from 10 to 21 weeks of age. Atherosclerotic lesions at the base of the aorta were assessed by cross sections of the aortic root (upper left: representative cross section of the aortic root in control mice). Shown at the upper right is the circumscribed lesion area of the aortic root cross section shown on the left panel (control mice), which was analyzed and quantified. The lower part of the figure is the summary bar graph showing that S18886 caused a significant (*, 25%) decrease in the lesion area, whereas aspirin had no significant effect (from Cayatte et al., 2000).

A_2 production during platelet aggregation to an even greater extent than S18886, and yet atherosclerotic lesions were unaffected (Cayatte et al., 2000). Although studies in another mouse model did identify antiatherogenic actions of aspirin that are compatible with a role of platelets in atherogenesis (Cyrus et al., 2002), our results strongly point to a mechanism of action of S18886 resulting in a decreased atherosclerotic lesion development distinct from inhibiting platelet aggregation. In support of a direct vascular effect of the TP antagonist, we found that serum-soluble intercellular adhesion molecule-1, which is shed by endothelial cells into the blood, was decreased by S18886, but not by aspirin. These results strongly suggest that at least some of the antiatherosclerotic action of S18886 is distinct from its antiplatelet activity.

In a subset of New Zealand white rabbits that lack TP receptors only in the vasculature, when compared to control rabbits, a cholesterol-enriched diet produces a less severe impairment of endothelium-dependent relaxations and the incidence of aortic lesions caused by the diet is diminished (Pfister, 2006). Pharmacological studies with TP receptor antagonists have confirmed the deleterious effect of TP receptor activation. In hypercholesterolemic rabbits, the inhibition of plaque formation by S18886 is accompanied by a decreased infiltration of macrophages (Worth et al., 2005). In rabbits with long-term atherosclerosis, the treatment with S18886 caused regression of the atherosclerotic lesions transforming them into a more stable phenotype (Viles-Gonzalez et al., 2005). The beneficial effect of blocking TP receptors on the development of atherosclerosis has been confirmed by the demonstration that apolipoprotein E mice that are genetically deficient in TP receptors also develop less atherosclerosis (Kobayashi et al., 2004). Although the platelet function of TP receptor deficient mice were inhibited as expected, studies in which TP receptor intact or deficient bone marrow was transplanted into TP receptor intact or deficient mice showed that the antiatherosclerotic protection was conferred by the lack of vascular TP receptors, but not by that of TP receptors on bone marrow-derived platelets or leukocytes (Zhuge et al., 2006). Additionally, in atherosclerotic plaques taken either from a murine model of atherosclerosis or from atherosclerotic patients, thromboxane synthase is expressed and is associated with thromboxane A_2 generation (Gabrielsen et al., 2010). These results confirm that a TP antagonist can inhibit atherosclerosis development independently of its antiplatelet effects.

The role of TP receptors in the marked deterioration of endothelial vasodilator function and inflammation associated with hyperglycemia and diabetes also predicts that TP receptor blockade might be particularly effective in combating the accelerated atherosclerosis associated with diabetes. Indeed, in apolipoprotein E deficient mice with type-1 diabetes induced with streptozotocin, atherosclerotic lesions were increased at least threefold (Fig. 5; Park et al., 1998; Zuccollo et al., 2005). Treatment with S18886 abrogated this dramatic enhancement of atherosclerosis caused by diabetes, indicating that TP receptors played an essential role (Zuccollo et al., 2005). The finding is all the more impressive, because there was no effect of the TP antagonist on the hyperlipidemia or hyperglycemia in the treated mice. The TP antagonist not only decreased lesions throughout the aorta but also prevented the decrease in eNOS expression and the increase in vascular VCAM-1 expression. In addition, the accumulation of nitrotyrosine and advanced glycation end products in the aorta were prevented. Nitrotyrosine accumulates in tissues and cells exposed chronically to oxidants and reactive nitrogen species including peroxynitrite, and advanced glycation end products are generated as a result of inflammation and oxidants, particularly in the setting of hyperglycemia. These results indicate that not only do *in vivo* studies confirm the results of studies

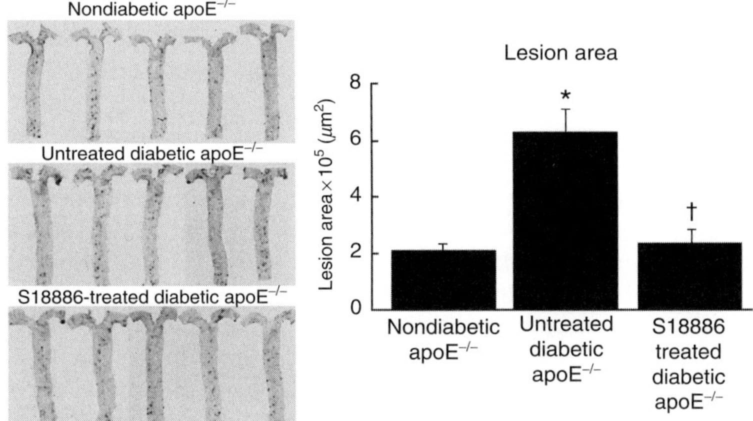

FIGURE 5 S18886 prevents the acceleration of atherosclerosis caused by diabetes. Apolipoprotein E deficient mice made diabetic with streptozotocin from 9 to 15 weeks of age were treated or not with S18886. Lipid laden atherosclerotic plaques were stained with Sudan IV and photographed (left). Lesions were visible on the aortic arch and at spinal branch points. Diabetic apolipoprotein E deficient mice experienced a threefold increase in lesions, but this increase was entirely prevented by treatment with S18886. * values significantly different from non diabetic apoE-/-; † value significantly different from untreated diabetic apoE-/- (from Zuccollo et al., 2005).

obtained *in vitro* in cultured endothelial cells exposed to hyperglycemic conditions (in which beneficial effects of S18886 on eNOS expression and inflammation were observed) but blocking TP receptors attenuates the deleterious tissue consequences of inflammation mediated by oxidants.

Of course, in diabetes, the tissular consequences of the metabolic dysfunction and inflammatory activation may be widespread. For example, the pathophysiology of diabetic nephropathy includes oxidant activation and eicosanoid generation which is stimulated by angiotensin II and prevented by ACEI (Candido et al., 2002). In apolipoprotein E deficient mice, we found that the induction of diabetes led to a more than 10-fold increase in microalbuminuria (Xu et al., 2006). Treatment with S18886 significantly prevented the albuminuria as well as the accompanying increases in TGFβ expression, collagen matrix deposition, and defects in glomerular morphology that were associated with the diabetic state (Xu et al., 2006). There were also dramatic increases in inflammatory enzymes in the kidney of diabetic apolipoprotein E deficient mice including the p47^{phox} subunit of NADPH oxidase, iNOS, and 12-lipoxygenase. The latter enzyme is the mouse homolog of human 15-lipoxygenase, both of which can elaborate HETEs that stimulate TP receptors. There was a fivefold increase in urinary 12-HETE, potentially explaining the beneficial effects of S18886 on albuminuria. In addition, there was an approximately 10-fold increase in 8-iso-PGF$_{1\alpha}$, one of the many isoprostanes produced by the actions of oxidants on arachidonic acid which also stimulate TP receptors. These studies point out that in diabetes not only is there increased production of multiple eicosanoids that stimulate TP receptors but also that treatment with the TP antagonist is associated with broad effects to decrease inflammation and activation of oxidant pathways. Thus, treatment with S18886 may positively reinforce the effect of blocking TP receptors by decreasing the production of TP stimulating eicosanoids. These effects go far beyond those of inhibiting platelet aggregation. This was demonstrated directly by the fact that diabetic animals treated with aspirin did not show the benefit observed in the diabetic animals treated with S18886 as regard either atherosclerotic lesions (unpublished studies) or nephropathy (Xu et al., 2006). The protective renal actions of S18886 have also been observed in the uninephrectomized obese Zucker rat (Sebekova et al., 2007), and the TP antagonist attenuated renal damage in the double transgenic hypertensive rat harboring the human renin and angiotensinogen genes (Sebekova et al., 2008). Thus, the protective renal actions of S18886 extend to both type 1 and type 2 diabetes and renovascular hypertension, illustrate the role of TP receptors in these pathologies, and are particularly advantageous in treating complex cardiovascular disease that is usually of a systemic nature.

VI. Implications for Clinical Usefulness of TP Antagonists

Antiplatelet treatment is a proven therapeutic modality in patients with arteriosclerotic cardiovascular disease. The benefits include the prevention of recurrent thromboembolic stroke and myocardial infarction (Maree & Fitzgerald, 2004). The mechanism underlying this therapeutic effect is largely attributed to preventing the formation of platelet thromboemboli. Our animal and cellular studies have revealed that treatment of vascular disease with a TP antagonist has actions far beyond its antithrombotic effect at the level of platelets, and can be attributed to direct effects on endothelial and vascular smooth muscle cells. These include effects on vascular adhesion molecules, eNOS expression and function, oxidant production, and accumulation of extracellular matrix and advanced glycation end products. In addition, these tissular effects appear to extend beyond the vasculature, particularly in diabetes in which hyperlipidemia and hyperglycemia induce tissue damage such as that observed in the kidney of diabetic hyperlipidemic mice. The importance of the TP receptors in vascular and other tissular pathologies, particularly in diabetes, may be due to the fact that not only thromboxane A_2 but also other eicosanoids (including HETEs and isoprostanes) are produced in diseased tissues to the extent that they activate TP receptors. Aspirin has no effect on the production of isoprostanes which are formed nonenzymatically from arachidonic acid, and aspirin can actually increase

HETE production by COX (Meade et al., 1993). Thus, the broad ability to block the actions of many eicosanoids that activate TP receptors may account for the added benefit of TP blockade. Of course, in addition, blocking platelet TP receptors inhibits platelet aggregation. Therefore, it is reasonable to anticipate that treatment of patients with a TP receptor antagonist will result in therapeutic benefits additional to those of aspirin. Furthermore, the addition of TP antagonism to COX-2 inhibitory activity may improve the cardiovascular risk profile of COX-2 inhibitors (Rovati et al., 2010).

Acknowledgments

Dr. Cohen is supported by National Institutes of Health Grants RO1 HL 31607, RO1 AG027080, PO1 HL081587, PO1 HL68758, and R21 DK084390. The work reviewed in this chapter was supported by a Strategic Alliance between Institut de Recherche Servier and Boston Medical Center as well as by Institut de Recherches Internationales Servier.

Conflict of Interest: Drs. Félétou and Verbeuren are currently employees of the Institut de Recherches Servier, a research organization that specializes in preclinical animal pharmacology, while Dr. Vanhoutte is a former employee of this institute.

Abbreviations

ACEI	angiotensin converting enzyme inhibitor
COX	cyclooxygenase
DOCA-salt rats	deoxycorticosterone acetate-salt hypertensive rats
EDCF	endothelium-derived contracting factor
EDHF	endothelium-derived hyperpolarizing factor
H_2O_2	hydrogen peroxide
HETE	hydroperoxyeicosatetraenoic acid
HMG CoA	3-hydroxy-3-methylglutaryl-coenzyme A
L-NAME	L-arginine-methylester
LPS	lipopolysaccharides
NO	nitric oxide
NOS	nitric oxide synthase
$O_2^{\bullet -}$	superoxide anion
$OONO^-$	peroxynitrite
PG	prostaglandin
PGH_2	prostaglandin H_2/endoperoxide
PGI_2	prostacyclin
SHR	spontaneously hypertensive rats
VCAM-1	vascular cell adhesion molecule 1
WKY rats	Wistar-Kyoto rats

References

Arehart, E., Stitham, J., Asselbergs, F. W., et al. (2008). Acceleration of cardiovascular disease by a dysfunctional prostacyclin receptor mutation: Potential implications for cyclooxygenase-2 inhibition. *Circulation Research*, *102*, 986–993.

Bachschmid, M., Thurau, S., Zou, M. H., et al. (2003). Endothelial cell activation by endotoxin involves superoxide/NO-mediated nitration of prostacyclin synthase and thromboxane receptor stimulation. *The FASEB Journal*, *17*, 914–916.

Bayat, H., Xu, S., Pimentel, D., et al. (2008). Activation of thromboxane receptor upregulates interleukin (IL)-1beta-induced VCAM-1 expression through JNK signaling. *Arteriosclerosis, Thrombosis, and Vascular Biology*, *28*, 127–134.

Behm, D. J., Ogbonna, A., Wu, C., et al. (2009). Epoxyeicosatrienoic acids function as selective, endogenous antagonists of native thromboxane receptors: Identification of a novel mechanism of vasodilation. *The Journal of Pharmacology and Experimental Therapeutics*, *328*, 231–239.

Belhassen, L., Pelle, G., Dubois-Rande, J. L., et al. (2003). Improved endothelial function by the thromboxane A2 receptor antagonist S 18886 in patients with coronary artery disease treated with aspirin. *Journal of the American College of Cardiology*, *41*, 1198–1204.

Boffa, J. J., Just, A., Coffman, T. M., et al. (2004). Thromboxane receptor mediates renal vasoconstriction and contributes to acute renal failure in endotoxemic mice. *Journal of the American Society of Nephrology*, *15*, 2358–2365.

Bulut, D., Liaghat, S., Hanefeld, C., et al. (2003). Selective cyclo-oxygenase-2 inhibition with parecoxib acutely impairs endothelium-dependent vasodilatation in patients with essential hypertension. *Journal of Hypertension*, *21*, 1663–1667.

Candido, R., Jandeleit-Dahm, K. A., Cao, Z., et al. (2002). Prevention of accelerated atherosclerosis by angiotensin-converting enzyme inhibition in diabetic apolipoprotein E-deficient mice. *Circulation*, *106*, 246–253.

Cayatte, A. J., Du, Y., Oliver-Krasinski, J., et al. (2000). The thromboxane receptor antagonist S 18886 but not aspirin inhibits atherogenesis in apo E-deficient mice evidence that eicosanoids other than thromboxane contribute to atherosclerosis. *Arteriosclerosis, Thrombosis, and Vascular Biology*, *20*, 1724–1728.

Chenevard, R., Hürlimann, D., Béchir, M., et al. (2003). Selective COX-2 inhibition improves endothelial function in coronary artery disease. *Circulation*, *107*, 405–409.

Cheng, Y., Austin, S. C., Rocca, B., et al. (2002). Role of prostacyclin in the cardiovascular response to thromboxane A2. *Science*, *296*, 539–541.

Cohen, R. A., Zitnay, K. M., Weisbrod, R. M., et al. (1988). Influence of the endothelium on tone and the response of isolated pig coronary artery to norepinephrine. *The Journal of Pharmacology and Experimental Therapeutics*, *244*, 550–555.

Cyrus, T., Sung, S., Zhao, L., et al. (2002). Effect of low-dose aspirin on vascular inflammation, plaque stability, and atherogenesis in low-density lipoprotein receptor-deficient mice. *Circulation*, *106*, 1282–1287.

Egan, K. M., Lawson, J. A., Fries, S., et al. (2004). COX-2-derived prostacyclin confers atheroprotection on female mice. *Science*, *306*, 1954–1957.

Fatehi-Hassanabad, Z., Chan, C. B., & Furman, B. L. (2010). Reactive oxygen species and endothelial function in diabetes. *European Journal of Pharmacology*, *636*, 8–17.

Félétou, M., Huang, Y., & Vanhoutte, P. M. (2010a). Vasoconsttrictor prostanoids. *Pflugers Archiv: European Journal of Physiology*, *459*, 941–950.

Félétou, M., & Vanhoutte, P. M. (2006a). Endothelial dysfunction: A multifaceted disorder (The Wiggers Award Lecture). *American Journal of Physiology. Heart and Circulatory Physiology*, *291*, H985–H1002.

Félétou, M., & Vanhoutte, P. M. (2006b). *EDHF: The Complete Story.* Boca Raton, USA: Taylor & Francis CRC Press.

Félétou, M., Vanhoutte, P. M., & Verbeuren, T. J. (2010b). The TP-receptor: The common villain. *Journal of Cardiovascular Pharmacology*, *55*, 317–332.

Fetalvero, K. M., Martin, K. A., & Hwa, J. (2007). Cardioprotective prostacyclin signaling in vascular smooth muscle. *Prostaglandins & Other Lipid Mediators*, *82*, 109–118.

Francois, H., Athirakul, K., Mao, L., et al. (2004). Role for thromboxane receptors in angiotensin-II-induced hypertension. *Hypertension*, *43*, 364–369.

Francois, H., Makhanova, N., Ruiz, P., et al. (2008). A role for the thromboxane receptor in L-NAME hypertension. *American Journal of Physiology: Renal Physiology*, *295*, F1096–F1102.

Frolich, J. C. (1990). Prostacyclin in hypertension. *Journal of Hypertension*, *8*(Suppl.), S73–S78.

Furchgott, R. F., & Zawadzki, J. V. (1980). The obligatory role of the endothelial cells in the relaxation of arterial smooth muscle by acetylcholine. *Nature*, *288*, 373–376.

Gabrielsen, A., Qiu, H., Bäck, M., et al. (2010). Thromboxane synthase expression and thromboxane A(2) production in the atherosclerotic lesion. *Journal of Molecular Medicine* 88, 795–806.

Gelosa, P., Ballerio, R., Banfi, C., et al. (2010). Terutroban, a TP receptor antagonist, increases survival in stroke-prone rats by preventing systemic inflammation and endothelial dysfunction: Comparison with aspirin and rosuvastatin. *The Journal of Pharmacology and Experimental Therapeutics*, *334*(1), 199–205.

Gluais, P., Lonchampt, M., Morrow, J. D., et al. (2005). Acetylcholine-induced endothelium-dependent contractions in the SHR aorta: The Janus face of prostacyclin. *British Journal of Pharmacology*, *146*, 834–845.

Harlan, J. M., & Callahan, K. S. (1984). Role of hydrogen peroxide in the neutrophil-mediated release of prostacyclin from cultured endothelial cells. *Journal of Clinical Investigation*, *74*, 442–448.

He, C., Choi, H. C., & Xie, Z. (2010). Enhanced tyrosine nitration of prostacyclin synthase is associated with increased inflammation in atherosclerotic carotid arteries from type 2 diabetic patients. *The American Journal of Pathology*, *176*, 2542–2549.

Ishimitsu, T., Uehara, Y., Iwai, J., et al. (1991). Vascular eicosanoid production in experimental hypertensive rats with different mechanisms. *Prostaglandins Leukotrienes and Essential Fatty Acids*, *43*, 179–184.

Ishizuka, T., Suzuki, K., Kawakami, M., et al. (1996). Thromboxane A_2 receptor blockade suppresses intercellular adhesion molecule-1 expression by stimulated vascular endothelial cells. *European Journal of Pharmacology*, *312*, 367–377.

Kobayashi, T., Tahara, Y., Matsumoto, M., et al. (2004). Roles of thromboxane A(2) and prostacyclin in the development of atherosclerosis in apoE-deficient mice. *Journal of Clinical Investigation*, *114*, 784–794.

Kuklinska, A. M., Mroczko, B., Musial, W. J., et al. (2009). Diagnostic biomarkers of essential arterial hypertension: The value of prostacyclin, nitric oxide, oxidized-LDL, and peroxide measurements. *International Heart Journal*, *50*, 341–351.

Liu, C. Q., Leung, F. P., Wong, S. L., et al. (2009). Thromboxane prostanoid receptor activation impairs endothelial nitric oxide-dependent vasorelaxations: The role of Rho kinase. *Biochemical Pharmacology*, *78*, 374–381.

Lüscher, T. F., & Vanhoutte, P. M. (1986). Endothelium-dependent contractions to acetylcholine in the aorta of the spontaneously hypertensive rat. *Hypertension*, *8*, 344–348.

Maree, A. O., & Fitzgerald, D. J. (2004). Aspirin and coronary artery disease. *Thrombosis and Haemostasis*, *92*, 1175–1181.

Martin, W., Villani, G. M., Jothianandan, D., et al. (1985). Selective blockade of endothelium-dependent and glyceryl trinitrate-induced relaxation by hemoglobin and by methylene blue in rabbit aorta. *The Journal of Pharmacology and Experimental Therapeutics*, *232*, 708–716.

Meade, E. A., Smith, W. L., & DeWitt, D. L. (1993). Differential inhibition of prostaglandin endoperoxide synthase (cyclooxygenase) isozymes by aspirin and other non-steroidal antiinflammatory drugs. *The Journal of Biological Chemistry*, *268*, 6610–6614.

Michel, F. S., Man, G. S., Man, R. Y. K., et al. (2008). Hypertension and the absence of EDHF-mediated responses favor endothelium-dependent contractions in renal arteries of the rat. *British Journal of Pharmacology*, *13*, 4198–4217.

Miggin, S. M., Lawler, O. A., & Kinsella, B. T. (2003). Palmitoylation of the human prostacycllin receptor. Functional implications of palmitoylation and isoprenylation. *The Journal of Biological Chemistry*, *278*, 6947–6958.

Moncada, S., & Vane, J. R. (1979). Pharmacology and endogenous roles of prostaglandin endoperoxides, thromboxane A2 and prostacyclin. *Pharmacological Reviews*, *30*, 293–331.

Murata, T., Ushikubi, F., Matsuoka, T., et al. (1997). Altered pain perception and inflammatory response in mice lacking prostacyclin receptor. *Nature*, *388*, 678–682.

Nakahata, N. (2008). Thromboxane A2: Physiology/pathophysiology, cellular signal transduction and pharmacology. *Pharmacology & Therapeutics*, *118*, 18–35.

Pagano, P. J., Lin, L., Sessa, W. C., et al. (1991). Arachidonic acid elicits endothelium-dependent release from the rabbit aorta of a constrictor prostanoid resembling prostaglandin endoperoxides. *Circulation Research*, *69*, 396–405.

Park, L., Raman, K. G., Lee, K. J., et al. (1998). Suppression of accelerated diabetic atherosclerosis by the soluble receptor for advanced glycation endproducts. *Nature Medicine*, *4*, 1025–1031.

Patrignani, P., Di febbo, C., Tacconelli, S., et al. (2008). Differential association between human prostacyclin receptor polymorphisms and the development of venous thrombosis and intimal hyperplasia: A clinical biomarker study. *Pharmacogenetics and Genomics*, *18*, 611–620.

Pfister, S. L. (2006). Aortic thromboxane receptor deficiency alters vascular reactivity in cholesterol-fed rabbits. *Atherosclerosis*, *189*, 358–363.

Rosenfeld, L., Grover, G. J., & Stier, C. T. Jr., (2001). Ifetroban sodium: An effective TxA_2/PGH_2 receptor antagonist. *Cardiovascular Drug Reviews*, *19*, 97–115.

Rovati, G. E., Sala, A., Capra, V., et al. (2010). Dual COXIB/TP antagonists: A possible new twist in NASID pharmacology and cardiovascular risk. *Trends in Pharmacological Sciences*, *31*, 102–107.

Schmidt, P., Youhnovski, N., Daiber, A., et al. (2003). Specific nitration at tyrosine-430 revealed by high resolution mass spectrometry as basis for redox regulation of bovine prostacyclin synthase. *The Journal of Biological Chemistry*, *278*, 12813–12819.

Sebekova, K., Eifert, T., Klassen, A., et al. (2007). Renal effects of S 18886 (Terutroban), a TP receptor antagonist, in an experimental model of type 2 diabetes. *Diabetes*, *56*, 968–974.

Sebekova, K., Ramuscak, A., Boor, P., et al. (2008). The selective TP receptor antagonist, S 18886 (terutroban), attenuates renal damage in the double transgenic rat model of hypertension. *American Journal of Nephrology*, *28*, 47–53.

Shi, Y., Félétou, M., Ku, D. D., et al. (2007). The calcium ionophore A23187 induces endothelium-dependent contractions in femoral arteries from rats with streptozotocin-induced diabetes. *British Journal of Pharmacology*, *150*, 624–632.

Stitham, J., Arehart, E. J., Gleim, S., et al. (2007). Arginine (CGC) codon targeting in the human prostacyclin receptor gene (PTGIR) and G-protein coupled receptors (GPCR). *Gene*, *396*, 180–187.

Stitham, J., Gleim, S. R., Douville, K., et al. (2006). Versatility and different roles of cysteine residues in human prostacyclin receptor structure and function. *The Journal of Biological Chemistry*, *281*, 37227–37236.

Taddei, S., Virdis, A., Ghiadoni, L., et al. (2000). Endothelial dysfunction in hypertension. *Journal of Nephrology*, *23*, 205–210.

Takayama, K., Yuhki, K., Ono, K., et al. (2005). Thromboxane A2 and prostaglandin F2alpha mediate inflammatory tachycardia. *Nature Medicine, 11, 562–566.*

Tang, E. H. C., Félétou, M., Huang, Y., et al. (2005). Acetylcholine and sodium nitroprusside cause long-term inhibition of EDCF-mediated contraction. *American Journal of Physiology: Heart and Circulatory Physiology, 289*, H2434–H2440.

Tang, E. H. C., Jensen, B. L., Skott, O., et al. (2008). The role of prostaglandin E and thromboxane-prostanoid receptors in the response to prostaglandin E2 in the aorta of Wistar Kyoto rats and spontaneously hypertensive rats. *Cardiovascular Research, 78*, 130–138.

Tang, E. H. C., & Vanhoutte, P. M. (2008). Gene expression changes of prostanoid synthases in endothelial cells and prostanoid receptors in vascular smooth muscle cells caused by aging and hypertension. *Physiological Genomics, 32*, 409–418.

Tesfamariam, B., Brown, M. L., & Cohen, R. A. (1991). Elevated glucose impairs endothelium-dependent relaxation by activating protein kinase C. *Journal of Clinical Investigation, 87*, 1643–1648.

Tesfamariam, B., Brown, M. L., & Cohen, R. A. (1995). 15-hydroxyeicosatetraenoic acid and diabetic endothelial dysfunction in rabbit aorta. *Journal of Cardiovascular Pharmacology, 25*, 748–755.

Tesfamariam, B., Brown, M. L., Deykin, D., et al. (1990). Elevated glucose promotes generation of endothelium-derived vasoconstrictor prostanoids in rabbit aorta. *Journal of Clinical Investigation, 85*, 929–932.

Tesfamariam, B., & Cohen, R. A. (1992a). Free radicals mediate endothelial cell dysfunction caused by elevated glucose. *The American Journal of Physiology, 263*, H321–H326.

Tesfamariam, B., & Cohen, R. A. (1992b). Role of superoxide anion and endothelium in vasoconstrictor action of prostaglandin endoperoxide. *The American Journal of Physiology, 262*, H1915–H1919.

Tesfamariam, B., Jakubowski, J. A., & Cohen, R. A. (1989). Contraction of diabetic rabbit aorta caused by endothelium-derived PGH_2-TxA_2. *The American Journal of Physiology, 257*, H1327–H1333.

Tesfamariam, B., & Ogletree, M. L. (1995). Dissociation of endothelial cell dysfunction and blood pressure in SHR. *American Journal of Physiology: Heart and Circulatory Physiology, 269*, H189–H194.

Thomas, D. W., Mannon, R. B., Mannon, P. J., et al. (1998). Coagulation defects and altered hemodynamic responses in mice lacking receptors for thromboxane A2. *Journal of Clinical Investigation, 102*, 1994–2001.

Tsuboi, K., Sugimoto, Y., & Ichikawa, A. (2002). Prostanoid receptor subtypes. *Prostaglandins & Other Lipid Mediators, 68–69, 535–556.*

Valentin, F., Field, M. C., & Tippins, J. R. (2004). The mechanism of oxidative stress stabilization of the thromboxane receptor in COS-7 cells. *The Journal of Biological Chemistry, 279*, 8316–8324.

Verbeuren, T. J. (2006). Terutroban and endothelial TP receptors in atherogenesis. *Medical Sciences (Paris), 22*, 437–443.

Verbeuren, T. J., Simonet, S., Descombes, J. J., et al. (1995). S 18204: A new powerful TXA2-receptor antagonist with a long duration of action. *Thrombosis and Haemostasis, 73*, 1324.

Verma, S., Raj, S. R., Shewchuk, L., et al. (2001). Cyclooxygenase-2 blockade does not impair endothelial vasodilator function in healthy volunteers: Randomized evaluation of rofecoxib versus naproxen on endothelium-dependent vasodilatation. *Circulation, 104*, 2879–2882.

Viles-Gonzalez, J. F., Fuster, V., Corti, R., et al. (2005). Atherosclerosis regression and TP receptor inhibition: Effect of S 18886 on plaque size and composition—A magnetic resonance imaging study. *European Heart Journal, 26*, 1557–1561.

Virdis, A., Ghiadoni, L., & Taddei, S. (2010). Human endothelial dysfunction: EDCFs. *Pflugers Archive: European Journal of Physiology*, *459*, 1015–1023.

Wang, H. D., Hope, S., Du, Y., et al. (1999). Paracrine role of adventitial superoxide anion in mediating spontaneous tone of the isolated rat aorta in angiotensin II-induced hypertension. *Hypertension*, *33*, 1225–1232.

Wang, H. D., Johns, D. G., Xu, S., et al. (2002). Role of superoxide anion in regulating the pressor and vascular hypertrophic response to angiotensin II. *American Journal of Physiology: Heart and Circulatory Physiology*, *292*, H1697–H1702.

Wang, H. D., Xu, S., Johns, D. G., et al. (2001). Role of NADPH oxidase in the vascular hypertrophic and oxidative stress response to angiotensin II in mice. *Circulation Research*, *88*, 947–953.

Wilson, S. J., Cavanagh, C. C., Lesher, A. M., et al. (2009). Activation-dependent stabilization of the human thromboxane receptor: Role of reactive oxygen species. *Journal of Lipid Research*, *50*, 1047–1056.

Wong, S. L., Leung, F. P., Lau, C. W., et al. (2009). Cyclooxygenase-2-derived prostaglandin F2alpha mediates endothelium-dependent contractions in the aortae of hamsters with increased impact during aging. *Circulation Research*, *104*, 228–235.

Worth, N. F., Berry, C. L., Thomas, A. C., et al. (2005). S 18886, a selective TP receptor antagonist, inhibits development of atherosclerosis in rabbits. *Atherosclerosis*, *183*, 65–73.

Xiao, C. Y., Hara, A., Yuhki, K., et al. (2001). Roles of prostaglandin I(2) and thromboxane A(2) in cardiac ischemia-reperfusion injury: A study using mice lacking their respective receptors. *Circulation*, *104*, 2210–2215.

Xu, S., Jiang, B., Maitland, K. A., et al. (2006). The thromboxane receptor antagonist S 18886 attenuates renal oxidant stress and proteinuria in diabetic apolipoprotein E-deficient mice. *Diabetes*, *55*, 110–119.

Yamada, T., Fujino, T., Yuhki, K., et al. (2003). Thromboxane A2 regulates vascular tone via its inhibitory effect on the expression of inducible nitric oxide synthase. *Circulation*, *108*, 2381–2386.

Yu, Y., Lucitt, M. B., Stubbe, J., et al. (2009). Prostaglandin F2alpha elevates blood pressure and promotes atherosclerosis. *Proceedings of the National Academy of Sciences of the United States of America*, *106*, 7985–7990.

Zhuge, X., Arai, H., Xu, Y., et al. (2006). Protection of atherogenesis in thromboxane A2 receptor-deficient mice is not associated with thromboxane A2 receptor in bone marrow-derived cells. *Biochemical and Biophysical Research Communications*, *351*, 865–871.

Zou, M. H., Leist, M., & Ullrich, V. (1999). Selective nitration of prostacyclin synthase and defective vasorelaxation in atherosclerotic bovine coronary arteries. *The American Journal of Pathology*, *154*, 1359–1365.

Zou, M. H., Shi, C., & Cohen, R. A. (2002a). High glucose via peroxynitrite causes tyrosine nitration and inactivation of prostacyclin synthase that is associated with thromboxane/prostaglandin H2 receptor-mediated apoptosis and adhesion molecule expression in cultured human aortic endothelial cells. *Diabetes*, *51*, 198–203.

Zou, M. H., Shi, C., & Cohen, R. A. (2002b). Oxidation of the zinc-thiolate complex and uncoupling of endothelial nitric oxide synthase by peroxynitrite. *Journal of Clinical Investigation*, *109*, 817–826.

Zuccollo, A., Shi, C., Mastroianni, R., et al. (2005). The thromboxane A2 receptor antagonist S 18886 prevents enhanced atherogenesis caused by diabetes mellitus. *Circulation*, *112*, 3001–3008.

David G. Harrison, Wei Chen, Sergey Dikalov, and Li Li

Department of Medicine and Division of Cardiology, Emory University School of Medicine and Atlanta Veterans Administration Medical Center, Decatur, Georgia, USA

Regulation of Endothelial Cell Tetrahydrobiopterin: Pathophysiological and Therapeutic Implications

Abstract

Tetrahydrobiopterin (BH_4) is a critical cofactor for the nitric oxide synthases. In the absence of BH_4, these enzymes become uncoupled, fail to produce nitric oxide, and begin to produce superoxide and other reactive oxygen species (ROS). BH_4 levels are modulated by a complex biosynthetic pathway, salvage enzymes, and by oxidative degradation. The enzyme GTP

Advances in Pharmacology, Volume 60

1054-3589/10 $35.00
10.1016/S1054-3589(10)60005-1

cyclohydrolase-1 catalyzes the first step in the *de novo* synthesis of BH_4 and new evidence shows that this enzyme is regulated by phosphorylation, which reduces its interaction with its feedback regulatory protein (GFRP). In the setting of a variety of common diseases, such as atherosclerosis, hypertension, and diabetes, reactive oxygen species promote oxidation of BH_4 and inhibit expression of the salvage enzyme dihydrofolate reductase (DHFR), promoting accumulation of BH_2 and NOS uncoupling. There is substantial interest in therapeutic approaches to increasing tissue levels of BH_4, largely by oral administration of this agent. BH_4 treatment has proved effective in decreasing atherosclerosis, reducing blood pressure, and preventing complications of diabetes in experimental animals. While these basic studies have been very promising, there are only a few studies showing any effect of BH_4 therapy in humans in treatment of these common problems. Whether BH_4 or related agents will be useful in treatment of human diseases needs additional study.

I. Introduction

One of the major developments in biology in the past 30 years has been the discovery of NO as a signaling molecule (Moncada et al., 1991). Nitric oxide is produced enzymatically in mammalian cells by a family of enzymes known as the nitric oxide synthases (NOSs), which include the neuronal isoform or NOS1, the inducible NOS also referred to as NOS2, and the endothelial enzyme or NOS3. NO produced by these enzymes serves myriad signaling properties in multiple organs and cells. The production of NO is generally beneficial, particularly when produced by the endothelial and neuronal isoforms. NO produced by these enzymes promotes vasodilatation, inhibits inflammation, stimulates long-term potentiation of neuronal transmission, and inhibits platelet aggregation (Khan et al., 1996; Kubes et al., 1991; Moncada et al., 1988). NO can also be damaging when produced in high concentrations, for example, upon induction of NOS2, which generates large amounts of NO constantly in inflammatory states. This toxicity is in part due to reactions of higher oxides of NO reacting with various cellular targets, including DNA, lipids, membranes, and various proteins (Burney et al., 1999; Zhang et al., 2003). While overt inflammation is associated with increased NO production (Fortin et al., 2010), more commonly, NO bioavailability is reduced in conditions such as vascular, muscular, and neuronal diseases and this contributes to pathophysiology in these conditions (Harrison & Cai, 2003; Thomas et al., 1998).

Given the profound effects of NO on target tissues, there has been interest in modulating either its delivery or synthesis pharmacologically. Various drugs that release NO, such as the organic nitrates, misoldamine, and sodium nitroprusside, have been employed for decades for treatment of cardiovascular diseases, even before endogenous production of NO was discovered (Miller & Megson, 2007). Organic nitrates are commonly used to treat coronary artery disease. When combined with hydralazine, isosorbide dinitrate improves outcome in subgroups of patients with congestive heart failure (Taylor et al., 2004), but otherwise the organic nitrates have not provided survival benefit and use of these agents is often limited by the development of tolerance (Kosmicki, 2009).

Because of this limited success with administering NO donors and nitrates over the long term, there has been interest in improving endogenous production of NO. One approach has been to supplement cofactors for the NO synthase enzyme. Among these, administration of the cofactor tetrahydrobiopterin (BH_4) seems to be an attractive option. Various diseases alter cellular levels of BH_4, and BH_4 administration or efforts to prevent its degradation seem to be effective in several diseases. In this review, we will discuss the chemistry of the pterins and the pathways involved in mammalian BH_4 synthesis and review a growing body of literature suggesting that BH_4 might prove to be an effective therapy for a variety of diseases.

II. Chemistry and Synthesis of Pterins

A. Structural Properties

Biopterin and its related molecules belong to a class of compounds that include pterins, folates, riboflavin, lumazines, and alloxazines. These all contain two or three heterocyclic six-membered rings and are ubiquitous in plants and animals. Pterins possess one pyrazine and one pyrimadine ring (Fig. 1), and were named so because they were first identified as a pigment in butterfly wings

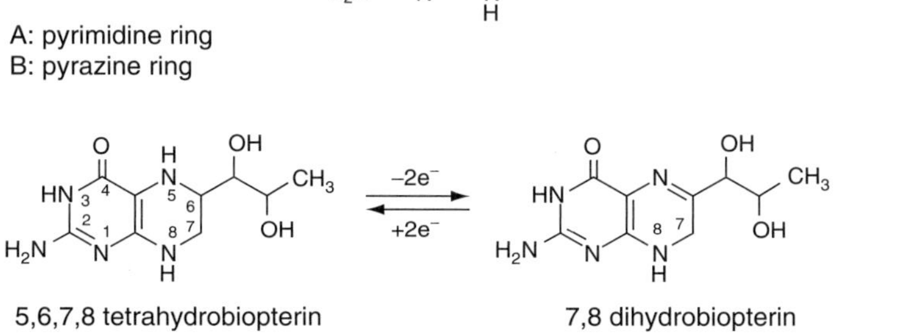

FIGURE 1 Structure of tetrahydrobiopterin and related compounds. Shown are representative examples of pyrimidine and pyrazine rings, tetrahydrofolic acid, 5,6,7,8-tetrahydrobiopterin, and 7,8-dihydrobiopterin.

(Gates, 1947). The major pterins in vertebrates are BH_4, which has hydrogens at the 5, 6, 7, and 8 positions, and its oxidized form, 7,8-dihydrobiopterin (BH_2) (Fig. 1). BH_4 is important in mammalian biology because it serves as a critical cofactor for the aromatic amino acid hydroxylases and the NOSs (Bigham et al., 1987). The three amino acid hydroxylases include phenylalanine hydroxylase, tyrosine hydroxylase, and tryptophan hydroxylase, ultimately leading to formation of tyrosine, dopamine, and serotonin. In the case of these enzymes, BH_4 is converted to the quinoid form of BH_2 (qBH_2) and must be converted back to BH_4 via a salvage pathway (discussed below). Genetic deficiencies of BH_4 are uncommon, but cause a form of phenylketonuria due to reduced conversion of phenylalanine to tyrosine (Tada et al., 1970).

B. Biosynthesis and Salvage Pathways for BH_4

1. De novo *Synthesis Pathway*

BH_4 is synthesized by the sequential actions of three enzymes, GTP cyclohydrolase-1 (GTPCH-1), 6-pyruvoyl-tetrahydropterin synthase (PTPS), and sepiapterin reductase (SR) (Fig. 2). This process has been reviewed in depth by Thony et al. (2000), and will only be briefly summarized here. The first step in this process, which requires the zinc-containing enzyme GTPCH-1, is generally rate limiting. This step involves initial opening of the imidazole and furanose rings, release of formate, an Amadori rearrangement, and ultimate formation of the pyrazine ring. The final product of the GTPCH-1-mediated reaction is the intermediate 7,8-dihydroneopterin triphosphate (7,8-DNTP) (Fig. 2). Monitoring neopterin production after dephosphorylation of 7,8-DNTP is often used to quantify GTPCH-1 activity. It is of interest that GTP cyclohydrolase-II, which is involved in production of riboflavin in prokaryotic organisms, involves a ring-opening step of GTP in a manner similar to GTPCH-1 (Ren et al., 2005).

The second reaction in the *de novo* pathway is conversion of 7,8-DNTP to 6-pyruvoyl-tetrahydropterin by the enzyme PTPS. This involves reduction of a double bond at the N5 position of the pyrazine ring, removal of the phosphate groups, and modification of the side chain (Fig. 3). The final step, catalyzed by SR, entails conversion of the keto-groups at the $1'$ and $2'$ positions on the side chain to hydroxyl groups yielding BH_4.

A variety of pathophysiological stimuli modulate levels of GTPCH-1 mRNA and protein in mammalian cells. Cytokines, such as IL-1β, IFN-γ, TNFα, and inflammatory mediators, such as lipopolysaccharide (LPS), markedly increase GTPCH-1 mRNA and protein levels in endothelial cells, vascular smooth muscle cells, fibroblasts, and macrophages (Katusic et al., 1998; Werner et al., 1998). Because of the potent stimulation of GTPCH-1 by inflammatory cytokines, accumulation of plasma neopterin is often employed as a measure of systemic inflammation (Dominguez-Rodriguez et al., 2009).

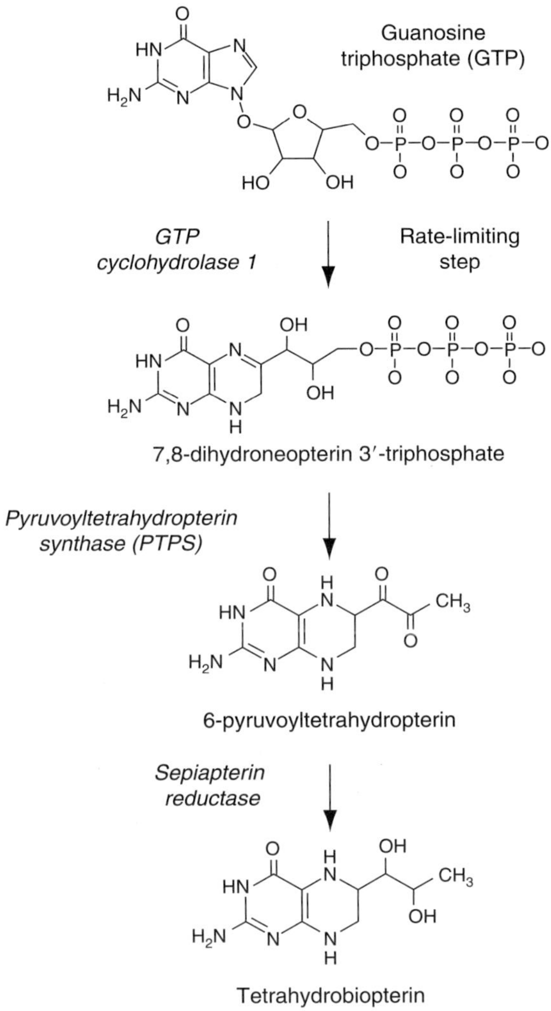

Guanosine triphosphate (GTP)

GTP cyclohydrolase 1

Rate-limiting step

7,8-dihydroneopterin 3′-triphosphate

Pyruvoyltetrahydropterin synthase (PTPS)

6-pyruvoyltetrahydropterin

Sepiapterin reductase

Tetrahydrobiopterin

FIGURE 2 *De novo* pathway of tetrahydrobiopterin synthesis from guanosine triphosphate.

Sun et al. (2009) have reported that 17-β estradiol stimulates GTPCH-1 mRNA transcription in pulmonary artery endothelial cells and that this response is dependent on protein kinase A and activation of the cAMP response element binding (CREB) protein. As might be expected, forskolin, which increases cellular cAMP, also increased GTPCH-1 expression in this study.

HMG-CoA reductase inhibitors, known as statins, enhance GTPCH-1 expression in cultured endothelial cells and in the aorta of streptozotocin-induced diabetic mice (Hattori et al., 2003; Wenzel et al., 2008).

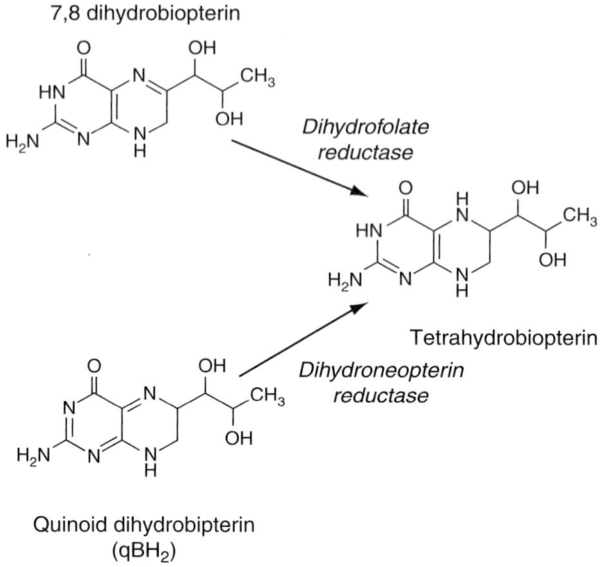

7,8 dihydrobiopterin

Dihydrofolate reductase

Tetrahydrobiopterin

Dihydroneopterin reductase

Quinoid dihydrobipterin (qBH_2)

FIGURE 3 Salvage pathways for recovery of BH_4 from dihydrobiopterin and quinoid BH_2 (qBH_2). Reactions of enzymes dihydrofolate reductase and dihydroneopterin reductase are shown.

Thus, induction of GTPCH-1 activity and increased BH_4 levels might reflect one of the pleiotropic effects of statin medications, beyond the benefit achieved by lipid lowering. Insulin also increases both mRNA and protein levels of GTPCH-1, and enhances endothelial BH_4 synthesis (Ishii et al., 2001). These effects of insulin are blocked by the phosphatidylinositol 3-kinase (PI3-kinase) pathway inhibitors Wortmannin and LY294002, suggesting a role of PI-3 kinase in this process. The precise mechanism by which PI-3 kinase affects GTPCH-1 transcription remains unclear.

2. *Salvage Pathways*

In addition to the *de novo* pathway discussed above, salvage pathways are important for maintaining cellular levels of BH_4 (Fig. 3). The ultimate product of the amino acid hydroxylases is the quinoid form of dihydrobipterin (q-BH_2), which has a double bond between the 4 and 5 positions in the pyrazine ring. This is reconverted to BH_4 by the enzyme dihydropteridine reductase (DHPR), using NADH as a substrate for reducing equivalents. In contrast, the nonenzymatic oxidation product of BH_4, 7,8-dihydrobipterin (which contains double bonds between the 6 and 7 positions) is reconverted to BH_4 by dihydrofolate reductase (DHFR) (Fig. 3). In endothelial cells, DHFR seems essential for maintenance of BH_4 levels. Chalupsky and

Cai (2005) showed that siRNA downregulation of DHFR reduces endothelial cell BH_4 levels and NO production (see discussion below). The authors have also shown that endogenously produced hydrogen peroxide inhibits DHFR protein expression in the endothelium. This could have pathophysiological relevance, as discussed in below, for diseases such as hypertension and diabetes. Similarly, Sugiyama et al. (2009) have shown that inhibition of DHFR expression by siRNA lowers the levels of BH_4 in bovine aortic endothelial cells in culture. Crabtree et al. (2009) have also demonstrated the importance of DHFR in a modulating BH_4 levels in transfected murine fibroblasts and mouse endothelial cells.

C. Redox Reactions of BH_4 and Related Pterins

BH_4 and its related compounds participate in redox reactions via electron donation and acceptance in the pyrazine ring. For example, BH_4 donates an electron during NOS catalysis (see below), or upon reaction with a variety of oxidants, such as sulfenyl radicals ($RS^{\cdot}$), the carbonate radical ($CO_3^{\cdot}$), nitrogen dioxide ($NO_2^{\cdot-}$), and the hydroxyl radical ($OH^{\cdot}$) (Patel et al., 2002). Oxidation of BH_4 has been recently reviewed in depth (Vasquez-Vivar, 2009), and will not be repeated extensively here. Briefly, however, BH_4 can undergo autooxidation involving initially electron loss, formation of the BH_4 radical, and addition of molecular oxygen to form a peroxy-species. This in turn can react with another molecule of BH_4 leading to an ongoing oxidation of successive molecules of BH_4 in a manner analogous to lipid chain peroxidation (Vasquez-Vivar, 2009). Data from our own laboratory have suggested that peroxynitrite is a particularly important oxidant for BH_4 in pathophysiological conditions (Laursen et al., 2001). Using electron spin resonance, we found that this reaction leads to formation of the BH_3 radical (Kuzkaya et al., 2003). In the presence of ascorbate, the $BH_3^{\cdot}$ radical is readily converted back to BH_4, while in its absence, $BH_3^{\cdot}$ degrades rapidly to BH_2 (Fig. 4). These findings might explain the ability of ascorbate to increase cellular BH_4 and to improve NOS catalysis in some studies (Heller et al., 2001; Huang et al., 2000; Kuzkaya et al., 2003; Nakai et al., 2003). Interestingly, peroxynitrite reacts with BH_4 more rapidly than with other cellular antioxidants, including cysteine, glutathione, or ascorbate (Kuzkaya et al., 2003). Thus, exposure of cells to peroxynitrite or similar oxidants predisposes to BH_4 oxidation.

The above-mentioned capacity of BH_4 to undergo autooxidation helps explain its instability in solution. Likewise, the chemical interplay with ascorbate outlined above underscores the ability of small amounts of ascorbate to stabilize solutions of BH_4. These properties should be kept in mind when BH_4 is used therapeutically or employed in biological solutions.

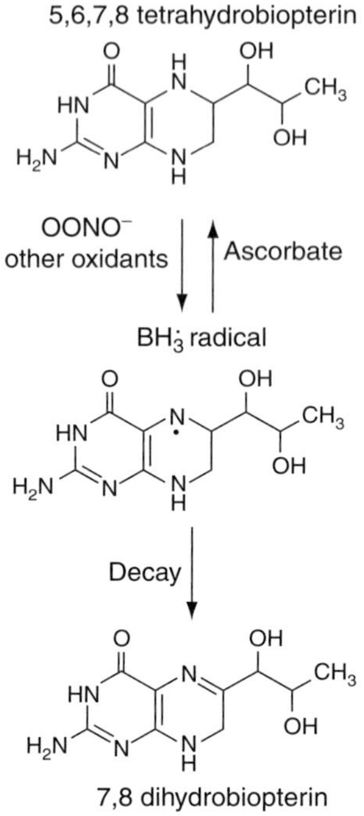

5,6,7,8 tetrahydrobiopterin

$OONO^-$
other oxidants

Ascorbate

$BH_3^{\cdot}$ radical

Decay

7,8 dihydrobiopterin

FIGURE 4 Oxidation of tetrahydrobiopterin to the $BH_3^{\cdot}$ radical and ultimately to 7,8-dihydrobiopterin. Tetrahydrobiopterin can be oxidized by various oxidants such as peroxynitrite, $RS^{\cdot}$, and others. These extract an electron from BH_4, leading to formation of the $BH_3^{\cdot}$ radical. Also shown is the effect of ascorbic acid on the $BH_3^{\cdot}$ radical.

III. The Role of BH_4 in NOS Function—Concept of NOS Uncoupling

The production of NO by the NOS enzymes is initiated when electrons from the reductase domains reduce a ferric heme in the oxygenase domain to the ferrous state, allowing binding of oxygen and ultimately formation of a ferric heme-superoxy-(FeIII-$O_2^{\cdot-}$) intermediate. BH_4, which is also bound in the oxygenase domain, plays a critical role by reacting with this intermediate leading to formation of a heme-oxy species that reacts with L-arginine and *N*-hydroxy-L-arginine, ultimately leading to production of NO and cirtulline (Wei et al., 2003, 2008). When BH_4 is absent, or oxidized, or in insufficient levels, reduction of the heme-superoxy intermediate does not occur and it dissociates to release superoxide and perhaps other reactive oxygen species. This phenomenon is commonly referred to as NOS uncoupling. There is also evidence that the NOS enzymes are stabilized in their functional dimeric form by BH_4 (Stuehr, 1997; Toth et al., 1997). Thus, BH_4 allosterically controls both function and structure of the NOS enzymes in a critical fashion.

Importantly, NOS uncoupling is often not complete, such that both NO and superoxide is produced, predisposing to formation of peroxynitrite (Berka et al., 2004). There is also evidence that L-arginine modulates hydrogen peroxide and superoxide production by some NOS isoforms (Tsai et al., 2005). Thus, the NOS enzymes can function as NO synthases, superoxide synthases, peroxynitrite synthases, or in some cases hydrogen peroxide synthases.

These diverse potentials of NOS enzymes have emphasized that BH_4 plays an important role in modulating endothelial function, vascular biology, and vascular disease. As discussed in part above, it has now been recognized that vascular and endothelial levels of BH_4 are affected by numerous factors, including changes in synthesis, changes in regeneration, and oxidative degradation. Moreover, studies in both experimental animals and humans have shown that NOS uncoupling occurs in a variety of common diseases, and can contribute to pathophysiology of these conditions.

IV. NOS Uncoupling in Diseases

In 1996, Higman and coworkers demonstrated that saphenous veins from smokers produced less NO when stimulated by the calcium ionophore A23187 than did veins from nonsmokers. This difference was not changed by incubation with L-arginine, but was minimized by pretreatment with BH_4 (Higman et al., 1996). In the following year, Stroes and coworkers showed that infusion of BH_4 could restore endothelial function in the forearms of humans with hypercholesterolemia (Stroes et al., 1997). These studies have prompted studies of the role of NOS uncoupling in a variety of common diseases that are discussed in detail below. A recurring pattern is that in many pathological conditions, there appears to be oxidation of BH_4 and accumulation of BH_2. This is important because in the presence of adequate amounts of BH_4, the addition of BH_2 can lead to superoxide production by the endothelial NOS, in a manner similar to a competitive antagonist (Vasquez-Vivar et al., 2002b). Thus, BH_4 does not need to be depleted to uncouple the NOS enzymes if its oxidized forms accumulate in sufficient concentrations. This concept, together with other experimental evidence presented below, has led to a paradigm, illustrated in Fig. 5. According to this scheme, reactive oxygen species from other sources ("kindling radicals") can lead to BH_4 oxidation, promoting NOS uncoupling and production of increasing amounts of superoxide and other radicals ("bonfire radicals") in a feed forward fashion. As mentioned above, there is evidence that hydrogen peroxide can inhibit expression of the BH_4 salvage enzyme DHFR (Chalupsky & Cai, 2005), leading to cellular accumulation of BH_2 and consequent NOS uncoupling (Fig. 5).

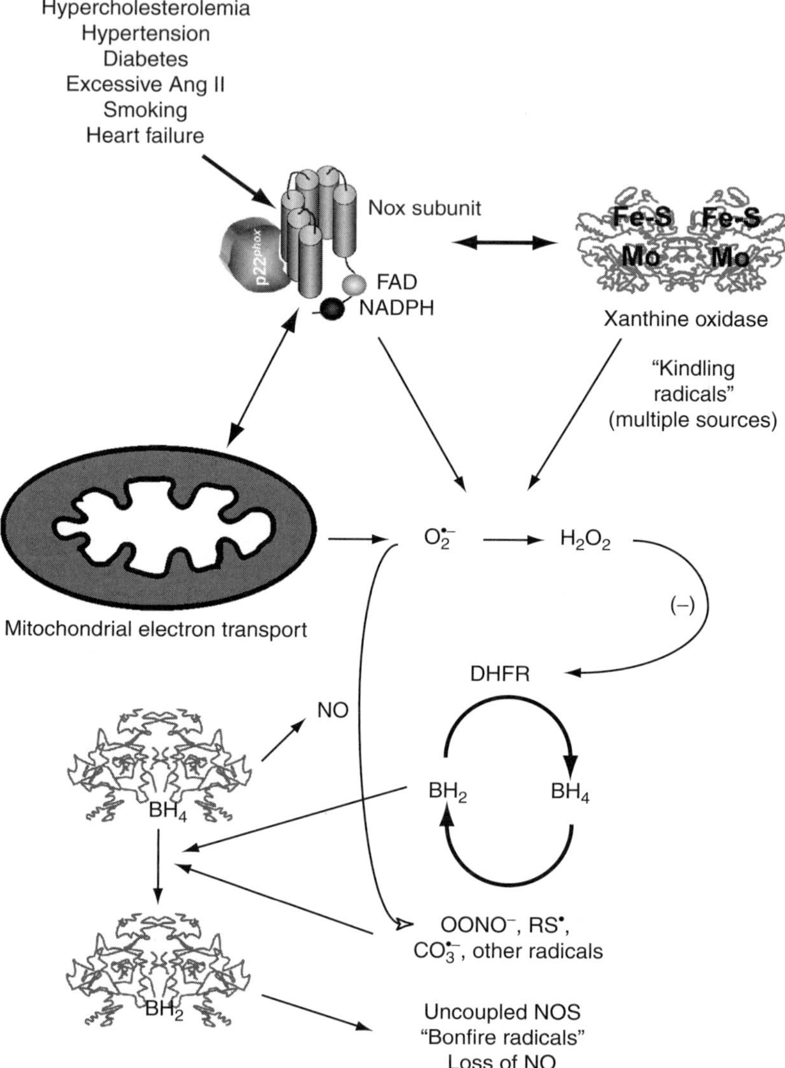

FIGURE 5 Effect of reactive oxygen species and diseases on NOS uncoupling. Various enzymes in mammalian cells, including the NADPH oxidases, mitochondrial electron transport, and xanthine oxidase produce oxidants in response to a variety of pathophysiological stimuli. These "kindling radicals" can oxidize BH_4 to BH_2, which occupies the binding site of the nitric oxide synthases, leading to impaired catalysis of NO and release of larger amounts or reactive oxygen species ("bonfire radicals") from these enzymes. Hydrogen peroxide also inhibits expression of the salvage enzyme dihydrofolate reductase (DHFR), which leads to accumulation of BH_2 and uncoupling of NOS.

A. Atherosclerosis and Hypercholesterolemia

For 25 years, it has been recognized that endothelial modulation of vascular tone is abnormal in atherosclerosis (Freiman et al., 1986; Ludmer et al., 1986). The mechanisms responsible for this abnormality have been studied in depth, and are likely multifactorial. Abnormalities that have been proposed included altered membrane signaling (Bossaller et al., 1987; Huraux et al., 1999), abnormalities of L-arginine availability (Boger et al., 1998; Creager et al., 1992), and oxidative inactivation of nitric oxide (Ohara et al., 1993). The latter hypothesis has been supported by the fact that the reaction of NO and radicals such as superoxide is extremely rapid (Beckman & Koppenol, 1996), by the observation that atherosclerosis and hypercholesterolemia increase vascular superoxide production (Ohara et al., 1993), and by studies in which scavenging of reactive oxygen species improves endothelium-dependent vasodilatation (Mugge et al., 1991). Stroes et al., 1997 provided the first hint that there might be a problem with BH_4 in hypercholesterolemia. These investigators showed that infusion of BH_4 in the forearm normalized serotonin-induced vasodilatation in hypercholesterolemic subjects while having no effect on normal subjects. The vasoconstrictor response to infusion of the NO synthase inhibitor L-NMMA was enhanced by BH_4, suggesting that it increased basal production of NO. Infusion of L-arginine did not add to the effect of BH_4, suggesting that L-arginine was not deficient in these subjects with hypercholesterolemia. Subsequently, Tiefenbacher et al. (2000) examined the effect of the BH_4 precursor sepiapterin in isolated coronary arterioles from atherosclerotic pigs and humans. The authors showed that sepiapterin improved endothelium-dependent vasodilatation to substance P and histamine, and reduced vasoconstrictions to serotonin (which is modulated by endothelium-derived NO) in coronary arterioles from pigs with diet-induced atherosclerosis. Constrictions to serotonin, which are normally blunted by NO, were reduced by sepiapterin treatment. In contrast, this agent had no effect in coronary arterioles from pigs fed a normal diet. Similar findings were obtained in coronary arterioles obtained from the right atrial appendage of humans with and without atherosclerosis undergoing cardiac surgery. Others have shown that treatment with BH_4 improves endothelium-dependent relaxations to acetylcholine in vessels from $ApoE^{-/-}$ mice and LDL-receptor-deficient mice (Jiang et al., 2000). In another study, increasing BH_4 via genetic means in $ApoE^{-/-}$ mice improved endothelium-dependent vasodilatation, reduced vascular superoxide production, and increased acetylcholine-stimulated vascular cGMP levels (Alp et al., 2004).

Studies such as these suggested that BH_4 might be deficient in the settings of atherosclerosis and hypercholesterolemia. Evidence to support this came from a study by Vasquez-Vivar et al. (2002a, 2002b), in which the authors found markedly diminished levels of BH_4 in aortas of rabbits with diet-induced hypercholesterolemia. The authors did not determine if this was due to reduced production of BH_4, or due to its oxidation. Studies from our own laboratory provided evidence that oxidation of BH_4 might indeed be responsible for a loss of BH_4 in atherosclerotic vessels (Laursen et al., 2001). We found that peroxynitrite rapidly oxidizes BH_4 *in vitro*, that addition of peroxynitrite to normal vascular segments causes NOS uncoupling and that atherosclerotic vessels produce excessive amounts of peroxynitrite. We further showed that scavenging of peroxynitrite improves endothelium-dependent vasodilatation of atherosclerotic vessels, in keeping with prevention of BH_4 oxidation. In accordance with this concept, Antoniades et al. showed that 5-methyltetrahydrofolate, which serves as an effective scavenger of peroxynitrite, reduces BH_4 oxidation in saphenous veins of atherosclerotic humans and improved endothelium-dependent vasodilatation. The authors provided evidence that this agent also recouples eNOS as evidenced by studies of vascular superoxide production in the presence and absence of NOS inhibitors.

Given the inhibitory role of NO on atherosclerotic lesion formation, and the propensity for superoxide and related radicals to enhance lipid oxidation and atherosclerosis, one might expect that measures to prevent NOS uncoupling would reduce atherosclerosis. Indeed, there is substantial evidence to support this concept. In the study by Alp et al. (2004), genetic enhancement of BH_4 in the endothelium reduced atherosclerosis in the aortic root. Hattori et al. (2007) showed that administration of BH_4 in the drinking water increases aortic levels of biopterin and reduces atherosclerotic lesion formation by half in $ApoE^{-/-}$ mice. A very recent study showed that oral BH_4 treatment in $ApoE^{-/-}$ mice not only reduces atherosclerotic lesion formation, but also decreases the content of inflammatory cells in these vessels, without altering plasma lipids (Schmidt et al., 2010). The mechanisms by which BH_4 could alter the inflammatory response in atherosclerosis remain undefined but could reflect modulation of T cell function or antigen formation in atherosclerosis.

B. Hypertension

In addition to atherosclerosis, there has been ample evidence that vascular BH_4 levels are altered in hypertension. Cosentino et al. (1998) studied vessels from spontaneously hypertensive rats (SHR) at 4 weeks of age, before the development of hypertension. The authors showed that these vessels produce less NO and more superoxide than vessels from control rats and that these differences were reduced by treatment with either superoxide dismutase or BH_4. In keeping with NOS uncoupling, the authors found that the increase in superoxide is prevented by NOS inhibition with L-NMMA. Subsequently, Kerr et al. (1999) showed that vessels from stroke-prone SHR produce excessive superoxide which is inhibitable by endothelial removal or the NOS antagonist L-nitroarginine methyl ester (L-NAME). The authors concluded that eNOS uncoupling occurs in this model of severe hypertension. Hong et al. (2001) showed that plasma levels of BH_4 are reduced in SHR, and that treatment with BH_4 improves endothelium-dependent vasodilatation to acetylcholine, reduces vascular superoxide production, and lowers blood pressure in these animals. Higashi and colleagues extended this research to humans, showing that BH_4 markedly enhances forearm blood flow responses to acetylcholine in hypertensive subjects. Interestingly, they also showed enhancement of endothelium-dependent responses in nonhypertensive controls, suggesting that normal subjects might also be relatively BH_4-deficient (Higashi et al., 2002). In an effort to further understand why BH_4 is reduced in hypertension, we measured levels of BH_4 and its oxidized forms in control mice and mice with DOCA-salt hypertension (Landmesser et al., 2003). We found that while total levels of biopterin are not altered in hypertensive animals, the level of oxidized biopterin is increased while the level of reduced BH_4 is similarly diminished (Fig. 6). These data are compatible with BH_4 oxidation in the setting of hypertension. We provided evidence that this is likely due to peroxynitrite produced by the NADPH oxidase and the endothelial NOS, because BH_4 oxidation was less obvious in hypertensive mice lacking the $p47^{phox}$ subunit of the NADPH oxidase or in mice lacking eNOS. This finding is in keeping with previous observations that peroxynitrite is capable of potently oxidizing BH_4 (Laursen et al., 2001). As in the case of studies by Hong et al. (2001), we found that BH_4 treatment reduces blood pressure in mice with DOCA-salt hypertension. Consistent with this finding, Kase et al. (2005) showed that BH_4 lowers blood pressure in mice during chronic angiotensin II infusion.

Chalupsky and Cai (2005) have uncovered a potentially important mechanism for uncoupling of NOS in hypertension. The authors showed that angiotensin II causes downregulation of the salvage enzyme DHFR and reduces endothelial BH_4 levels in cultured endothelial cells. This is blocked by scavenging of hydrogen peroxide with a catalase analog. The authors proposed a pathway in which hydrogen peroxide, produced by the NADPH oxidase in response to angiotensin II, inhibits expression of DHFR, leading to loss of BH_4 and accumulation of BH_2 leading to NOS uncoupling as illustrated in Fig. 5.

More recently, we found that oral treatment with BH_4 lowers blood pressure in humans with hypertension. We found that oral BH_4 also dramatically improves endothelial function, as monitored by flow-mediated vasodilatation (Porkert et al., 2008). The effect of BH_4 in this study was very promising, however studies involving larger numbers of patients are needed to determine if this will be a useful therapy for human hypertension.

The above studies have focused on how BH_4 regulates endothelial function in hypertension. There is also evidence that BH_4 affects NO production in the kidney and that this is altered in hypertension. Taylor et al. (2006)

found that there is oxidation of BH_4 in the outer renal medulla of hypertensive Dahl salt-sensitive rats. Sodium reabsorption in this portion of the kidney is largely mediated by the Na/K/Cl cotransporter, which in turn is modulated by local production of NO and superoxide. Superoxide stimulates the Na/K/Cl cotransporter and enhances sodium reabsorption in the medullary thick ascending limb. Thus, uncoupling of NOS in these cells caused by reduced BH_4 could contribute to hypertension.

Despite the availability of numerous drugs to treat hypertension, many patients require multiple agents to reduce their blood pressure and treatment is often only partly successful. Efforts to increase BH_4 in the vessel or perhaps the kidney might therefore provide an alternative therapeutic option that could also correct underlying pathophysiology in this disease.

C. Insulin Resistance and Diabetes

Endothelial dysfunction is common in both type 1 and type 2 diabetes. While multiple mechanisms likely contribute to this abnormality, substantial evidence indicates that this is in part due to NOS uncoupling and perturbations of BH_4 metabolism. Interestingly, one of the earliest studies focused on brain BH_4 levels in rats with streptozocin-induced diabetes. The authors found oral treatment with streptozocin caused a striking increase in BH_2 and a concomitant increase in BH_4 whole brain homogenates from these rats, but did not affect the liver content of the pterins (Hamon et al., 1989). The authors also found a marked reduction in the BH_2 salvage enzyme dihydropterin reductase. Such a perturbation of BH_4 could affect neuronal production of NO, which has important roles in long-term potentiation and memory development in the brains of diabetic animals and humans. Also, given the importance of BH_4 in production of neurotransmitters such as serotonin, dopamine, and norepinephrine, this might have major effects on central nervous system function. In 1997, Pieper demonstrated that an analog of BH_4 improved endothelium-dependent vasodilatation to acetylcholine in rings of vessels from diabetic rats (Pieper, 1997). This intervention also enhanced peak relaxations to nitroglycerin, possibly by reducing vascular levels of superoxide. In a very interesting subsequent study, Shinozaki et al. (1999) produced insulin resistance in rats by fructose feeding. Fructose feeding produced a modest increase in weight, and while baseline glucose levels were not altered, plasma insulin levels were doubled. Fructose feeding also elevated systolic pressure to 140 mmHg. Thus, these animals had many parameters of the metabolic syndrome, with modest obesity, hypertension, and insulin resistance. The vascular production of superoxide was markedly increased in these animals, and this was eliminated by endothelial removal or by treatment of rings with L-NAME, SOD, or BH_4. Levels of BH_2 were increased in these vessels, and this was ameliorated by insulin treatment. The authors also found that fructose feeding reduced activity of GTPCH-1, the rate-limiting enzyme in BH_4 *de novo* synthesis. In a subsequent study, Meininger et al. (2000) isolated endothelial cells from diabetic BioBreeding (BB) rats, which have insulin-dependent diabetes. These cells showed a reduction in GTPCH-1 protein by Western blot, diminished BH_4 levels, and markedly reduced NO production, as detected by monitoring nitrate and nitrite release. The reduced BH_4 and NO levels were partly restored by addition of sepiapterin, a precursor to BH_4. This study demonstrated that endothelial BH_4 levels also decline in a genetic model of diabetes.

Hink et al. (2001) provided evidence that protein kinase C (PKC) signaling might be responsible for NOS uncoupling in diabetes. The authors showed that various PKC inhibitors reduced parameters of NOS uncoupling in streptozocin-induced diabetes. NADPH oxidase activity was also markedly increased in diabetic vessels. This study again links activation of the NADPH oxidase to NOS uncoupling and perturbations similar to that observed in hypertension by Landmesser et al. (2003).

Using mice similar to those employed in their study of hypercholesterolemia (Alp et al., 2004), Alp et al. (2003) showed that targeted overexpression of GTPCH-1 in the endothelium reduces vascular superoxide production caused by streptozocin-induced diabetes. The authors also showed that this genetic manipulation prevents the development of abnormal endothelium-dependent vasodilatation maintains intracellular glutathione levels in these animals. Okumura et al. (2006) provided evidence that the proteinuria caused by diabetes could in part be due to BH_4 deficiency. Oak and Cai (2007) showed that angiotensin II signaling played a role in diabetes-induced NOS uncoupling (Oak & Cai, 2007). The authors demonstrated that streptozocin-induced diabetes caused a decline in vascular levels of the salvage protein DHFR in mice, and that this was prevented by treatment with the angiotensin I-converting enzyme captopril or the angiotensin II receptor antagonist candesartan. These agents also increased blood biopterin content, reduced vascular superoxide content, increased NO production, and improved endothelium-dependent vasodilatation. Similar to the prior report from this laboratory mentioned above (Chalupsky & Cai, 2005), and in keeping with findings by Hink et al. (2001), the authors showed that this effect on DHFR was likely due to increased NADPH oxidase activity in diabetes.

While the above findings have largely been made in experimental animals and cultured cells, Heitzer et al. (2000b) provided evidence that BH_4 deficiency likely contributes to endothelial dysfunction in diabetic humans. The authors showed that acetylcholine-mediated vasodilatation was blunted in subjects with type-II diabetes and was improved by BH_4 administration. In contrast, BH_4 treatment had no effect on normal subjects. BH_4 had no effect on the vasodilatation evoked by nitroprusside infusion.

Beyond vasomotor regulation, there is also evidence that some of the tragic consequences of diabetes might be due to BH_4 deficiency. One such problem is defective wound healing. Recently, Tie et al. (2009) provided evidence that streptocotocin-induced diabetes caused NOS uncoupling in the skin, and that this impaired closure of experimentally induced wounds. The authors showed that BH_4 levels were reduced in the skin of these animals, and that both BH_4 levels and wound closure were normalized by genetic overexpression of GTPCH-1. Another terrible complication of diabetes is delayed gastric emptying, which leads to excessive vomiting and an inability to digest food. Gastric emptying is mediated by release of NO from myenteric plexus neurons. Very recently, Gangula et al. (2010) demonstrated that BH_4 plays an important role in gastric relaxations. The authors showed that inhibition of the enzyme GTPCH-1 reduced relaxation of the pylorus caused by nerve stimulation. They further showed that BH_4 levels and NOS activity are depressed in the pylorus of rats with streptocotocin-induced diabetes and that this was associated with impaired gastric emptying. Treatment of the animals with BH_4 for either 3 or 9 weeks normalized gastric emptying in diabetic animals. These findings suggest that BH_4 treatment might prove beneficial in this otherwise very-difficult-to-treat problem in diabetes.

D. Miscellaneous Conditions

Atherosclerosis, hypertension, and diabetes are not the only conditions in which BH_4 deficiency has been implicated. Others include cigarette smoking (Heitzer et al., 2000a), pulmonary hypertension (Nandi et al., 2005), aging, cardiac fibrosis, and hypertrophy in aortic stenosis (Moens et al., 2008). Recently, we found that diastolic dysfunction, a common problem associated with hypertension, is associated with altered BH_2/BH_4 ratios in the heart and is improved by BH_4 treatment (Silberman et al., 2010).

V. Structural Characteristics of GTPCH-1 and Posttranslational Modulation

A. GTPCH-1 Structure

There is substantial interest in how the GTPCH-1 enzyme functions and it is modulated posttranslationally. Crystal-structure analysis has shown that GTPCH-1 exists as a homodecamer, composed of two homopentamers interfaced, simulating a bagel cut sagitally and then placed back together (Fig. 6A and B). The GTP-binding site is located on the outer surface of each subunit. Importantly, a regulatory protein known as the GTPCH-1 feedback regulatory protein (GFRP) modulates GTPCH-1 activity in a posttranscriptional fashion. GFRP exists as a homopentamer, and two of these interface

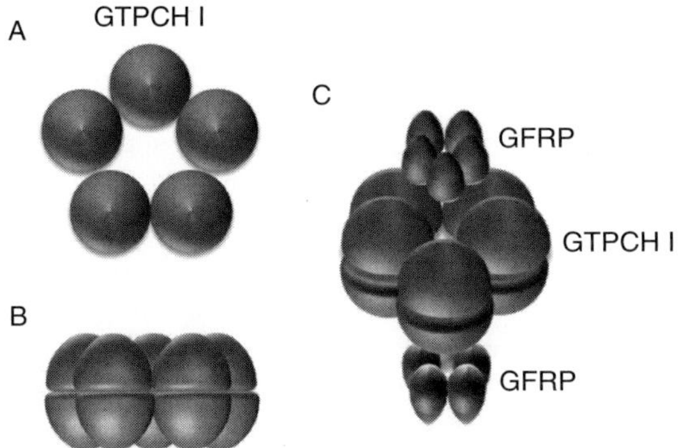

FIGURE 6 Schematic representation of the GTPCH-1 enzyme and its interaction with its regulatory protein. Panel A shows a "top down" view of the enzyme, illustrating the "bagel" configuration of the subunits grouped together. In panel B, a "side view" is shown showing the homodecamer conformation. Panel C shows association with the GTPCH-1 feed back regulatory protein (GFRP).

with either side of the "bagel" structure of GTPCH-1 as illustrated in Fig. 6C. The activity of GTPCH-1 is regulated both negatively and positively by BH_4 and phenylalanine, which are bound by pockets at the interface of these two proteins. Thus, binding of BH_4 at this site inhibits, while binding of phenylalanine activates GTPCH-1 activity. This property of GTPCH-1 is thought to provide allosteric negative feedback inhibition, such that when BH_4 levels decrease, activity of the enzyme is increased and when BH_4 levels are sufficient, the enzyme is inhibited. As discussed below, we have found that phosphorylation of human GTPCH-1 at serine 81 affects its binding and inhibition by GFRP. This amino acid exists in a regulatory arm that extends from each subunit (not shown in the simplified image in Fig. 6).

B. Regulation of Endothelial BH_4 by Mechanical Stimuli

Mechanical forces have enormous impact on endothelial cell biology. Laminar shear, the tangential force of flow over the surface of endothelial cells, provides anti-atherogenic effects by promoting NO production, inhibiting expression of adhesion molecules, and preventing leukocyte adhesion to the endothelial cell surface (Chiu et al., 2009). In particular, regions of the circulation exposed to well-developed laminar shear are protected against development of atherosclerosis. In contrast, disturbed flow patterns, such as oscillatory shear, have untoward effects on endothelial cell function, by promoting production of reactive oxygen species, enhancing proinflammatory genes and reducing nitric oxide production. For this reason, portions of the circulation exposed to oscillatory shear, such as the carotid bulb, the proximal coronary arteries and the distal aorta, have a predilection for atherosclerotic lesion development. The mechanisms underlying the varying responses to these mechanical forces remain a matter of intense investigation; however, it is thought that mechanical signals are transduced from cell surface proteins, including intracellular proteins such as VE-cadherin and PECAM to intracellular signaling kinases to induce both acute and chronic changes in endothelial function (Tzima et al., 2005).

Recently, our laboratory provided new information regarding regulation of BH_4 biosynthesis in the human endothelium. The purpose of this work was to understand how various shear stresses affect BH_4 biosynthesis, but the research has also provided new insight into function of GTPCH-1 and its interaction with GFRP. Previously, we and others found that laminar shear stress markedly increases both activity and subsequently, expression of the endothelial NOS (Noris et al., 1995; Uematsu et al., 1995). Because the newly synthesized NOS enzyme could not function without additional BH_4, we postulated that shear would also increase BH_4 biosynthesis. We, therefore, exposed human endothelial cells to either static conditions, laminar, or oscillatory shear stress using a cone-in-plate viscometer and examined the effect of these varying physical forces on BH_4 levels (Widder et al., 2007). As reported in other cells, we found, under static conditions, that GTPCH-1 activity is substantially lower than activity of the downstream enzymes PTPS or SR, in keeping with GTPCH-1 being rate limiting in this pathway (Widder et al., 2007). A striking finding was that laminar shear stress increased endothelial cell BH_4 levels by 30-fold (Fig. 7A). This was accompanied by a 30-fold increase in the activity of GTPCH-1, but no change in activities of PTPS or SR. Interestingly, oscillatory shear stress only modestly affected these parameters. In subsequent experiments, we found that shear did not increase GTPCH-1 protein levels, but promoted phosphorylation of the enzyme by the alpha prime isoform of casein kinase II, and that this markedly stimulated its activity. Using a panel of anti-phosphoprotein antibodies, we found that serine 81 on GTPCH-1 was phosphorylated in response to laminar, but not oscillatory shear stress. These studies identified a novel phosphorylation site on GTPCH-1 and provided evidence that it dramatically modulated function of this enzyme.

In a subsequent study, we provided additional insight into how phosphorylation of GTPCH-1 on serine 81 could have such a dramatic impact on its activity (Li et al., 2010). Using site-directed mutagenesis, we replaced serine 81 of human GTPCH-1 with an alanine (to inhibit phosphorylation, referred to as S81A) and an aspartate (to mimic phosphorylation, referred to as S81D). We found that the S81D mutant produced high amounts of BH_4 when transfected into endothelial cells, even in the absence of shear stress. In contrast, the S81A mutant produced low amounts of BH_4 and did not respond to shear. In subsequent studies *in vitro*, we showed that S81D activity was higher than that of the wild-type GTPCH-1, but more importantly, S81D was completely resistant to inhibition by GFRP. In these studies,

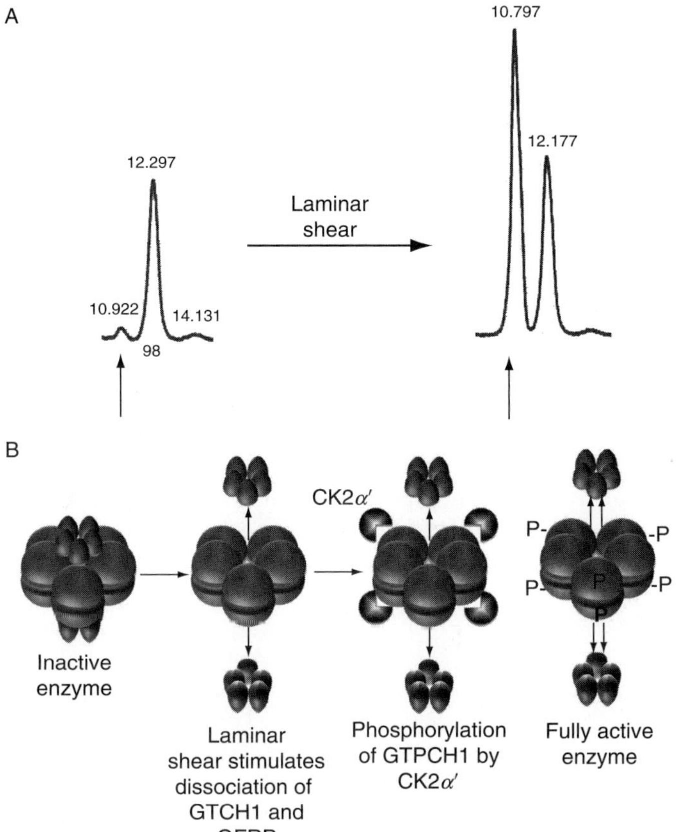

FIGURE 7 Effect of shear stress on human endothelial cell tetrahydrobiopterin levels and GTPCH-1 activity. Panel A shows HPLC traces with representative peaks for total pterins eluting at approximately 11 min following injection of endothelial cell samples (arrow). Panel B shows the proposed mechanism for phosphorylation and modulation of GTPCH-1 activity by shear stress. Laminar shear promotes dissociation of GTPCH-1 from the regulatory protein GFRP, which allows GTPCH-1 phosphorylation by casein kinase 2α' (CK2). The fully phosphorylated enzyme has diminished binding affinity to GFRP and is not inhibited by GFRP, leading to a fully functioning enzyme.

we defined a new pathway in which shear causes dissociation of GTPCH-1 from GFRP. This allows phosphorylation of GTPCH-1 at serine 81. Phosphorylated GTPCH-1 is resistant to GFRP binding and inhibition. This interplay between GTPCH-1 phosphorylation, its binding and regulation by GFRP and its production of BH_4 is illustrated in Fig. 7B. In experiments in mice, we showed that GTPCH-1 phosphorylation in response to shear also occurs *in vivo*, and that this ultimately affects tissue BH_4 levels and NO production.

VI. Conclusion

In this review, we have summarized a large number of studies showing that BH_4 biosynthesis, recovery, and oxidation can be altered in many diseases. There is a wealth of information from studies of experimental animals suggesting that BH_4 therapy can improve a variety of conditions including atherosclerosis, hypertension, diabetes, and others. Despite this, it is unclear if BH_4 therapy will be effective in treatment of human diseases. A major concern is that most of the recent experimental studies have involved early treatment or even concomitant administration of BH_4 with the onset of the imposed disease. Therapy in human diseases is often initiated many years after disease onset. It is unclear if interventions such as BH_4 therapy will be effective at these latter stages. It is also possible that BH_4 treatment could exacerbate inflammatory processes by promoting function of the inducible NOS. Despite these considerations, therapeutic efforts to enhance BH_4 levels are currently an attractive therapeutic option that needs further exploration. Given the impact that diseases like diabetes, atherosclerosis, and hypertension have on human health, in-depth study of the potential benefits of BH_4 therapy is warranted.

Acknowledgments

This work was supported by Veterans Administration Merit Funding, NIH grants R01 HL390006, P01 HL05800, and P01 HL095070. Ms. Li was supported by a postgraduate fellowship from the American Heart Association.

Conflict of Interest Statement: The authors have no conflict of interests to declare.

Abbreviations

GTP	guanosine triphosphate
GFRP	GTP cyclohydrolase-1 feedback regulatory protein
NOS	nitric oxide synthase
BH_4	tetrahydrobiopterin
BH_2	dihydrobiopterin
q-BH_2	qunoid form BH_2
$RS^{\cdot}$	thiyl radical
$CO_3^{\cdot}$	carbonate radical
NO_2^-	nitrogen dioxide
$OH^{\cdot}$	hydroxyl radical
$BH_3^{\cdot}$	trihdyrobiopterin radical
$OONO^-$	peroxynitrite
PTPS	6-pyruvoyl-tetrahydropterin synthase

SR	sepiapterin reductase
IL-1β	interleukin-1β
IFN-γ	gamma interferon
TNFα	tumor necrosis factor alpha
LPS	lipopolysaccharide
cAMP	cyclic adenosine monophosphate
HMG-CoA	3-hydroxy-3-methyl-glutaryl-CoA
PI3-kinase	phosphatidylinositol 3-kinase
PECAM	platelet endothelial cell adhesion molecule
DHPR	dihydropteridine reductase
siRNA	small interfering RNA
FeIII-O_2^-	ferric heme-superoxy intermediate
DHFR	dihydrofolate reductase
L-NMMA	L-*N*-monomethyl arginine
L-NAME	L-nitroarginine methyl ester
WKY	Wistar Kyoto rats
DOCA	deoxycoricosterone
SOD	superoxide dismutase
PKC	protein kinase C

References

Alp, N. J., McAteer, M. A., Khoo, J., Choudhury, R. P., & Channon, K. M. (2004). Increased endothelial tetrahydrobiopterin synthesis by targeted transgenic GTP-cyclohydrolase I overexpression reduces endothelial dysfunction and atherosclerosis in ApoE-knockout mice. *Arteriosclerosis, Thrombosis, and Vascular Biology*, *24*(3), 445–450.

Alp, N. J., Mussa, S., Khoo, J., Cai, S., Guzik, T., Jefferson, A., et al. (2003). Tetrahydrobiopterin-dependent preservation of nitric oxide-mediated endothelial function in diabetes by targeted transgenic GTP-cyclohydrolase I overexpression. *Journal of Clinical Investigation*, *112*(5), 725–735.

Beckman, J. S., & Koppenol, W. H. (1996). Nitric oxide, superoxide, and peroxynitrite: The good, the bad, and ugly. *The American Journal of Physiology*, *271*(5 Pt 1), C1424–C1437.

Berka, V., Wu, G., Yeh, H. C., Palmer, G., & Tsai, A. L. (2004). Three different oxygen-induced radical species in endothelial nitric-oxide synthase oxygenase domain under regulation by L-arginine and tetrahydrobiopterin. *The Journal of Biological Chemistry*, *279*(31), 32243–32251.

Bigham, E. C., Smith, G. K., Reinhard, J. F., Jr., Mallory, W. R., Nichol, C. A., & Morrison R. W. Jr., (1987). Synthetic analogues of tetrahydrobiopterin with cofactor activity for aromatic amino acid hydroxylases. *Journal of Medicinal Chemistry*, *30*(1), 40–45.

Boger, R. H., Bode-Boger, S. M., Szuba, A., Tsao, P. S., Chan, J. R., Tangphao, O., et al. (1998). Asymmetric dimethylarginine (ADMA): A novel risk factor for endothelial dysfunction: Its role in hypercholesterolemia. *Circulation*, *98*(18), 1842–1847.

Bossaller, C., Habib, G. B., Yamamoto, H., Williams, C., Wells, S., & Henry, P. D. (1987). Impaired muscarinic endothelium-dependent relaxation and cyclic guanosine 5′-monophosphate formation in atherosclerotic human coronary artery and rabbit aorta. *Journal of Clinical Investigation*, *79*(1), 170–174.

Burney, S., Caulfield, J. L., Niles, J. C., Wishnok, J. S., & Tannenbaum, S. R. (1999). The chemistry of DNA damage from nitric oxide and peroxynitrite. *Mutation Research*, *424*(1–2), 37–49.

Chalupsky, K., & Cai, H. (2005). Endothelial dihydrofolate reductase: Critical for nitric oxide bioavailability and role in angiotensin II uncoupling of endothelial nitric oxide synthase. *Proceedings of the National Academy of Sciences of the United States of America*, *102*(25), 9056–9061.

Chiu, J. J., Usami, S., & Chien, S. (2009). Vascular endothelial responses to altered shear stress: Pathologic implications for atherosclerosis. *Annali Medici*, *41*(1), 19–28.

Cosentino, F., Patton, S., d'Uscio, L. V., Werner, E. R., Werner-Felmayer, G., Moreau, P., et al. (1998). Tetrahydrobiopterin alters superoxide and nitric oxide release in prehypertensive rats. *Journal of Clinical Investigation*, *101*(7), 1530–1537.

Crabtree, M. J., Tatham, A. L., Hale, A. B., Alp, N. J., & Channon, K. M. (2009). Critical role for tetrahydrobiopterin recycling by dihydrofolate reductase in regulation of endothelial nitric-oxide synthase coupling: Relative importance of the de novo biopterin synthesis versus salvage pathways. *The Journal of Biological Chemistry*, *284*(41), 28128–28136.

Creager, M. A., Gallagher, S. J., Girerd, X. J., Coleman, S. M., Dzau, V. J., & Cooke, J. P. (1992). L-arginine improves endothelium-dependent vasodilation in hypercholesterolemic humans. *Journal of Clinical Investigation*, *90*(4), 1248–1253.

Dominguez-Rodriguez, A., Abreu-Gonzalez, P., & Kaski, J. C. (2009). Inflammatory systemic biomarkers in setting acute coronary syndromes–effects of the diurnal variation. *Current Drug Targets*, *10*(10), 1001–1008.

Fortin, C. F., McDonald, P. P., Fulop, T., & Lesur, O. (2010). Sepsis, leukocytes, and nitric oxide (NO): An intricate affair. *Shock*, *33*(4), 344–352.

Freiman, P. C., Mitchell, G. G., Heistad, D. D., Armstrong, M. L., & Harrison, D. G. (1986). Atherosclerosis impairs endothelium-dependent vascular relaxation to acetylcholine and thrombin in primates. *Circulation Research*, *58*(6), 783–789.

Gangula, P. R., Mukhopadhyay, S., Ravella, K., Cai, S., Channon, K. M., Garfield, R. E., et al. (2010). Tetrahydrobiopterin (BH4), a cofactor for nNOS, restores gastric emptying and nNOS expression in female diabetic rats. *American Journal of Physiology. Gastrointestinal and Liver Physiology*, *298*(5), G692–G699.

Gates, M. (1947). The chemistry of the pteridines. *Chemical Reviews*, *41*(1), 63–95.

Hamon, C. G., Cutler, P., & Blair, J. A. (1989). Tetrahydrobiopterin metabolism in the streptozotocin induced diabetic state in rats. *Clinica Chimica Acta*, *181*(3), 249–253.

Harrison, D. G., & Cai, H. (2003). Endothelial control of vasomotion and nitric oxide production. *Cardiology Clinics*, *21*(3), 289–302.

Hattori, Y., Hattori, S., Wang, X., Satoh, H., Nakanishi, N., & Kasai, K. (2007). Oral administration of tetrahydrobiopterin slows the progression of atherosclerosis in apolipoprotein E-knockout mice. *Arteriosclerosis, Thrombosis, and Vascular Biology*, *27*(4), 865–870.

Hattori, Y., Nakanishi, N., Akimoto, K., Yoshida, M., & Kasai, K. (2003). HMG-CoA reductase inhibitor increases GTP cyclohydrolase I mRNA and tetrahydrobiopterin in vascular endothelial cells. *Arteriosclerosis, Thrombosis, and Vascular Biology*, *23*(2), 176–182.

Heitzer, T., Brockhoff, C., Mayer, B., Warnholtz, A., Mollnau, H., Henne, S., et al. (2000a). Tetrahydrobiopterin improves endothelium-dependent vasodilation in chronic smokers: Evidence for a dysfunctional nitric oxide synthase. *Circulation Research*, *86*(2), E36–E41.

Heitzer, T., Krohn, K., Albers, S., & Meinertz, T. (2000b). Tetrahydrobiopterin improves endothelium-dependent vasodilation by increasing nitric oxide activity in patients with Type II diabetes mellitus. *Diabetologia*, *43*(11), 1435–1438.

Heller, R., Unbehaun, A., Schellenberg, B., Mayer, B., Werner-Felmayer, G., & Werner, E. R. (2001). L-ascorbic acid potentiates endothelial nitric oxide synthesis via a chemical stabilization of tetrahydrobiopterin. *The Journal of Biological Chemistry*, *276*(1), 40–47.

Higashi, Y., Sasaki, S., Nakagawa, K., Fukuda, Y., Matsuura, H., Oshima, T., et al. (2002). Tetrahydrobiopterin enhances forearm vascular response to acetylcholine in both normotensive and hypertensive individuals. *American Journal of Hypertension*, *15*(4 Pt 1), 326–332.

Higman, D. J., Strachan, A. M., Buttery, L., Hicks, R. C., Springall, D. R., Greenhalgh, R. M., et al. (1996). Smoking impairs the activity of endothelial nitric oxide synthase in saphenous vein. *Arteriosclerosis, Thrombosis, and Vascular Biology*, *16*(4), 546–552.

Hink, U., Li, H., Mollnau, H., Oelze, M., Matheis, E., Hartmann, M., et al. (2001). Mechanisms underlying endothelial dysfunction in diabetes mellitus. *Circulation Research*, *88*(2), E14–E22.

Hong, H. J., Hsiao, G., Cheng, T. H., & Yen, M. H. (2001). Supplemention with tetrahydrobiopterin suppresses the development of hypertension in spontaneously hypertensive rats. *Hypertension*, *38*(5), 1044–1048.

Huang, A., Vita, J. A., Venema, R. C., & Keaney, J. F. Jr., (2000). Ascorbic acid enhances endothelial nitric-oxide synthase activity by increasing intracellular tetrahydrobiopterin. *The Journal of Biological Chemistry*, *275*(23), 17399–17406.

Huraux, C., Makita, T., Kurz, S., Yamaguchi, K., Szlam, F., Tarpey, M. M., et al. (1999). Superoxide production, risk factors, and endothelium-dependent relaxations in human internal mammary arteries. *Circulation*, *99*(1), 53–59.

Ishii, M., Shimizu, S., Nagai, T., Shiota, K., Kiuchi, Y., & Yamamoto, T. (2001). Stimulation of tetrahydrobiopterin synthesis induced by insulin: Possible involvement of phosphatidylinositol 3-kinase. *The International Journal of Biochemistry & Cell Biology*, *33*(1), 65–73.

Jiang, J., Valen, G., Tokuno, S., Thoren, P., & Pernow, J. (2000). Endothelial dysfunction in atherosclerotic mice: Improved relaxation by combined supplementation with L-arginine-tetrahydrobiopterin and enhanced vasoconstriction by endothelin. *British Journal of Pharmacology*, *131*(7), 1255–1261.

Kase, H., Hashikabe, Y., Uchida, K., Nakanishi, N., & Hattori, Y. (2005). Supplementation with tetrahydrobiopterin prevents the cardiovascular effects of angiotensin II-induced oxidative and nitrosative stress. *Journal of Hypertension*, *23*(7), 1375–1382.

Katusic, Z. S., Stelter, A., & Milstien, S. (1998). Cytokines stimulate GTP cyclohydrolase I gene expression in cultured human umbilical vein endothelial cells. *Arteriosclerosis, Thrombosis, and Vascular Biology*, *18*(1), 27–32.

Kerr, S., Brosnan, M. J., McIntyre, M., Reid, J. L., Dominiczak, A. F., & Hamilton, C. A. (1999). Superoxide anion production is increased in a model of genetic hypertension: Role of the endothelium. *Hypertension*, *33*(6), 1353–1358.

Khan, B. V., Harrison, D. G., Olbrych, M. T., Alexander, R. W., & Medford, R. M. (1996). Nitric oxide regulates vascular cell adhesion molecule 1 gene expression and redox-sensitive transcriptional events in human vascular endothelial cells. *Proceedings of the National Academy of Sciences of the United States of America*, *93*(17), 9114–9119.

Kosmicki, M. A. (2009). Long-term use of short- and long-acting nitrates in stable angina pectoris. *Current Clinical Pharmacology*, *4*(2), 132–141.

Kubes, P., Suzuki, M., & Granger, D. N. (1991). Nitric oxide: An endogenous modulator of leukocyte adhesion. *Proceedings of the National Academy of Sciences of the United States of America*, *88*(11), 4651–4655.

Kuzkaya, N., Weissmann, N., Harrison, D. G., & Dikalov, S. (2003). Interactions of peroxynitrite, tetrahydrobiopterin, ascorbic acid, and thiols: Implications for uncoupling endothelial nitric-oxide synthase. *The Journal of Biological Chemistry*, *278*(25), 22546–22554.

Landmesser, U., Dikalov, S., Price, S. R., McCann, L., Fukai, T., Holland, S. M., et al. (2003). Oxidation of tetrahydrobiopterin leads to uncoupling of endothelial cell nitric oxide synthase in hypertension. *Journal of Clinical Investigation*, *111*(8), 1201–1209.

Laursen, J. B., Somers, M., Kurz, S., McCann, L., Warnholtz, A., Freeman, B. A., et al. (2001). Endothelial regulation of vasomotion in apoE-deficient mice: Implications for interactions between peroxynitrite and tetrahydrobiopterin. *Circulation*, *103*(9), 1282–1288.

Li, L., Rezvan, A., Salerno, J. C., Husain, A., Kwon, K., Jo, H., et al. (2010). GTP cyclohydrolase I phosphorylation and interaction with GTP cyclohydrolase feedback regulatory protein provide novel regulation of endothelial tetrahydrobiopterin and nitric oxide. *Circulation Research*, *106*(2), 328–336.

Ludmer, P. L., Selwyn, A. P., Shook, T. L., Wayne, R. R., Mudge, G. H., Alexander, R. W., et al. (1986). Paradoxical vasoconstriction induced by acetylcholine in atherosclerotic coronary arteries. *The New England Journal of Medicine*, *315*(17), 1046–1051.

Meininger, C. J., Marinos, R. S., Hatakeyama, K., Martinez-Zaguilan, R., Rojas, J. D., Kelly, K. A., et al. (2000). Impaired nitric oxide production in coronary endothelial cells of the spontaneously diabetic BB rat is due to tetrahydrobiopterin deficiency. *The Biochemical Journal*, *349*(Pt 1), 353–356.

Miller, M. R., & Megson, I. L. (2007). Recent developments in nitric oxide donor drugs. *British Journal of Pharmacology*, *151*(3), 305–321.

Moens, A. L., Takimoto, E., Tocchetti, C. G., Chakir, K., Bedja, D., Cormaci, G., et al. (2008). Reversal of cardiac hypertrophy and fibrosis from pressure overload by tetrahydrobiopterin: Efficacy of recoupling nitric oxide synthase as a therapeutic strategy. *Circulation*, *117*(20), 2626–2636.

Moncada, S., Palmer, R. M., & Higgs, E. A. (1988). The discovery of nitric oxide as the endogenous nitrovasodilator. *Hypertension*, *12*(4), 365–372.

Moncada, S., Palmer, R. M., & Higgs, E. A. (1991). Nitric oxide: Physiology, pathophysiology, and pharmacology. *Pharmacological Reviews*, *43*(2), 109–142.

Mugge, A., Elwell, J. H., Peterson, T. E., Hofmeyer, T. G., Heistad, D. D., & Harrison, D. G. (1991). Chronic treatment with polyethylene-glycolated superoxide dismutase partially restores endothelium-dependent vascular relaxations in cholesterol-fed rabbits. *Circulation Research*, *69*(5), 1293–1300.

Nakai, K., Urushihara, M., Kubota, Y., & Kosaka, H. (2003). Ascorbate enhances iNOS activity by increasing tetrahydrobiopterin in RAW 264.7 cells. *Free Radical Biology & Medicine*, *35*(8), 929–937.

Nandi, M., Miller, A., Stidwill, R., Jacques, T. S., Lam, A. A., Haworth, S., et al. (2005). Pulmonary hypertension in a GTP-cyclohydrolase 1-deficient mouse. *Circulation*, *111*(16), 2086–2090.

Noris, M., Morigi, M., Donadelli, R., Aiello, S., Foppolo, M., Todeschini, M., et al. (1995). Nitric oxide synthesis by cultured endothelial cells is modulated by flow conditions. *Circulation Research*, *76*(4), 536–543.

Oak, J. H., & Cai, H. (2007). Attenuation of angiotensin II signaling recouples eNOS and inhibits nonendothelial NOX activity in diabetic mice. *Diabetes*, *56*(1), 118–126.

Ohara, Y., Peterson, T. E., & Harrison, D. G. (1993). Hypercholesterolemia increases endothelial superoxide anion production. *Journal of Clinical Investigation*, *91*(6), 2546–2551.

Okumura, M., Masada, M., Yoshida, Y., Shintaku, H., Hosoi, M., Okada, N., et al. (2006). Decrease in tetrahydrobiopterin as a possible cause of nephropathy in type II diabetic rats. *Kidney International*, *70*(3), 471–476.

Patel, K. B., Stratford, M. R., Wardman, P., & Everett, S. A. (2002). Oxidation of tetrahydrobiopterin by biological radicals and scavenging of the trihydrobiopterin radical by ascorbate. *Free Radical Biology & Medicine*, *32*(3), 203–211.

Pieper, G. M. (1997). Acute amelioration of diabetic endothelial dysfunction with a derivative of the nitric oxide synthase cofactor, tetrahydrobiopterin. *Journal of Cardiovascular Pharmacology*, *29*(1), 8–15.

Porkert, M., Sher, S., Reddy, U., Cheema, F., Niessner, C., Kolm, P., et al. (2008). Tetrahydrobiopterin: A novel antihypertensive therapy. *Journal of Human Hypertension*, *22*(6), 401–407.

Ren, J., Kotaka, M., Lockyer, M., Lamb, H. K., Hawkins, A. R., & Stammers, D. K. (2005). GTP cyclohydrolase II structure and mechanism. *The Journal of Biological Chemistry*, *280* (44), 36912–36919.

Schmidt, T. S., McNeill, E., Douglas, G., Crabtree, M. J., Hale, A. B., Khoo, J., et al. (2010). Tetrahydrobiopterin supplementation reduces atherosclerosis and vascular inflammation in apolipoprotein E-knockout mice. *Clinical Science (London)*, *19*(3), 131–142.

Shinozaki, K., Kashiwagi, A., Nishio, Y., Okamura, T., Yoshida, Y., Masada, M., et al. (1999). Abnormal biopterin metabolism is a major cause of impaired endothelium-dependent relaxation through nitric oxide/O2-imbalance in insulin-resistant rat aorta. *Diabetes*, *48* (12), 2437–2445.

Silberman, G. A., Fan, T. H., Liu, H., Jiao, Z., Xiao, H. D., Lovelock, J. D., et al. (2010). Uncoupled cardiac nitric oxide synthase mediates diastolic dysfunction. *Circulation*, *121*(4), 519–528.

Stroes, E., Kastelein, J., Cosentino, F., Erkelens, W., Wever, R., Koomans, H., et al. (1997). Tetrahydrobiopterin restores endothelial function in hypercholesterolemia. *Journal of Clinical Investigation*, *99*(1), 41–46.

Stuehr, D. J. (1997). Structure–function aspects in the nitric oxide synthases. *Annual Review of Pharmacology and Toxicology*, *37*, 339–359.

Sugiyama, T., Levy, B. D., & Michel, T. (2009). Tetrahydrobiopterin recycling, a key determinant of endothelial nitric-oxide synthase-dependent signaling pathways in cultured vascular endothelial cells. *The Journal of Biological Chemistry*, *284*(19), 12691–12700.

Sun, X., Kumar, S., Tian, J., & Black, S. M. (2009). Estradiol increases guanosine 5′-triphosphate cyclohydrolase expression via the nitric oxide-mediated activation of cyclic adenosine 5′-monophosphate response element binding protein. *Endocrinology*, *150*(8), 3742–3752.

Tada, K., Yoshida, T., Mochizuki, K., Konno, T., & Nakagawa, H. (1970). Two siblings of hyperphenylalaninemia: Suggestion to a genetic variant of phenylketonuria. *The Tohoku Journal of Experimental Medicine*, *100*(3), 249–253.

Taylor, N. E., Maier, K. G., Roman, R. J., & Cowley, A. W. Jr., (2006). NO synthase uncoupling in the kidney of Dahl S rats: Role of dihydrobiopterin. *Hypertension*, *48*(6), 1066–1071.

Taylor, A. L., Ziesche, S., Yancy, C., Carson, P., D'Agostino, R., Jr., Ferdinand, K., et al. (2004). Combination of isosorbide dinitrate and hydralazine in blacks with heart failure. *The New England Journal of Medicine*, *351*(20), 2049–2057.

Thomas, G. D., Sander, M., Lau, K. S., Huang, P. L., Stull, J. T., & Victor, R. G. (1998). Impaired metabolic modulation of alpha-adrenergic vasoconstriction in dystrophin-deficient skeletal muscle. *Proceedings of the National Academy of Sciences of the United States of America*, *95*(25), 15090–15095.

Thony, B., Auerbach, G., & Blau, N. (2000). Tetrahydrobiopterin biosynthesis, regeneration and functions. *The Biochemical Journal*, *347*(Pt 1), 1–16.

Tie, L., Li, X. J., Wang, X., Channon, K. M., & Chen, A. F. (2009). Endothelium-specific GTP cyclohydrolase I overexpression accelerates refractory wound healing by suppressing oxidative stress in diabetes. *American Journal of Physiology. Endocrinology and Metabolism*, *296*(6), E1423–E1429.

Tiefenbacher, C. P., Bleeke, T., Vahl, C., Amann, K., Vogt, A., & Kubler, W. (2000). Endothelial dysfunction of coronary resistance arteries is improved by tetrahydrobiopterin in atherosclerosis. *Circulation*, *102*(18), 2172–2179.

Toth, M., Kukor, Z., & Sahin-Toth, M. (1997). Activation and dimerization of type III nitric oxide synthase by submicromolar concentrations of tetrahydrobiopterin in microsomal preparations from human primordial placenta. *Placenta*, *18*(2–3), 189–196.

Tsai, P., Weaver, J., Cao, G. L., Pou, S., Roman, L. J., Starkov, A. A., et al. (2005). L-arginine regulates neuronal nitric oxide synthase production of superoxide and hydrogen peroxide. *Biochemical Pharmacology*, *69*(6), 971–979.

Tzima, E., Irani-Tehrani, M., Kiosses, W. B., Dejana, E., Schultz, D. A., Engelhardt, B., et al. (2005). A mechanosensory complex that mediates the endothelial cell response to fluid shear stress. *Nature*, *437*(7057), 426–431.

Uematsu, M., Ohara, Y., Navas, J. P., Nishida, K., Murphy, T. J., Alexander, R. W., et al. (1995). Regulation of endothelial cell nitric oxide synthase mRNA expression by shear stress. *The American Journal of Physiology*, *269*(6 Pt 1), C1371–C1378.

Vasquez-Vivar, J. (2009). Tetrahydrobiopterin, superoxide, and vascular dysfunction. *Free Radical Biology & Medicine*, *47*(8), 1108–1119.

Vasquez-Vivar, J., Duquaine, D., Whitsett, J., Kalyanaraman, B., & Rajagopalan, S. (2002a). Altered tetrahydrobiopterin metabolism in atherosclerosis: Implications for use of oxidized tetrahydrobiopterin analogues and thiol antioxidants. *Arteriosclerosis, Thrombosis, and Vascular Biology*, *22*(10), 1655–1661.

Vasquez-Vivar, J., Martasek, P., Whitsett, J., Joseph, J., & Kalyanaraman, B. (2002b). The ratio between tetrahydrobiopterin and oxidized tetrahydrobiopterin analogues controls superoxide release from endothelial nitric oxide synthase: An EPR spin trapping study. *The Biochemical Journal*, *362*(Pt 3), 733–739.

Wei, C. C., Crane, B. R., & Stuehr, D. J. (2003). Tetrahydrobiopterin radical enzymology. *Chemical Reviews*, *103*(6), 2365–2383.

Wei, C. C., Wang, Z. Q., Tejero, J., Yang, Y. P., Hemann, C., Hille, R., et al. (2008). Catalytic reduction of a tetrahydrobiopterin radical within nitric-oxide synthase. *The Journal of Biological Chemistry*, *283*(17), 11734–11742.

Wenzel, P., Daiber, A., Oelze, M., Brandt, M., Closs, E., Xu, J., et al. (2008). Mechanisms underlying recoupling of eNOS by HMG-CoA reductase inhibition in a rat model of streptozotocin-induced diabetes mellitus. *Atherosclerosis*, *198*(1), 65–76.

Werner, E. R., Werner-Felmayer, G., & Mayer, B. (1998). Tetrahydrobiopterin, cytokines, and nitric oxide synthesis. *Proceedings of the Society for Experimental Biology and Medicine*, *219*(3), 171–182.

Widder, J. D., Chen, W., Li, L., Dikalov, S., Thony, B., Hatakeyama, K., et al. (2007). Regulation of tetrahydrobiopterin biosynthesis by shear stress. *Circulation Research*, *101*(8), 830–838.

Zhang, H., Bhargava, K., Keszler, A., Feix, J., Hogg, N., Joseph, J., et al. (2003). Transmembrane nitration of hydrophobic tyrosyl peptides. Localization, characterization, mechanism of nitration, and biological implications. *The Journal of Biological Chemistry*, *278*(11), 8969–8978.

Valérie B. Schini-Kerth, Cyril Auger,
Nelly Étienne-Selloum, and Thierry Chataigneau

Laboratoire de Biophotonique et Pharmacologie, UMR 7213 CNRS, Université de Strasbourg, Faculté de Pharmacie, Illkirch, France

Polyphenol-Induced Endothelium-Dependent Relaxations: Role of NO and EDHF

Abstract

The Mediterranean diet has been associated with greater longevity and quality of life in epidemiological studies. Indeed, because of the abundance of fruits and vegetables and a moderate consumption of wine, the Mediterranean diet provides high amounts of polyphenols thought to be essential bioactive compounds that might provide health benefits in terms of cardiovascular diseases and mortality. Several polyphenol-rich sources, such as grape-derived products, cocoa, and tea, have been shown to decrease mean

Advances in Pharmacology, Volume 60

1054-3589/10 $35.00
10.1016/S1054-3589(10)60004-X

blood pressure in patients with hypertension. The improvement of the endothelial function is likely to be one of the mechanisms by which polyphenols may confer cardiovascular protection. Indeed, polyphenols are able to induce nitric oxide (NO)-mediated endothelium-dependent relaxations in a large number of arteries including the coronary artery; they can also induce endothelium-derived hyperpolarizing factor (EDHF)-mediated relaxations in some of these arteries. Altogether, these mechanisms might contribute to explain the antihypertensive and cardio-protective effects of polyphenols *in vivo*. The aim of this review was to provide a nonexhaustive analysis of the effect of several polyphenol-rich sources and isolated compounds on the endothelium in *in vitro*, *ex vivo*, and *in vivo* models as well as in humans.

I. Introduction: The Epidemiological Evidence

The Mediterranean diet has been associated with greater longevity and quality of life in epidemiological studies. Recent meta-analyses found that the Mediterranean diet had a favorable effect on lipid profile, endothelium-dependent vasodilatation, insulin resistance, metabolic syndrome, antioxidant capacity, myocardial, and cardiovascular mortality (Mead et al., 2006; Serra-Majem et al., 2006). Indeed, greater adherence to a Mediterranean diet is associated with a significant reduction (9%) in mortality from cardiovascular diseases (CVDs) (Sofi et al., 2008). Interestingly, because of the abundance of fruits and vegetables as well as a moderate consumption of wine, the Mediterranean diet provides high amounts of polyphenols thought to be essential bioactive compounds that might provide health benefit. This phenomenon is well supported by numerous epidemiological studies suggesting a protective effect against CVDs of various rich sources of polyphenols such as vegetables and fruits (Dauchet et al., 2005, 2006; He et al., 2006, 2007; Law & Morris, 1998), chocolate and cocoa (Ding et al., 2006; Hooper et al., 2008), red wine (Di Castelnuovo et al., 2002; Renaud and de Lorgeril, 1992; Rimm et al., 1999; St Leger et al., 1979), and green tea (Arab et al., 2009; Arts et al., 2001; Kuriyama, 2008; Peters et al., 2001).

Impaired endothelial function, assessed by flow-mediated dilation (FMD), has been shown to be an independent predictor of cardiovascular outcome in subjects with cardiovascular risk factors or established CVDs (Yeboah et al., 2007). Therefore, to assess the protective effect of polyphenols on the cardiovascular system, the effect of dietary intake of polyphenols has often been evaluated on FMD as well as on blood pressure in humans. Chocolate and cocoa showed significant improvement of FMD, both acutely and chronically, and, in addition, chronic intake of chocolate and cocoa also had a beneficial effect on systolic and diastolic blood pressure (Balzer et al., 2008; Desch et al., 2010; Grassi et al., 2008; Hooper et al., 2008; Njike et al., 2009). Moreover, red wine restored normal endothelial function in hypercholesterolemic patients with impaired FMD (Andrade et al., 2009; Coimbra et al., 2005). Other grape-derived sources of polyphenols also improved FMD under basal conditions and after a high-fat meal in healthy subjects and in hypercholesterolemic patients (Chaves et al., 2009; Coimbra et al., 2005), and in adolescents with metabolic syndrome (Hashemi et al., 2010).

Clinical studies investigating vascular protection by tea and its major components, catechins, also revealed a significant increase in FMD in healthy subjects (Jochmann et al., 2008) and in patients with coronary heart disease (Duffy et al., 2001; Widlansky et al., 2007). In addition, black tea, green tea, benifuuki tea, and (−)-epigallocatechin gallate (EGCG) decreased systolic and diastolic blood pressure in healthy subjects (Brown et al., 2009; Grassi et al., 2009; Kurita et al., 2010).

II. Polyphenols Induce Endothelium-Dependent Responses *In Vitro* and *Ex Vivo*

A. Grape-Derived Polyphenols and Endothelial Function

1. Role of Nitric Oxide

a. Polyphenols Induce Endothelium-Dependent Nitric Oxide-Mediated Relaxations in Arteries The possibility that grape-derived products affect vascular tone was first assessed by Fitzpatrick et al. (1993) in rat aortic rings suspended in organ chambers. Following precontraction of the rings, the addition of increasing volumes of several wines and grape juice induced pronounced relaxations but had little effect in rings in which the endothelium had been mechanically removed. The endothelium-dependent relaxation was associated with increased guanosine 3′,5′-cyclic monophosphate (cyclic GMP) levels in the intact aorta and both the relaxation and the increase in cyclic GMP were prevented by competitive inhibitors of nitric oxide (NO) synthase such as N^G-monomethyl-L-arginine (L-NMMA) and N^G-nitro-L-arginine (L-NA) suggesting an increased formation of NO (Fitzpatrick et al., 1993). A red wine polyphenolic extract (RWP) from a Cabernet-Sauvignon grape variety also caused full relaxation of intact precontacted rat aortic rings, whereas a 1000-fold higher concentration was needed to relax those without endothelium (Andriambeloson et al., 1997). The endothelium-dependent but not the endothelium-independent relaxation was abolished by N^G-nitro-L-arginine-methyl-ester (L-NAME; Andriambeloson et al., 1997). The direct evidence that the RWPs cause endothelial formation of NO was obtained using electron paramagnetic resonance spectroscopy with Fe^{2+}-diethyldithiocarbamate as a NO spin trap (Andriambeloson et al., 1997). Indeed, RWPs increased NO level in aortic rings by about twofold and this response was strictly dependent on the presence of the endothelium and abolished by

L-NAME. Thereafter, the ability of grape-derived products to induce endothelium-dependent NO-mediated relaxations has been observed in several conductance arteries including the rat and rabbit aorta (Andriambeloson et al., 1997; Cishek et al., 1997; Fitzpatrick et al., 1993), the rat mesenteric artery (Dal-Ros et al., 2009), the porcine coronary artery (Ndiaye et al., 2005), the human coronary, and internal mammary artery (Flesch et al., 1998; Rakici et al., 2005). NO-mediated relaxations to grape-derived products have also been observed in resistance arteries such as the rat mesenteric bed and cerebral arterioles (Soares de Moura et al., 2002; Chan et al., 2008) and porcine retinal arterioles (Nagaoka et al., 2007), and in veins such as the human saphenous vein (Rakici et al., 2005). The possibility that NO-mediated relaxations to grape-derived products are dependent on gender has been assessed using aortic rings from female and male rats (Kane et al., 2009). These studies indicate that RWPs cause NO-mediated relaxations in aortic rings from both female and male rats but that the response is more pronounced in females (Kane et al., 2009). As acetylcholine-induced NO-mediated relaxations are also greater in female compared to male aortic rings, the enhanced vasorelaxation in females is most likely the consequence of the twofold higher expression level of eNOS in females (Kane et al., 2009; Weiner et al., 1994). Thus, blood vessels from both female and male may benefit from the ability of polyphenols to improve the NO-mediated vascular protection.

Experiments comparing the effect of numerous wines have indicated that some red wines induce pronounced endothelium-dependent relaxations of isolated blood vessels, whereas others have little effect (Burns et al., 2000; Cishek et al., 1997; Flesch et al., 1998). The difference in vasodilator activity of red wines is strongly correlated to their phenolic content (Burns et al., 2000). The major role of polyphenols is further supported by the fact that polyphenolic-rich extracts from red wines are strong inducers of endothelium-dependent relaxations (Andriambeloson et al., 1997; Ndiaye et al., 2005). Red wine contains large amounts of phenolic compounds (1–4 g/l of polyphenols expressed as gallic acid equivalents) from grapes, particularly the skins and seeds, and also wood- and yeast-derived phenolics (Hooper et al., 2008; Singleton, 1981). Indeed, grape skin and grape seed extracts induce strong endothelium-dependent relaxations (Fitzpatrick et al., 1993, 2000; Madeira et al., 2009; Mendes et al., 2003; Soares de Moura et al., 2002). In addition, several types of white wines, which contain about 10% of the phenolic content of red wines, cause little or no endothelium-dependent relaxations (Flesch et al., 1998). Besides red wines, potent endothelium-dependent relaxations of isolated blood vessels have also been observed in response to several grape juices and in particular purple grape juice, a rich source of polyphenols (2.3 g/l; Anselm et al., 2007; Fitzpatrick et al., 1993).

Wines are complex mixture of several hundreds of polyphenolic compounds (German & Walzem, 2000). Fractionation of RWPs has indicated that a great variety of compounds are able to stimulate the endothelial formation of NO and in particular, procyanidins and anthocyanidins (Andriambeloson et al., 1998; Auger et al., 2010a). In particular, the anthocyanin petunidin-3-O-(6″coumaroyl)-glucoside was identified as a potent activator of eNOS, whereas its related compound petunidin-3-O-glucoside and petunidin aglycone were inactive (Auger et al., 2010a).

b. Mechanisms Involved in the Polyphenol-Induced Activation of eNOS The investigation of the signaling pathway mediating the stimulatory effect of grape-derived polyphenols on eNOS activation has surprisingly indicated a key role for an intracellular redox-sensitive mechanism (Fig. 1; Ndiaye et al., 2005). Indeed, NO-mediated relaxations to RWPs in coronary artery rings were markedly reduced by membrane-permeant analogs of superoxide dismutase (SOD), Mn(III)tetrakis(1-methyl-4-pyridyl)porphyrin (MnTMPyP) and polyethyleneglycol–SOD (PEG-SOD), and of catalase (PEG-catalase) but remained unaltered by native SOD and native catalase (Ndiaye et al., 2005). In addition, Mn(III)tetrakis(1-methyl-4-pyridyl)porphyrin (MnTMPyP), PEG-catalase, and the antioxidant *N*-acetylcysteine also prevented the RWPs-induced formation of NO in cultured endothelial cells (Ndiaye et al., 2005). Similar findings were also observed with Concord grape juice and a grape skin extract (Anselm et al., 2007; Madeira et al., 2009). In contrast,

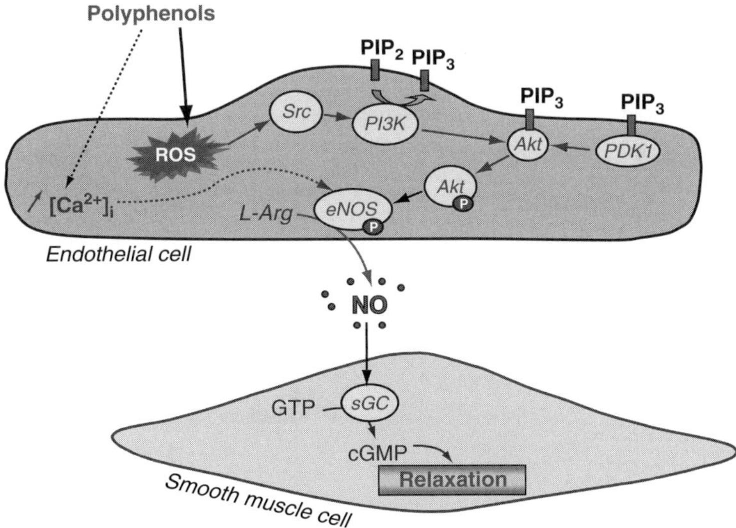

FIGURE 1 Polyphenols are potent inducers of the endothelial formation of NO via the Src/PI3-kinase/Akt pathway. eNOS, endothelial NO synthase; L-Arg, L-arginine; NO, nitric oxide; PDK1, phosphoinositide-dependent kinase 1; PI3K, phosphatidylinositol 3-kinase; ROS, reactive oxygen species; sGC, soluble guanylyl cyclase.

NO-mediated relaxations to a physiological agonist bradykinin were not affected by MnTMPyP indicating that grape-derived polyphenols and bradykinin activate distinct signaling pathways leading to eNOS activation (Ndiaye et al., 2005). More direct evidence that grape-derived polyphenols are able to cause the intracellular formation of superoxide anions was obtained in cultured coronary artery endothelial cells using the redox-sensitive probe dihydroethidine (Anselm et al., 2007; Madeira et al., 2009; Ndiaye et al., 2003). Although the endothelial source of superoxide anions remains unclear, it does not seem to involve classic enzymatic sources such as NADPH oxidase, xanthine oxidase, cytochrome P450, or the mitochondrial respiratory chain but rather appears to be dependent on the polyphenol hydroxyl moieties suggesting their auto-oxidation, a mechanism known to generate semi-quinones and then quinones with the concomitant formation of superoxide anions (Fig. 2; Auger et al., 2010b; Madeira et al., 2009; Ndiaye et al., 2005). Such a hypothesis is consistent with the fact that methylation of the hydroxyl groups of a grape seed extract resulted in the loss of its biological activity (Edirisinghe et al., 2008a). The pro-oxidant response in endothelial cells triggers the Src/PI3-kinase/Akt pathway, which ultimately causes eNOS

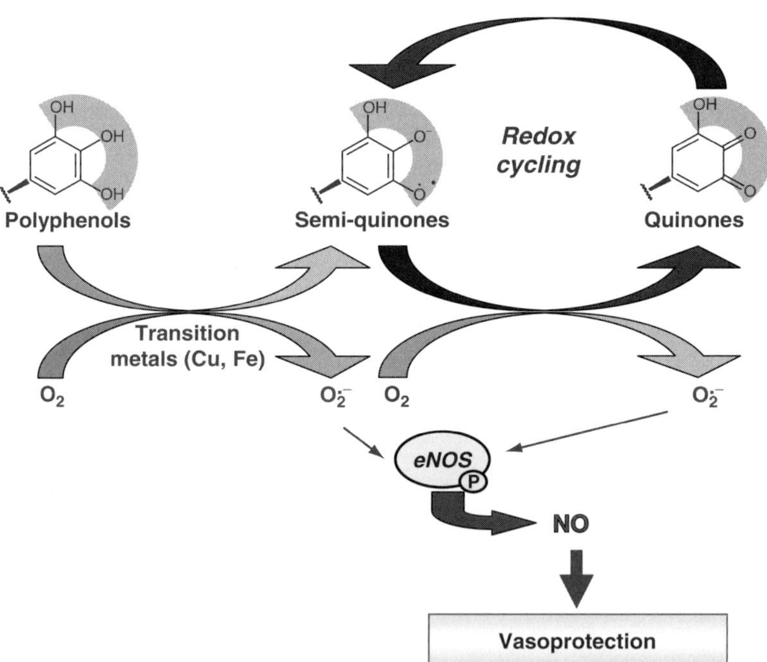

FIGURE 2 Proposed mechanism of auto-oxidation of polyphenols, a process where a group of two *ortho* hydroxyl functions provides through a two-step oxidation process first a semi-quinone and, second, a quinone, with the concomitant generation of superoxide anions.

activation by increasing the phosphorylation level of Ser 1177, a positive regulatory site, and the dephosphorylation of Thr 495, a negative regulatory site (Fig. 1; Anselm et al., 2009; Edirisinghe et al., 2008a, 2008b; Madeira et al., 2009; Ndiaye et al., 2005). Interestingly, the polyphenol-induced changes in eNOS phosphorylation level persisted for more than 30 min, which may explain their ability to stimulate NO formation for several hours, whereas that induced by bradykinin is a transient event, which vanishes within 5–10 min (Anselm et al., 2009; Ndiaye et al., 2005).

Besides changes in phosphorylation levels, eNOS activity can also be enhanced in response to an increase in the calcium signal via the interaction of the calcium–calmodulin complex with the calmodulin-binding domain within eNOS (Fleming, 2010). Experiments with cultured endothelial cells have indicated that the RWP-induced formation of NO requires calcium in the extracellular medium, and that this effect is associated with an increase in the intracellular concentration of calcium (Martin et al., 2002). However, since the calcium signal is relatively small (the peak increase is about 100 nM) and short-lasting (about 2–3 min), the calcium signal most likely acts in synergy with the Src/PI3-kinase/Akt pathway to rapidly initiate a significant eNOS-dependent NO formation, whereas changes in eNOS phosphorylation levels may control the sustained NO formation.

As estrogens have been shown to activate eNOS via the PI3-kinase/Akt pathway and polyphenols can interact with estrogen receptors (Chambliss & Shaul, 2002), several studies have investigated the involvement of estrogen receptors in the stimulatory effect of polyphenols on eNOS. A red wine extract Provinol® caused small endothelium-dependent relaxations in aortic rings from estrogen receptor alpha wild-type mice but not in estrogen receptor alpha knockout mice (Chalopin et al., 2010). Moreover, the Provinol®-induced formation of NO and phosphorylation of eNOS on Ser 1177 in EaHy926 cells, a human endothelial cell line, were abolished by fulvestrant, a selective estrogen receptor alpha antagonist, and by siRNA directed against estrogen receptor alpha (Chalopin et al., 2010). In contrast, fulvestrant did not affect endothelium-dependent NO-mediated relaxations to RWPs and two major red wine polyphenols, kaempferol and rutin, in the rat aorta, and also not the RWPs-induced eNOS phosphorylation on Ser1177 in porcine coronary artery endothelial cells studied at first passage (Kane et al., 2009; Padilla et al., 2005). Thus, these findings suggest that the estrogen receptor alpha may be involved in the endothelial formation of NO in response to some red wine polyphenols and in certain types of endothelial cells.

2. Role of Endothelium-Derived Hyperpolarizing Factor

a. Endothelium-Derived Hyperpolarizing Factor-Mediated Relaxations Endothelium-derived hyperpolarizing factor (EDHF)-mediated relaxations are usually characterized as the component of arterial endothelium-dependent relaxations which are resistant to inhibitors of NO synthase and cyclooxygenases. EDHF is proposed to be an electrical signal that is propagated from endothelial cells via myoendothelial gap junctions to the underlying smooth muscle (Griffith et al., 2004) and/or a substance, or even potassium ions released from endothelial cells, and which action is to hyperpolarize the vascular smooth muscle (Edwards et al., 1998; Fisslthaler et al., 1999; Matoba et al., 2000). It is now accepted that these mechanisms may operate in parallel in many arteries and, for instance, in maximally contracted rat mesenteric artery, myoendothelial gap junctions and endothelium-derived potassium ions are both responsible for the EDHF phenomenon (Dora et al., 2003; Edwards et al., 1998).

EDHF has been the focus of several recent reviews (Campbell & Fleming, 2010; de Wit & Griffith, 2010; Edwards et al., 2010; Shimokawa, 2010).

b. Polyphenols Induce Endothelium-Dependent EDHF-Mediated Responses in Arteries The first demonstration of a participation of EDHF to endothelium-dependent responses to polyphenols comes from a study performed with isolated porcine coronary arteries (Fig. 3; Ndiaye et al., 2003). In this study, it was shown that RWPs are able to induce concentration-dependent EDHF-mediated relaxations associated with hyperpolarization of vascular smooth muscle cells. EDHF-mediated endothelium-dependent relaxations have also

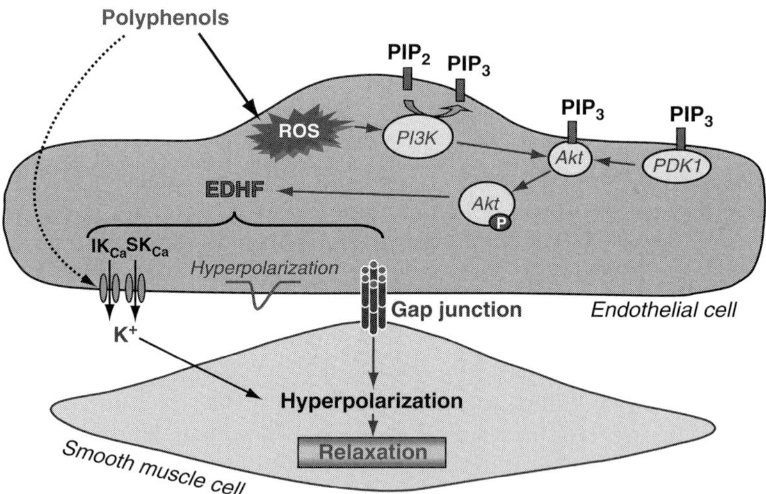

FIGURE 3 Polyphenols are potent inducers of the endothelial formation of EDHF via the PI3-kinase/Akt pathway. Some polyphenols such as resveratrol might also directly activate IK_{Ca} channels in endothelial cells. EDHF, endothelium-derived hyperpolarizing factor; IK_{Ca}, intermediate conductance calcium-activated potassium channels; PDK1, phosphoinositide-dependent kinase 1; PI3K, phosphatidylinositol 3-kinase; ROS, reactive oxygen species; SK_{Ca}, small conductance calcium-activated potassium channels.

been characterized in the isolated, perfused or not, mesenteric arterial bed in response to an aqueous extract of the roots of Siberian ginseng (Kwan et al., 2004), an extract of *Eucommia ulmoides* Oliv. bark, a traditional Chinese medicinal herb containing polyphenols (Kwan et al., 2004), açai (*Euterpe oleracea* Mart.; Rocha et al., 2007), an alcohol-free lyophilized Brazilian red wine (de Moura et al., 2004), and RWPs (Dal-Ros et al., 2009). It is noteworthy that Concord grape juice, a nonalcoholic rich source of grape-derived polyphenols, is also able to induce EDHF-mediated relaxations of porcine coronary arteries (Anselm et al., 2007).

c. Mechanisms Involved in Polyphenol-Induced EDHF-Mediated Responses The mechanisms involved in the polyphenol-induced EDHF-mediated relaxation are still poorly understood. However, studies aiming to characterize the endothelial mechanisms involved in RWPs-induced EDHF-mediated responses clearly indicated that relaxation and hyperpolarization are dependent on the endothelial intracellular formation of superoxide anions in porcine coronary arteries (Fig. 3; Ndiaye et al., 2003). Indeed, RWPs-induced EDHF-mediated relaxations which are examined in the presence of L-NA and indomethacin, to rule out the formation of NO and vasoactive prostanoids, respectively, are reduced by antioxidants, such as *N*-acetylcysteine, and membrane-permeant analogs of SOD, such as MnTMPyP and PEG-SOD. In addition, exposure of cultured porcine coronary artery endothelial cells to RWPs induced the MnTMPyP-sensitive formation of superoxide anions (Ndiaye et al., 2003). It has been shown thereafter that the RWPs-induced EDHF-mediated relaxation of porcine coronary arteries involves the redox-sensitive activation of PI3-kinase leading to Akt phosphorylation in endothelial cells (Ndiaye et al., 2004). The possibility that the PI3-kinase/Akt pathway modulates potassium channel activity and/or myoendothelial gap junctions in response to polyphenols remains to be examined (Fig. 4).

d. Effects of Polyphenols on Ionic Channels in Endothelial Cells The capacity of the combination of charybdotoxin plus apamin to inhibit EDHF-mediated responses indicates that polyphenols are able to open calcium-activated potassium (K_{Ca}) channels (Ndiaye et al., 2003). Indeed, it is known from electrophysiological evaluation in voltage-clamped *Xenopus laevis* oocytes injected with *m*Slo mRNA that flavonoids are effective openers of large conductance K_{Ca} (BK_{Ca}) channels (Li et al., 1997). From this study, it appears that meta-OH substitution at position A7 of the flavonoid structure is important for activity. In addition, it has also been shown, in human umbilical vein endothelial cells, that resveratrol, a polyphenolic phytoalexin found in grape skin, red wine, and peanuts, is able to activate BK_{Ca} channels. The resveratrol-induced increase in channel activity is independent of internal calcium (Li et al., 2000). Furthermore, in the MS1 cell line, a pancreatic islet

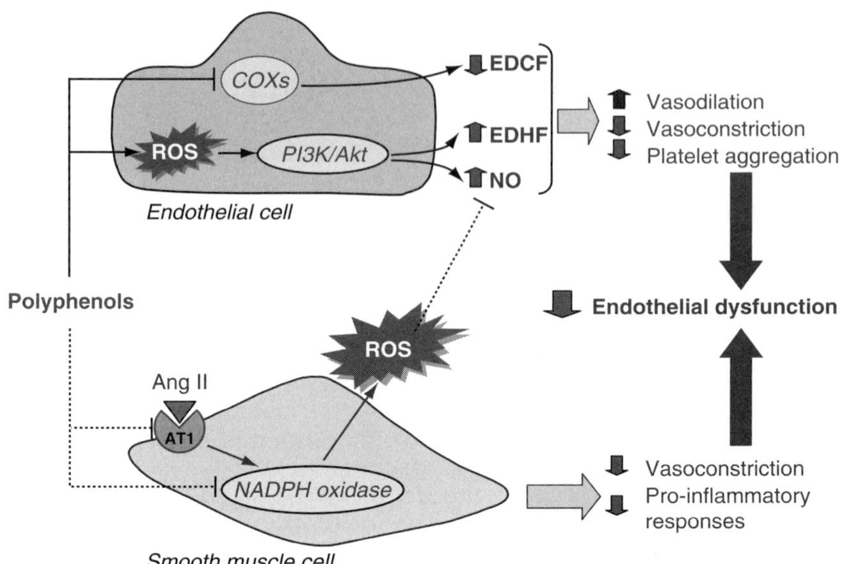

FIGURE 4 Schematic summarizing the protective effect of polyphenols on the arterial wall. Polyphenols are able to strongly stimulate the redox-sensitive endothelial formation of NO and EDHF-mediated relaxation, two potent vasoprotective mechanisms, and to prevent the endothelial formation of cyclooxygenase-derived metabolites of arachidonic acid that act on thromboxan receptors in vascular smooth muscle cells to cause vasoconstriction. In addition, polyphenols also prevent the excessive vascular formation of reactive oxygen species mostly by normalizing the expression of NADPH oxidase and angiotensin II AT1 receptor in the arterial wall. Both the effect in endothelial cells and vascular smooth muscle cells will promote vasodilation, and prevent inflammatory responses as well as activation of platelets. Ang II, angiotensin II; COXs, cyclooxygenases; EDCF, endothelium-derived contracting factors; EDHF, endothelium-derived hyperpolarizing factor; NO, nitric oxide; PI3K, phosphatidylinositol 3-kinase; ROS, reactive oxygen species.

endothelial cell line, resveratrol stimulated intermediate K_{Ca} (IK_{Ca}) channels (24 pS) by increasing their open probability (Li et al., 2000). Altogether, these results suggest that activation of IK_{Ca} channels in endothelial cells may contribute to EDHF-mediated responses to polyphenols in arteries such as the porcine coronary artery and the rat mesenteric artery. In these types of blood vessels, the consecutive endothelial hyperpolarization may then be transferred to the underlying vascular smooth muscle cells via gap junctions.

e. *Effects of Polyphenols on Gap Junctions* Intercellular communication through vascular gap junctions has been shown to be involved in the RWPs-induced EDHF-mediated relaxations in the isolated rat mesenteric artery using pharmacological inhibitors such as carbenoxolone and 18 α-glycyrrhetinic acid (Dal-Ros et al., 2009). Similarly, gap junctional communication is involved in endothelium-dependent vasodilatation of the perfused rat mesenteric artery induced by an aqueous extract prepared from the leaves of *E. ulmoides* Oliv., which contains polyphenols such as chlorogenic acid, caffeic acid, rutin, quercetin, and kaempferol (Jin et al., 2008; Luo et al., 2004). However, the mechanism of activation of vascular gap junctions by polyphenols remains to be investigated. Decreased EDHF-mediated responses to red wine polyphenols have been observed in angiotensin II-induced hypertensive rats (Dal-Ros et al., 2009) and in rats submitted to common bile duct ligation (Dal-Ros et al., 2010), which were associated with a reduced vascular expression of connexins Cx37, Cx40, and Cx43.

B. A Great Variety of Polyphenol-Rich Sources Induce the Endothelial Formation of NO

Over the last few years, several sources of phenolics compounds have been shown to induce the endothelial formation of NO in various *in vitro* and *ex vivo* models. Since polyphenols are plant secondary metabolites, these studies have examined predominantly products derived from plants used as traditional remedies predominantly against hypertension, and also polyphenol-rich food items such as red wine, cocoa, or tea.

1. *Plant-Derived Polyphenols*

Anselm et al. have shown that the procyanidin-rich extract of Crataegus (hawthorn) WS1442® induced the redox-sensitive activation of Akt and eNOS in endothelial cells, and endothelium-dependent NO-mediated relaxations in porcine coronary artery rings (Anselm et al., 2009). Moreover, the same extract also induced activation of eNOS and formation of NO in human coronary artery endothelial cells as well as endothelium-dependent NO-mediated relaxations in rat aortic rings and human mammary artery from coronary bypass patients (Brixius et al., 2006) (see Table I).

Other rich sources of polyphenols include maritime pine bark extracts, which are rich in procyanidins. Pycnogenol® and Flavangenol®, two commercially available pine bark extracts, have been shown to induce endothelium-dependent NO-mediated relaxations in rat and mouse aortic rings and rat mesenteric artery rings (Fitzpatrick et al., 1998; Kwak et al., 2009).

Moreover, studies with traditional plant remedies for the treatment of hypertension have indicated that among 11 plants used for traditional treatment of hypertension in sub-Saharan Benin, endothelium-dependent relaxations in porcine coronary artery rings were observed only with three plant extracts, namely *Croton celtidifolius*, *Gardenia ternifolia*, and *Parkia biglobosa*, the later being the most potent (Tokoudagba et al., 2010). Endothelium-dependent NO-mediated relaxations have also been reported for an extract of *Hibiscus sabdariffa* (Sarr et al., 2009).

TABLE I Nonexhaustive Review of Studies on Endothelial Effects of Polyphenol-Containing Plants

Plant	*Model*	*Reported effects*	*Dose used*	*References*
In vitro				
Crataegus extract WS1442®	Porcine coronary artery endothelial cells	eNOS activation		Anselm et al. (2009)
Gingko biloba extract	EaHy926 endothelial cells	Induction of eNOS protein expression, eNOS phosphorylation, and activity		Koltermann et al. (2007)
Kaempferia parviflora extract	Human umbilical vein endothelial cells	NO formation		Wattanapitayakul et al. (2007)
Lysimachia clethroides extract	Bovine aortic endothelial cells	eNOS activation		Lee et al. (2010)
Ex vivo				
Açaì stone extract	Rat mesenteric vascular bed	Endothelium-dependent NO-mediated relaxation		Rocha et al. (2007)
Crataegus extract WS1442®	Rat aortic rings and human internal mammary artery rings from coronary bypass patients	Endothelium-dependent NO-mediated relaxations in rat aortic and human mammary artery rings		Brixius et al. (2006)
Crataegus extract WS1442®	Porcine coronary artery rings	Endothelium-dependent NO-mediated relaxation		Anselm et al. (2009)
Gardenia ternifolia leaf extract	Porcine coronary artery rings	Endothelium-dependent relaxation		Tokoudagba et al. (2010)
Hawthorn extract	Rat mesenteric artery rings	Endothelium-dependent NO-mediated relaxation		Chen et al. (1998)
Hibiscus sabdariffa extract	Rat aortic rings	Endothelium-dependent NO-mediated relaxation		Sarr et al. (2009)
Lysimachia clethroides extract	Rat aortic rings	Endothelium-dependent NO-mediated relaxation		Lee et al. (2010)
Parkia biglobosa leaf extract	Porcine coronary artery rings	Endothelium-dependent relaxation		Tokoudagba et al. (2010)
Procyanidin-rich extract of *Croton celtidifolius*	Rat mesenteric vascular bed	NO-mediated relaxation		DalBo et al. (2008)

Pycnogenol®	Rat aortic rings	Endothelium-dependent NO-mediated relaxation		Fitzpatrick et al. (1998)
Siberan Ginseng extract	Dog carotid arterial rings, rat aortic, and mesenteric artery rings	Endothelium-dependent NO-mediated relaxation		Kwan et al. (2004)
Spondia mombin leaf extract	Porcine coronary artery rings	Endothelium-dependent relaxation		Tokoudagba et al. (2010)
In vivo				
Citrus unshiu Marc extract	Diabetic and control rats	3% and 10% extract in diet restored acetylcholine-induced relaxation in aortic rings	1%, 3%, and 10% in diet for 10 weeks	Kamata et al. (2005)
Flavangenol® (maritime pine bark extract)	DOCA-salt rats	Reduction of systolic blood pressure and increased acetylcholine-induced relaxation	0.1% in the diet for 5 weeks	Kwak et al. (2009)
Flavangenol® (maritime pine bark extract)	Rat aortic rings and mesenteric vascular bed	Endothelium-dependent NO-mediated relaxation in aortic strips and endothelium-dependent relaxation in mesenteric vascular bed	0.1% in the diet for 5 weeks	Kwak et al. (2009)
Gingko biloba extract	Sprague–Dawley rats	NO-dependent reduction of blood pressure	5 mg intravenous administration	Koltermann et al. (2007)
Mammea africana stem extract	L-NAME-treated rats	Reduction of systolic blood pressure	200 mg/kg/day for 4 weeks	Nguelefack-Mbuyo et al. (2008)
Solanum torvum fruit extract	L-NAME-treated rats	Reduction of systolic blood pressure	200 mg/kg/day for 4 weeks	Nguelefack et al. (2009)
In human				
Gingko biloba extract	Patients with coronary artery disease	Increased in NO formation in left anterior coronary artery	87.5 mg/day intravenous administration for 2 weeks	Wu et al. (2008)
Pycnogenol®	Healthy young men	NO-mediated increase of forearm blood flow response to acetylcholine	180 mg/day for 2 weeks	Nishioka et al. (2007)

2. Nutritional-Derived Polyphenols

Besides grape-derived products, several other dietary polyphenol-rich sources such as berries have been shown to induce the endothelial formation of NO (Table II). Indeed, endothelium-dependent relaxations in isolated arteries have been observed in response to blackcurrant (Nakamura et al., 2002), chokeberry, bilberry (Bell & Gochenaur, 2006), cranberry (Maher et al., 2000), raspberry, and strawberry extracts (Edirisinghe et al., 2008a, 2008b).

C. Several Authentic Polyphenols Induce the Endothelial Formation of NO

Polyphenol-rich food sources are complex mixtures also containing, besides a great variety of polyphenols, nonphenolic matrix including peptides, sugars, organic acids, and other natural compounds. It has been estimated that red wines contain several hundreds of different phenolic compounds (German & Walzem, 2000). Therefore, the study of purified phenolic compounds is necessary to demonstrate unambiguously that polyphenols are able to stimulate the endothelial formation of NO and also to better understand the structure–activity relationship for their vasorelaxant activity.

Phenolics compounds can be separated into nonflavonoids and flavonoids compounds (Tables III and IV). Among nonflavonoids, *trans*-resveratrol, a stilbene, has been reported to have several beneficial effects on the cardiovascular system (Penumathsa & Maulik, 2009). *trans*-Resveratrol stimulated eNOS protein expression in cultured endothelial cells (Nicholson et al., 2008; Wallerath et al., 2005) and caused endothelium-dependent and -independent relaxations in rat aortic rings (Leblais et al., 2008; Rush et al., 2007). In addition, eNOS protein expression in cultured EaHy926 endothelial cells was increased in response to the nonflavonoids *p*-coumaric acid, vanillic acid, benzoic acid, and caffeic acid (Wallerath et al., 2005).

Similarly, several studies have reported that various flavonoids can induce activation of eNOS *in vitro* and *ex vivo* (Table IV). Anthocyanins are flavonoid pigments responsible for the blue–red colors of many fruits and vegetables such as berries, apple skins, and red and purple grape skins. The anthocyanins, cyanidin and cyanidine-3-glucoside, have been shown to induce eNOS protein expression and activation in cultured endothelial cells (Wallerath et al., 2005; Xu et al., 2004a, 2004b), and delphinidin induced endothelium-dependent NO-mediated relaxations in isolated arteries (Andriambeloson et al., 1998; Leblais et al., 2008; Stoclet et al., 1999). Recently, we have isolated by fractionation of a red wine extract, petunidin-3-coumaroylglucoside, a potent activator of eNOS, and stimulator of NO formation in cultured endothelial cells from coronary arteries (Auger et al., 2010a).

TABLE II Nonexhaustive Review of Studies on Endothelial Effects of Nutritional Polyphenol-Rich Sources

Alimentary source	*Type of study*	*Model*	*Reported effects*	*Dose used (in vivo)*	*References*
Cocoa					
Cocoa	*Clinical*	Diabetic patients	Dose-dependent increases of flow-mediated dilatation		Balzer et al. (2008)
			Increased flow-mediated dilatation	3 × 321 mg flavonols/days for 30 days	Balzer et al. (2008)
Cocoa extract CocoanOX®	*In vivo*	Spontaneously hypertensive rats	Reduction of blood pressure	100 and 300 mg/kg per gavage	Cienfuegos-Jovellanos et al. (2009)
Dark chocolate	*Clinical*	Healthy adult subjects	Improvement of endothelium-dependent flow-mediated dilatation	46 g	Engler et al. (2004)
		Healthy adult smokers	Increased flow-mediated dilatation 2 h after intake and effect last for 8 h	40 g	Hermann et al. (2006)
		Hypertensive patients	Reduction of systolic and diastolic blood pressure	100 g/day for 14 days	Taubert et al. (2003)
Dark chocolate and cocoa	*Clinical*	Overweight adults	Increased flow-mediated dilatation and decreased blood pressure	22 g	Faridi et al. (2008)
Tea					
Green tea and black tea polyphenol mix	*In vivo*	Stroke-prone spontaneously hypertensive rats	Reduction of systolic and diastolic blood pressure	5 g/l	Negishi et al. (2004)
Black tea	*Clinical*	Patients with coronary artery disease	Increased flow-mediated dilatation	450 ml tea	Duffy et al. (2001)
			Increased flow-mediated dilatation	900 ml/day for 8 weeks	Duffy et al. (2001)
Green tea extract Teavigo®	*Clinical*	Patients with coronary artery disease	Increased flow-mediated dilatation	300 mg	Widlansky et al. (2007)

(*continued*)

Table II *(continued)*

Alimentary source	*Type of study*	*Model*	*Reported effects*	*Dose used (in vivo)*	*References*
Grape juice					
Purple grape juice	*Ex vivo*	Porcine coronary artery rings	Endothelium-dependent NO- and EDHF-mediated relaxation		Anselm et al. (2007)
	Clinical	Hypertensive patients	Reduction of systolic and diastolic blood pressure	5.5 mg/kg/day for 8 weeks	Park et al. (2004)
		Patients with coronary artery disease	Increased flow-mediated dilatation	4 and 8 ml/kg for 4 weeks	Chou et al. (2001)
			Increased flow-mediated dilatation	7.7 ml/kg/days for 14 days	Stein et al. (1999)
		Hypercholesterolemic patients	Increased flow-mediated dilatation	500 ml/day for 14 days	Coimbra et al. (2005)
Grape products					
Grape marc extract	*Ex vivo*	Sprague–Dawley rat aortic rings	Endothelium-dependent relaxation		Auger et al. (2004)
Grape marc extract	*In vivo*	Fructose-fed Sprague–Dawley rats	Reduction of blood pressure	21 mg/kg for 6 weeks	Al-Awwadi et al. (2005)
Grape procyanidins extract	*Ex vivo*	Human internal mammary artery rings	Endothelium-dependent NO-mediated relaxation		Aldini et al. (2003)
Grape seed extract	*In vitro*	Human umbilical vein endothelial cells	Phosphorylation of eNOS		Edirisinghe et al. (2008a)
	Ex vivo	Sprague–Dawley rat aortic rings	Endothelium-dependent relaxation		Auger et al. (2004)
		New Zealand rabbit aortic rings	Endothelium-dependent relaxation		Edirisinghe et al. (2008a)
		Wistar rat aortic rings	Endothelium-dependent NO-mediated relaxation		Mendes et al. (2003)

		Sprague–Dawley rat aortic rings and mesenteric vascular bed	Endothelium-dependent relaxation		Fitzpatrick et al. (2000)
	In vivo	Fructose-fed Sprague–Dawley rats	Reduction of blood pressure	21 mg/kg for 6 weeks	Al-Awwadi et al. (2005)
Grape skin extract	*In vitro*	Porcine coronary artery endothelial cells	Phosphorylation of eNOS		Madeira et al. (2009)
	Ex vivo	Porcine coronary artery rings	Endothelium-dependent NO-mediated relaxation		Madeira et al. (2009)
		Rat aortic rings	Endothelium-dependent NO-mediated relaxation		Fitzpatrick et al. (1993)
	In vivo	Ovariectomized spontaneously hypertensive rats	Reduction of blood pressure	0.5% in diet for 10 weeks	Peng et al. (2005)
Procyanidin-rich red wine fractions	*Ex vivo*	Wistar rat aortic rings	Endothelium-dependent relaxation		Andriambeloson et al. (1998)
Provinols®	*In vitro*	EaHy926 endothelial cells	ERalpha-dependent phosphorylation of eNOS		Chalopin et al. (2010)
	Ex vivo	Wistar rat femoral artery rings	NO-mediated relaxation and NO formation		Zenebe et al. (2003)
	In vivo	ERalpha$^{(-/-)}$ mice	Endothelium-dependent ERalpha-dependent relaxation		Chalopin et al. (2010)
			Induction of ERalpha-dependent formation of NO in aorta		Chalopin et al. (2010)
		L-NAME-treated NO deficient rats	Reduction of systolic blood pressure, improved acetylcholine-induced relaxation, and NO synthase activity	40 mg/kg/day for 3 weeks	Bernatova et al. (2002)
			Reduction of blood pressure and increased eNOS activity	40 mg/kg/day for 4 weeks	Pechanova et al. (2004)

(*continued*)

Table II *(continued)*

Alimentary source	*Type of study*	*Model*	*Reported effects*	*Dose used (in vivo)*	*References*
Red grape extract	*Clinical*	Patients with coronary artery disease	Increased flow-mediated dilatation	600 mg	Lekakis et al. (2005)
Wines					
Red and white wines	*Clinical*	Patients with coronary artery disease	Increased flow-mediated dilatation	4 ml/kg	Whelan et al. (2004)
Spanish wines	*Ex vivo*	Wistar rat aortic rings	Endothelium-dependent relaxation		Padilla et al. (2005)
Red wine	*In vitro*	EaHy926 endothelial cells	Increased eNOS expression and eNOS activity		Wallerath et al. (2005)
		EaHy926 endothelial cells	Induction of eNOS protein expression and NO formation		Wallerath et al. (2003)
	Ex vivo	Rat aortic rings	Endothelium-dependent NO-mediated relaxation		Andriambeloson et al. (1997), Fitzpatrick et al. (1993)
		New Zealand rabbit aortic rings	NO-mediated relaxation		Cishek et al. (1997)
		Human coronary artery rings	Endothelium-dependent relaxation		Flesch et al. (1998)
		Wistar rat perfused mesenteric bed	Endothelium-dependent NO- and EDHF-mediated relaxation		de Moura et al. (2004)
	In vivo	L-NAME-treated NO deficient rats	Reduction of blood pressure	100 mg/kg/day for 10 days	de Moura et al. (2004)
	Clinical	Healthy adults	Increased flow-mediated dilatation		Boban et al. (2006)
		Healthy adult men	Increased flow-mediated dilatation	240 ml/day for 30 days	Cuevas et al. (2000)

		Hypercholesterolemic patients	Increased flow-mediated dilatation	250 ml/day for 14 days	Coimbra et al. (2005)
Red wine and alcohol-free red wine	*Ex vivo*	Rat aortic rings	Red wine with or without alcohol-induced relaxation		Boban et al. (2006)
	In vivo	Sprague–Dawley rats	NO-mediated increase of bleeding time and decrease of platelet adhesion to collagen	8.4 ml/day for 10 days	Wollny et al. (1999)
	Clinical	Habitual smokers	Reduction of blood pressure		Papamichael et al. (2006)
Red wine without alcohol	*Ex vivo*	Sprague–Dawley rats	Increased eNOS activity and NO formation	35% in diet for 10 days	Benito et al. (2002)
	Clinical	Healthy adults	Increased flow-mediated dilatation	250 ml	Agewall et al. (2000)
		Healthy adult men	Increased flow-mediated dilatation	500 ml	Hashimoto et al. (2001)
		Patients with coronary artery disease	Increased flow-mediated dilatation	250 ml	Karatzi et al. (2004)
Red wine fraction	*Ex vivo*	Rat mesenteric arterial bed	NO-mediated relaxation		Schuldt et al. (2005)
Anthocyanins-rich red wine fraction	*Ex vivo*	Wistar rats	Endothelium-dependent relaxation		Andriambeloson et al. (1998)
Red wine extract					
Red wine extract	*In vitro*	Porcine coronary artery endothelial cells	eNOS activation		Auger et al. (2010a)
			eNOS activation		Ndiaye et al. (2005)
		EaHy926 and human umbilical vein endothelial cells	Induction of eNOS protein expression and NO formation		Leikert et al. (2002)
	Ex vivo	Rat mesenteric artery rings	Endothelium-dependent NO-mediated relaxation		Duarte et al. (2004)
		Sprague–Dawley rats	Endothelium-dependent relaxation		Stoclet et al. (1999)
		Hypertensive commun bile duct ligation rats	Endothelium-dependent NO-mediated relaxation		Dal-Ros et al. (2010)

(*continued*)

Table II *(continued)*

Alimentary source	*Type of study*	*Model*	*Reported effects*	*Dose used (in vivo)*	*References*
		Guinea pig aortic rings	Endothelium-dependent NO-mediated relaxation		Brizic et al. (2009)
		Wistar rats	Endothelium-dependent relaxation		Andriambeloson et al. (1998)
		Wistar rats	Endothelium-dependent NO-mediated relaxation		Leblais et al. (2008)
		Porcine coronary artery rings	Endothelium-dependent NO-mediated relaxation		Ndiaye et al. (2005)
		Angiotensin II-induced hypertension in rats	NO- and EDHF-mediated relaxation		Dal-Ros et al. (2009) Kane et al. (2010)
		Porcine coronary artery rings	Endothelium-dependent EDHF-mediated relaxation		Ndiaye et al. (2004)
	In vivo	Spontaneously hypertensive rats	Improved endothelium-dependent and pressure-induced dilatation in cerebral arterioles	100 mg/kg/day	Chan et al. (2008)
		Ovariectomized spontaneously hypertensive rats	Reduction of systolic blood pressure	40 mg/kg/day for 5 weeks	Lopez-Sepulveda et al. (2008)
		DOCA-salt rats	Reduction of systolic blood pressure	40 mg/kg/day for 5 weeks	Jimenez et al. (2007)
		Angiotensin II-induced hypertension in rats	Reduction of systolic blood pressure	150 mg/kg/day for 3 weeks	Sarr et al. (2006)
		Spontaneously hypertensive rats	Reduction of systolic blood pressure	1% in drinking water for 8 weeks	Mizutani et al. (1999)
Beans					
Soy isoflavones	*Clinical*	Postmenopausal women	Increased flow-mediated dilatation and plasmatic NO metabolites	80 mg in low-fat meal	Hall et al. (2008)
Polyphenol-containing azuki bean extract (*Vigna angularis*)	*In vivo*	Spontaneously hypertensive rats	Reduction of systolic blood pressure, increased NOx excretion in urine	0.9% in diet for 8 weeks	Mukai and Sato (2009)

Berries					
Blackcurrant extract	*Ex vivo*	Sprague–Dawley aortic rings	Endothelium-dependent NO-mediated relaxation		Nakamura et al. (2002)
Blueberry extract	*In vivo*	Spontaneously hypertensive rats	Reduction of systolic blood pressure at 4 and 6 weeks, but not at 7 and 8 weeks	3% freeze-dried blueberry in diet for 8 weeks	Shaughnessy et al. (2009)
Chokeberry and bilberry anthocyanins extracts	*Ex vivo*	Porcine coronary artery rings	Endothelium-dependent relaxation		Bell and Gochenaur (2006)
Cranberry juice	*Ex vivo*	Sprague–Dawley aortic rings	Endothelium-dependent NO-mediated relaxation		Maher et al. (2000)
	In vivo	Sprague–Dawley anaesthetized rats	Reduction of blood pressure		Maher et al. (2000)
Pomegranate concentrated juice	*In vivo*	Female obese Zucker rats	Endothelium-dependent relaxation	6.25 ml/l in drinking water	de Nigris et al. (2007)
Pomegranate extract	*In vivo*	Female obese Zucker rats	Endothelium-dependent relaxation	6.25 ml/l in drinking water	de Nigris et al. (2007)
Pomegranate juice	*Clinical*	Patients with severe carotid artery stenosis	Reduction of systolic blood pressure	50 ml per day for 1 year	Aviram et al. (2004)
		Hypertensive patients	Reduction of systolic blood pressure	50 ml per day for 2 weeks	Aviram and Dornfeld (2001)
Procyanidin-rich apple extract	*Ex vivo*	Rat aortic rings	Endothelium-dependent NO-mediated relaxation		Matsui et al. (2009)
Raspberry extract and fractions	*Ex vivo*	New Zealand rabbit aortic rings	Vasorelaxant effect associated with fractions enriched in lambertianin C and sanguiin H-6		Mullen et al. (2002)
Strawberry extract	*In vitro*	Human umbilical vein endothelial cells	Phosphorylation of eNOS		Edirisinghe et al. (2008b)
	Ex vivo	New Zealand rabbit aortic rings	Endothelium-dependent NO-mediated relaxation		Edirisinghe et al. (2008b)

TABLE III Nonexhaustive Review of Studies on Endothelial Effects of Isolated Nonflavonoid Compounds

Molecule	*Type of study*	*Model*	*Reported effects*	*Dose used*	*References*
Stilbenes					
trans-Resveratrol	*In vitro*	EaHy926 endothelial cells	Increased eNOS protein expression		Wallerath et al. (2005)
		Human umbilical vein endothelial cells	Increased eNOS protein expression		Nicholson, et al. (2008)
	Ex vivo	Wistar rat aortic rings	Endothelium-independent relaxation		Novakovic et al. (2006)
		Spontaneously hypertensive rats	Improved endothelium-dependent relaxation	0.448 and 4.48 mg/l	Rush et al. (2007)
		Wistar rat aortic rings	NO-mediated relaxation		Leblais et al. (2008)
	In vivo	Obese Zucker rats	Reduced systolic blood pressure and increased eNOS protein expression	10 mg/kg/day for 8 weeks	Rivera et al. (2009)
Phenolic acids					
Benzoic acid	*In vitro*	EaHy926 endothelial cells	Increased eNOS protein expression		Wallerath et al. (2005)
Caffeic acid	*In vitro*	EaHy926 endothelial cells	Increased eNOS protein expression		Wallerath et al. (2005)
p-Coumaric acid	*In vitro*	EaHy926 endothelial cells	Increased eNOS protein expression		Wallerath et al. (2005)
Vanillic acid	*In vitro*	EaHy926 endothelial cells	Increased eNOS protein expression		Wallerath et al. (2005)

Chlorogenic acid	*In vivo*	Spontaneously hypertensive rats	Reduction of systolic blood pressure	0.5% of diet (ca. 300 mg/kg) for 8 weeks	Suzuki et al. (2006)
Other nonflavonoid compounds					
1-α-O-Galloylpunicalagin	*In vitro*	Bovine aortic endothelial cells	Activation of eNOS and NO formation		Chen et al. (2008)
Curcumin	*Ex vivo*	Porcine coronary artery rings	Endothelium-dependent NO-mediated relaxation		Xu et al. (2007)
Dioclein	*Ex vivo*	Wistar rat aortic rings	Endothelium-dependent NO-mediated relaxation		Lemos et al. (1999)
Vitisin C	*Ex vivo*	Rabbit aortic rings	Endothelium-dependent NO-mediated relaxation		Seya et al. (2003)

TABLE IV Nonexhaustive Review of Studies on Endothelial Effects of Isolated Flavonoid Compounds

Molecule	*Type of study*	*Model*	*Reported effects*	*Dose used*	*References*
Anthocyans					
Cyanidin	*In vitro*	EaHy926 endothelial cells	Increased eNOS protein expression		Wallerath et al. (2005)
Cyanidin-3-glucoside	*In vitro*	Bovine aortic endothelial cells	eNOS activation		Xu et al. (2004b)
		Bovine aortic endothelial cells	eNOS protein expression and NO formation		Xu et al. (2004a)
Delphinidin	*Ex vivo*	Rat aortic rings	Endothelium-dependent relaxation		Stoclet et al. (1999)
		Rat aortic rings	Endothelium-dependent relaxation		Andriambeloson et al. (1998)
		Rat aortic rings	NO-mediated relaxation		Leblais et al. (2008)
Petunidin-3 (6″coumaroyl)-glucoside	*In vitro*	Porcine coronary artery endothelial cells	eNOS activation and NO formation		Auger et al. (2010a)
Flavanols					
Catechin	*In vivo*	Sprague–Dawley rats	eNOS activation and NO formation in aorta	0.3% in diet for 10 days	Benito et al. (2002)
EGCg	*In vitro*	Bovine aortic endothelial cells	eNOS activation by a PI3-kinase-, PKA-, and Akt-dependent pathways	100 μM	Lorenz et al. (2004)
		Porcine coronary artery endothelial cells	eNOS activation		Auger et al. (2010b)
		Bovine aortic endothelial cells	eNOS activation		Kim et al. (2007)
	Ex vivo	Rat aortic rings	Endothelium- NO-dependent relaxation in aortic rings	1–50 μM	Lorenz et al. (2004)
		Bovine ophtalmic artery rings	Endothelium-dependent NO-mediated relaxation in bovine ophtalmic artery		Romano and Lograno (2009)

		Mesenteric vascular bed from spontaneously hypertensive rats	Endothelium-dependent NO-mediated relaxation		Potenza et al. (2007)
		Porcine coronary artery rings	Endothelium-dependent NO- and EDHF-mediated relaxation		Auger et al. (2010b)
		Wistar rat mesenteric artery rings	Endothelium-dependent NO-mediated relaxation		Kim et al. (2007)
	In vivo	Streptozotocin-diabetic rats	Prevention of endothelial dysfunction in aortic rings	25 mg/kg/day per os for 8 weeks	Roghani and Baluchnejadmojarad (2009)
		Spontaneously hypertensive rats	Reduction of blood pressure	200 mg/kg/day for 3 weeks	Potenza et al. (2007)
Epicatechin	*Ex vivo*	Sprague–Dawley rat mesenteric artery rings	Endothelium-dependent NO-mediated relaxation		Huang et al. (1999)
	Clinical	Healthy subjects	Induction of flow-mediated dilatation	1 or 2 mg/kg	Schroeter et al. (2006)
Procyanidin dimers	*In vitro*	Porcine coronary artery endothelial cells	eNOS activation		Auger et al. (2010a)
Procyanidin oligomers	*Ex vivo*	Sprague–Dawley rat aortic rings and mesenteric vascular bed	Endothelium-dependent relaxation		Fitzpatrick et al. (2000)
Flavones					
Baicalein	*In vivo*	Spontaneously hypertensive rats	Improved endothelium-dependent relaxation in aortic rings	10 mg/kg for 4 weeks	Machha et al. (2007a)
Baicalein	*Ex vivo*	Rat mesenteric artery rings	Endothelium-dependent NO-mediated relaxation		Chen et al. (1999)
Chrysin	*Ex vivo*	Rat aortic rings	Endothelium-dependent relaxation		Duarte et al. (2001a)
Flavonols					
Quercetin	*In vivo*	Spontaneously hypertensive rats	Prevention of hypertension and of the increased eNOS expression	10 mg/kg for 13 weeks	Sanchez et al. (2006)

(*continued*)

Table IV *(continued)*

Molecule	*Type of study*	*Model*	*Reported effects*	*Dose used*	*References*
		Diabetic rat aortic rings	Restoration of endothelium-dependent relaxation in diabetic rats	10 mg/kg for 6 weeks	Machha et al. (2007b)
		2K1C Goldblatt hypertensive rats	Prevention of hypertension and of endothelial dysfunction	10 mg/kg for 5 weeks	Garcia-Saura et al. (2005)
		DOCA-salt rats	Reduction of systolic blood pressure similarly to 20 mg/kg verapamil, prevention of endothelial dysfunction	10 mg/kg for 5 weeks	Galisteo et al. (2004)
		L-NAME-treated NO deficient rats	Prevention of hypertension	10 mg/kg for 6 weeks	Duarte et al. (2002)
		Normal and streptozotocin-diabetic rats	Endothelium-dependent relaxations		Roghani et al. (2004)
		Spontaneously hypertensive rats	Reduction of blood pressure	10 mg/kg/day for 5 weeks	Duarte et al. (2001b)
		Spontaneously hypertensive rats	Reduction of blood pressure, improved endothelium-dependent and independent vasorelaxation in aortic rings	10 mg/kg/day for 4 weeks	Machha and Mustafa (2005)
		Dahl–sensitive hypertensive rats	Reduction of blood pressure	10 mg/kg/day for 4 weeks	Mackraj et al. (2008)
		Obese Zucker rats	Reduction of systolic blood pressure	2 and 10 mg/kg/day for 10 weeks	Rivera et al. (2008)
		Sprague–Dawley rats	Increased eNOS activity and NO formation in the aorta	0.3% in diet for 10 days	Benito et al. (2002)
Quercetin and isorhamnetin	*Ex vivo*	Wistar rat aortic rings	Inhibition of endothelial dysfunction induced by ET-1	1–10 μM	Romero et al. (2009)

		Wistar rat aortic rings	Prevention of angiotensin II-induced endothelial dysfunction	10 μM	Sanchez et al. (2007)
Enzymatically modified isoquercitrin	*In vivo*	Spontaneously hypertensive rats	Reduction of systolic blood pressure at 3 mg/kg for up to 50 days	2 mg/kg/day for	Emura et al. (2007)
Baicalein, flavone' and quercetin	*In vivo*	Spontaneously hypertensive rats	Improved endothelium-dependent relaxation in aortic rings, quercetin, and flavone reduced systolic blood pressure	10 mg/kg for 4 weeks	Machha and Mustafa (2005)
Isoflavones					
Equol	*In vitro*	Human umbilical vein endothelial cells	Induction of Akt-mediated eNOS activation		Joy et al. (2006)
	Ex vivo	Wistar rat aortic rings	Induction of NO-mediated relaxation		Joy et al. (2006)
Genistein and daidzein	*Ex vivo*	Spontaneously hypertensive rat aortic rings	Improved endothelium-dependent relaxation	10 μM	Vera et al. (2005)
Genistein	*In vivo*	Spontaneously hypertensive rats	Reduction of blood pressure, improved endothelium-dependent relaxation, increased eNOS activity, and decreased oxidative stress	10 mg/kg for 5 weeks	Vera et al. (2007)
		Streptozotocin-diabetic rats	Prevention of endothelial dysfunction in aortic rings	1 mg/kg/day IP for 4 weeks	Baluchnejadmojarad and Roghani (2008)
Flavanones					
Hesperetin	*In vitro*	Human umbilical vein endothelial cells	Induction of ER-dependent NO formation and eNOS expression		Liu et al. (2008)
	In vivo	Spontaneously hypertensive rats	Reduction of systolic blood pressure	50 mg/kg IP	Yamamoto et al. (2008)

Flavanol monomers such as EGCG also induced activation of eNOS in endothelial cells and endothelium-dependent relaxations in artery rings (Auger et al., 2010b; Kim et al., 2007; Lorenz et al., 2004). Similar effects have been reported with flavanol oligomers (Auger et al., 2010b; Fitzpatrick et al., 2002).

Studies showing that a great variety of polyphenols are able to activate eNOS suggest that a common structure within these polyphenols may account for their biological activity. The fractionation of a red wine extract let to the isolation of two closely related anthocyanins, petunidin-3-coumaroylglucoside, and malvidin-3-coumaroylglucoside (Auger et al., 2010a). These anthocyanins differ only by a single substituent: a hydroxyl function on position 5′ for the petunidin-3-coumaroylglucoside and a methoxyl function on position 5′ for the malvidin-3-coumaroylglucoside. Interestingly, petunidin-3-coumaroylglucoside strongly activated eNOS by phosphorylation on serine 1177, whereas malvidin-3-coumaroylglucoside did not have such an effect. These results indicate a critical role for the hydroxyl function on position 5′ of the B ring for eNOS activation by anthocyanins. Moreover, the comparison of petunidin-3-coumaroylglucoside with closely related petunidin derivatives revealed that the nature of the substitution on position C3 is also critical for the induction of eNOS activation. Indeed, petunidin-3-glucoside and petunidin aglycon did not induce activation of eNOS, suggesting that the presence of the coumaroyl moiety on the glucose is necessary for the activation of eNOS by petunidin-3-coumaroylglucoside.

Similarly, the study of several catechins has indicated that EGCG and epicatechin gallate induced eNOS activation in cultured endothelial cells, whereas epicatechin and epigallocatechin were without effect (Auger et al., 2010b), indicating that the gallate moiety on position C3 is critical for the stimulatory effect of tea catechins on eNOS. In addition, the degree of hydroxylation of catechins appears to be another important factor regulating eNOS activation. Indeed, the methoxylation of all hydroxyl functions except those on the gallate moiety of EGCg resulted in a partial reduction of eNOS activation, whereas the replacement of all hydroxyl functions by methoxyl functions resulted in the total loss of the biological activity. Edirisinghe et al. (2008a) have also reported that methoxylation of all hydroxyl functions of phenolic compounds in a grape seed extract abolished its ability to activate eNOS. Altogether, these findings indicate that hydroxyl functions on the polyphenolic structure are required to activate eNOS in response to polyphenols.

In addition, the methoxylation of all hydroxyl functions on EGCg is associated with the loss of their ability to simulate ROS formation in endothelial cells (Auger et al., 2010b). Thus, it has been suggested that hydroxyl functions are necessary for the production of ROS in endothelial cells possibly due to the auto-oxidation of the phenolic structures, a process where a group of two *ortho* hydroxyl functions provides through a two-step oxidation process first a semi-quinone and, then, a quinone (Fig. 2; Sang et al., 2007). Each step of oxidation will generate superoxide anions (Hou et al., 2005). Auto-oxidation of EGCg has been reported to occur *in vitro*, both in cultured cells and in some physiological solutions, generating either an EGCg quinone or various dimeric structures like theasinensins (Hong et al., 2002; Hou et al., 2005; Sang et al., 2007).

III. Polyphenols and Endothelial Function *In Vivo*

One of the major limitations of *in vitro* and *ex vivo* studies is the fact that most of the phenolic compounds applied on cells and artery rings are not the circulating form that have been identified after oral absorption. Indeed, most polyphenols found in food sources are not circulating in blood in their intact form but rather in the form of metabolites or catabolites. For example, in intestinal and hepatic cells, anthocyanins are metabolized into glucuronides, methyl, and sulfo-conjugates (Aprikian et al., 2003; Mullen et al., 2008). However, some polyphenols may also be absorbed in the intact form like procyanidin oligomers that have been identified in blood and urine in intact form following ingestion of a procyanidin-rich grape seed extract by humans and rats (Sano et al., 2003; Tsang et al., 2005). Moreover, they have been shown to reach the nanomolar range in human plasma after consumption of procyanidins-rich cocoa and grape seeds (Holt et al., 2002; Sano et al., 2003). Similarly, anthocyanins seem also to have a low absorption and bioavailability (Manach et al., 2005).

Recent studies indicate that compounds unabsorbed in the small intestine may reach the large intestine where they can be catabolized by the microflora into small ring fission products and that these small phenolic acids can then be directly absorbed (Del Rio et al., 2009). In addition, fermentation studies showed that monomeric flavan-3-ols, procyanidin dimers, and anthocyanins may produce various small phenolic acids, namely hydroxyphenylacetic and hydroxycarboxylic acids, when exposed to human fecal microflora action in fermentation reactors (Appeldoorn et al., 2009; Aura et al., 2005; Deprez et al., 2000; Fleschhut et al., 2006; Tsuda et al., 1999; van't Slot & Humpf, 2009). Indeed, Tsuda et al. have shown that protocatechuic acid is found in plasma after oral absorption of cyanidin-3-glucoside by rats (Tsuda et al., 1999).

Altogether, these data suggest that despite the apparent low absorption of flavonoids, the absorption of smaller phenolic acid catabolites may increase strongly the bioavailability of flavonoid compounds. However, the biological activity of the circulating flavonoid metabolites and catabolites remains to be assessed, especially regarding their ability to activate eNOS.

A. Experimental Animal Models

Since polyphenols can induce activation of the formation of NO and EDHF-mediated responses, they may exert beneficial effects on the endothelial function *in vivo*. As endothelial dysfunction is associated with hypertension, the evaluation of the protective effect of polyphenols has mostly been assessed in experimental models of hypertension.

Several studies have reported reduction of blood pressure after intake of polyphenol-rich products in various experimental models of hypertension including the spontaneously hypertensive rat, the N^G-nitro L-arginine-induced hypertension, the DOCA salt-induced hypertension, and the angiotensin II-induced hypertension in rats (Tables II–IV). As an example, blood pressure reduction has been reported after ingestion by spontaneously hypertensive rats of polyphenols in the form of grape seed extract (Peng et al., 2005), red wine (Machha & Mustafa, 2005), green and black tea extracts (Negishi et al., 2004), azuki bean extract (Mukai & Sato, 2009), and blueberry (Shaughnessy et al., 2009). Moreover, Cienfuegos-Jovellanos et al. (2009) recently demonstrated that a polyphenol-rich cocoa powder (up to 300 mg/kg body weight) in the spontaneously hypertensive rat reduced blood pressure similarly to 50 mg/kg of captopril, an angiotensin converting enzyme inhibitor (Cienfuegos-Jovellanos et al., 2009). A similar blood pressure lowering effect has been reported with red wine in experimental models of hypertension including the DOCA salt-induced hypertension (Jimenez et al., 2007), the N^G-nitro L-arginine-induced hypertension (Bernatova et al., 2002; de Moura et al., 2004; Pechanova et al., 2004), and the angiotensin II-induced hypertension in rats (Sarr et al., 2006).

Several blood-lowering effects have also been reported after chronic ingestion by spontaneously hypertensive rats of isolated polyphenolic compounds such as 200 mg/kg/day of EGCg (Potenza et al., 2007), 10 mg/kg/day of baicalein (Machha et al., 2005), 0.5% of chlorogenic acid (Suzuki et al., 2006), 10 mg/kg/day of quercetin (Sanchez et al., 2006), and 10 mg/kg/day of genistein (Vera et al., 2007).

Altogether, these experimental studies indicate that ingestion of several polyphenol-rich products is associated with a beneficial effect in several experimental models of hypertension.

B. Humans

In humans, impaired endothelial function assessed by FMD has been shown to be an independent predictor of cardiovascular outcomes in subjects with cardiovascular risk factors or established CVDs (Yeboah et al., 2007). Therefore, the effect of dietary intake of polyphenols has been often evaluated on FMD as well as on blood pressure in humans.

In healthy subjects, the basal FMD has been increased after consumption of 3 ml/kg or two glasses of red wine with or without alcohol (Agewall et al., 2000; Djousse et al., 1999; Karatzi et al., 2004). The same effect has been reported after consumption of a single dose of 46 g of dark chocolate (Engler et al., 2004). Schroeter et al. (2006) suggested that the effect of dark chocolate is related to its content in epicatechin, and demonstrated that ingestion of a low dose of purified epicatechin (1 or 2 mg/kg body weight) similarly increased basal FMD (Schroeter et al., 2006). It also has been reported that chronic consumption of the procyanidin-rich maritime pine bark extract Pycnogenol® (180 mg/day for 2 weeks) increased NO-mediated forearm blood flow to acetylcholine in healthy subjects (Nishioka et al., 2007).

More interestingly, polyphenol-rich sources may also exert beneficial effects on endothelial function under pathophysiological conditions associated with endothelial dysfunction (Tables I and II). For example, Duffy et al. (2001) indicated that in patients with coronary artery disease, FMD is increased either after consumption of a single 450 ml dose of black tea or after chronic daily ingestion of 900 ml of black tea for 8 weeks. A similar effect has been observed after ingestion of a single dose of either 300 mg of the EGCg-rich green tea extract Teavigo® (Widlansky et al., 2007) or 600 mg of a red grape extract (Lekakis et al., 2005). Chronic consumption of pomegranate juice (50 ml daily for a year) also reduced systolic blood and the intima-media thickness in patients with severe carotid artery stenosis (Aviram et al., 2004). Increased FMD has also been reported after dark chocolate ingestion in habitual smokers and overweight subjects, and after red wine consumption in hypercholesterolemic patients (Coimbra et al., 2005; Faridi et al., 2008; Hermann et al., 2006). Hall et al. (2008) also reported that supplementation of a low-fat meal with 80 mg of soybean isoflavones increased FMD in postmenopausal women, a population with increased risk of CVDs.

Moreover, consumption of polyphenol-rich sources may decrease blood pressure in hypertensive patients. Indeed, Taubert et al. (2003) showed that daily ingestion of 100 g of dark chocolate for 2 weeks reduced systolic and diastolic blood pressure in mildly hypertensive patients. Park et al. (2004) also showed improvement of blood pressure in hypertensive patients after intake of 5.5 ml/kg/day of purple grape juice, roughly equivalent to two glasses, for 8 weeks. In addition, Aviram et al. showed that daily consumption of 50 ml of pomegranate juice for 2 weeks by hypertensive patients reduced systolic blood pressure by 5% (Aviram & Dornfeld, 2001).

Taken together, these studies suggest that chronic consumption of polyphenol-rich sources has a beneficial effect on the endothelial function both in physiological and in pathophysiological conditions.

IV. Conclusion

The aim of this review was to present the experimental and clinical evidence indicating that several rich natural sources of polyphenols are able to enhance endothelial vasoprotective mechanisms including NO formation and EDHF-mediated relaxations both *in vitro* and *in vivo* as well as in normal and pathological conditions. However, additional investigations are crucially warranted to identify polyphenol-rich sources with a high vasoprotective activity, to determine the bioavailability and metabolism of polyphenols, as well as to characterize the molecular mechanisms involved in the endothelial vasoprotective mechanisms and the structure/activity relationship. Such information will help to better understand the potential of polyphenols to protect the vascular system in health and disease.

Conflict of Interest: The authors have no conflicts of interest to declare.

Abbreviations

EGCG	epigallocatechin gallate
PEG-catalase	polyethylene glycol catalase
PEG-SOD	polyethylene glycol–SOD
RWPs	red wine polyphenols

References

Agewall, S., Wright, S., Doughty, R. N., et al. (2000). Does a glass of red wine improve endothelial function? *European Heart Journal*, *21*, 74–78.

Al-Awwadi, N. A., Araiz, C., Bornet, A., et al. (2005). Extracts enriched in different polyphenolic families normalize increased cardiac NADPH oxidase expression while having differential effects on insulin resistance, hypertension, and cardiac hypertrophy in high-fructose-fed rats. *Journal of Agricultural and Food Chemistry*, *53*, 151–157.

Aldini, G., Carini, M., Piccoli, A., et al. (2003). Procyanidins from grape seeds protect endothelial cells from peroxynitrite damage and enhance endothelium-dependent relaxation in human artery: New evidences for cardio-protection. *Life Sciences*, *73*, 2883–2898.

Andrade, A. C., Cesena, F. H., Consolim-Colombo, F. M., et al. (2009). Short-term red wine consumption promotes differential effects on plasma levels of high-density lipoprotein cholesterol, sympathetic activity, and endothelial function in hypercholesterolemic, hypertensive, and healthy subjects. *Clinics*, *64*, 435–442.

Andriambeloson, E., Kleschyov, A. L., Muller, B., et al. (1997). Nitric oxide production and endothelium-dependent vasorelaxation induced by wine polyphenols in rat aorta. *British Journal of Pharmacology*, *120*, 1053–1058.

Andriambeloson, E., Magnier, C., Haan-Archipoff, G., et al. (1998). Natural dietary polyphenolic compounds cause endothelium-dependent vasorelaxation in rat thoracic aorta. *The Journal of Nutrition*, *128*, 2324–2333.

Anselm, E., Chataigneau, M., Ndiaye, M., et al. (2007). Grape juice causes endothelium-dependent relaxation via a redox-sensitive Src- and Akt-dependent activation of eNOS. *Cardiovascular Research*, *73*, 404–413.

Anselm, E., Socorro, V. F., Dal-Ros, S., et al. (2009). Crataegus special extract WS 1442 causes endothelium-dependent relaxation via a redox-sensitive Src- and Akt-dependent activation of endothelial NO synthase but not via activation of estrogen receptors. *Journal of Cardiovascular Pharmacology*, *53*, 253–260.

Appeldoorn, M. M., Venema, D. P., Peters, T. H. F., et al. (2009). Some phenolic compounds increase the nitric oxide level in endothelial cells in vitro. *Journal of Agricultural and Food Chemistry*, *57*, 7693–7699.

Aprikian, O., Duclos, V., Guyot, S., et al. (2003). Apple pectin and a polyphenol-rich apple concentrate are more effective together than separately on cecal fermentations and plasma lipids in rats. *The Journal of Nutrition*, *133*, 1860–1865.

Arab, L., Liu, W., & Elashoff, D. (2009). Green and black tea consumption and risk of stroke: A meta-analysis. *Stroke*, *40*, 1786–1792.

Arts, I. C., Hollman, P. C., Feskens, E. J., Bueno de Mesquita, H. B., Kromhout, D., et al. (2001). Catechin intake might explain the inverse relation between tea consumption and ischemic heart disease: The Zutphen Elderly Study. *The American Journal of Clinical Nutrition*, *74*, 227–232.

Auger, C., Chaabi, M., Anselm, E., et al. (2010a). The red wine extract-induced activation of endothelial nitric oxide synthase is mediated by a great variety of polyphenolic compounds. *Molecular Nutrition & Food Research*, *54*, S171–S183.

Auger, C., Gerain, P., Laurent-Bichon, F., et al. (2004). Phenolics from commercialized grape extracts prevent early atherosclerotic lesions in hamsters by mechanisms other than antioxidant effect. *Journal of Agricultural and Food Chemistry*, *52*, 5297–5302.

Auger, C., Kim, J. H., Chabert, P., et al. (2010b). The EGCg-induced redox-sensitive activation of endothelial nitric oxide synthase and relaxation are critically dependent on hydroxyl moieties. *Biochemical and Biophysical Research Communications*, *393*, 162–167.

Aura, A. M., Martin-Lopez, P., O'Leary, K. A., et al. (2005). In vitro metabolism of anthocyanins by human gut microflora. *European Journal of Nutrition*, *44*, 133–142.

Aviram, M., & Dornfeld, L. (2001). Pomegranate juice consumption inhibits serum angiotensin converting enzyme activity and reduces systolic blood pressure. *Atherosclerosis*, *158*, 195–198.

Aviram, M., Rosenblat, M., Gaitini, D., et al. (2004). Pomegranate juice consumption for 3 years by patients with carotid artery stenosis reduces common carotid intima-media thickness, blood pressure and LDL oxidation. *Clinical Nutrition*, *23*, 423–433.

Baluchnejadmojarad, T., & Roghani, M. (2008). Chronic administration of genistein improves aortic reactivity of streptozotocin-diabetic rats: Mode of action. *Vascular Pharmacology*, *49*, 1–5.

Balzer, J., Rassaf, T., Heiss, C., et al. (2008). Sustained benefits in vascular function through flavanol-containing cocoa in medicated diabetic patients a double-masked, randomized, controlled trial. *Journal of the American College of Cardiology*, *51*, 2141–2149.

Bell, D. R., & Gochenaur, K. (2006). Direct vasoactive and vasoprotective properties of anthocyanin-rich extracts. *Journal of Applied Physiology*, *100*, 1164–1170.

Benito, S., Lopez, D., Saiz, M. P., et al. (2002). A flavonoid-rich diet increases nitric oxide production in rat aorta. *British Journal of Pharmacology*, *135*, 910–916.

Bernatova, I., Pechanova, O., Babal, P., et al. (2002). Wine polyphenols improve cardiovascular remodeling and vascular function in NO-deficient hypertension. *American Journal of Physiology. Heart and Circulatory Physiology*, *282*, H942–H948.

Boban, M., Modun, D., Music, I., et al. (2006). Red wine induced modulation of vascular function: Separating the role of polyphenols, ethanol, and urates. *Journal of Cardiovascular Pharmacology*, *47*, 695–701.

Brixius, K., Willms, S., Napp, A., et al. (2006). Crataegus special extract WS 1442 induces an endothelium-dependent, NO-mediated vasorelaxation via eNOS-phosphorylation at serine 1177. *Cardiovascular Drugs and Therapy*, *20*, 177–184.

Brizic, I., Modun, D., Vukovic, J., et al. (2009). Differences in vasodilatory response to red wine in rat and guinea pig aorta. *Journal of Cardiovascular Pharmacology*, *53*, 116–120.

Brown, A. L., Lane, J., Coverly, J., et al. (2009). Effects of dietary supplementation with the green tea polyphenol epigallocatechin-3-gallate on insulin resistance and associated metabolic risk factors: Randomized controlled trial. *The British Journal of Nutrition*, *101*, 886–894.

Burns, J., Gardner, P. T., O'Neil, J., et al. (2000). Relationship among antioxidant activity, vasodilation capacity, and phenolic content of red wines. *Journal of Agricultural and Food Chemistry*, *48*, 220–230.

Campbell, W. B., & Fleming, I. (2010). Epoxyeicosatrienoic acids and endothelium-dependent responses. *Pflugers Archiv*, *459*, 881–895.

Chalopin, M., Tesse, A., Martinez, M. C., et al. (2010). Estrogen receptor alpha as a key target of red wine polyphenols action on the endothelium. *PLoS ONE*, *5*, e8554.

Chambliss, K. L., & Shaul, P. W. (2002). Estrogen modulation of endothelial nitric oxide synthase. *Endocrine Reviews*, *23*, 665–686.

Chan, S. L., Tabellion, A., Bagrel, D., et al. (2008). Impact of chronic treatment with red wine polyphenols (RWP) on cerebral arterioles in the spontaneous hypertensive rat. *Journal of Cardiovascular Pharmacology*, *51*, 304–310.

Chaves, A. A., Joshi, M. S., Coyle, C. M., et al. (2009). Vasoprotective endothelial effects of a standardized grape product in humans. *Vascular Pharmacology*, *50*, 20–26.

Chen, L. G., Liu, Y. C., Hsieh, C. W., et al. (2008). Tannin 1-alpha-O-galloylpunicalagin induces the calcium-dependent activation of endothelial nitric-oxide synthase via the phosphatidylinositol 3-kinase/Akt pathway in endothelial cells. *Molecular Nutrition & Food Research*, *52*, 1162–1171.

Chen, Z. Y., Su, Y. L., Lau, C. W., et al. (1999). Endothelium-dependent contraction and direct relaxation induced by baicalein in rat mesenteric artery. *European Journal of Pharmacology*, *374*, 41–47.

Chen, Z. Y., Zhang, Z. S., Kwan, K. Y., et al. (1998). Endothelium-dependent relaxation induced by hawthorn extract in rat mesenteric artery. *Life Sciences*, *63*, 1983–1991.

Chou, E. J., Keevil, J. G., Aeschlimann, S., et al. (2001). Effect of ingestion of purple grape juice on endothelial function in patients with coronary heart disease. *The American Journal of Cardiology*, *88*, 553–555.

Cienfuegos-Jovellanos, E., Quinones, M. M., Muguerza, B., et al. (2009). Antihypertensive effect of a polyphenol-rich cocoa powder industrially processed to preserve the original flavonoids of the cocoa beans. *Journal of Agricultural and Food Chemistry*, *57*, 6156–6162.

Cishek, M. B., Galloway, M. T., Karim, M., et al. (1997). Effect of red wine on endothelium-dependent relaxation in rabbits. *Clinical Science*, *93*, 507–511.

Coimbra, S. R., Lage, S. H., Brandizzi, L., et al. (2005). The action of red wine and purple grape juice on vascular reactivity is independent of plasma lipids in hypercholesterolemic patients. *Brazilian Journal of Medical and Biological Research*, *38*, 1339–1347.

Cuevas, A. M., Guasch, V., Castillo, O., et al. (2000). A high-fat diet induces and red wine counteracts endothelial dysfunction in human volunteers. *Lipids*, *35*, 143–148.

DalBo, S., Moreira, E. G., Brandao, F. C., et al. (2008). Mechanisms underlying the vasorelaxant effect induced by proanthocyanidin-rich fraction from *Croton celtidifolius* in rat small resistance arteries. *Journal of Pharmacological Sciences*, *106*, 234–241.

Dal-Ros, S., Bronner, C., Schott, C., et al. (2009). Angiotensin II-induced hypertension is associated with a selective inhibition of endothelium-derived hyperpolarizing factor-mediated responses in the rat mesenteric artery. *The Journal of Pharmacology and Experimental Therapeutics*, *328*, 478–486.

Dal-Ros, S., Oswald-Mammosser, M., Pestrikova, T., et al. (2010). Losartan prevents portal hypertension-induced, redox-mediated endothelial dysfunction in the mesenteric artery in rats. *Gastroenterology*, *138*, 1574–1584.

Dauchet, L., Amouyel, P., & Dallongeville, J. (2005). Fruit and vegetable consumption and risk of stroke: A meta-analysis of cohort studies. *Neurology*, *65*, 1193–1197.

Dauchet, L., Amouyel, P., Hercberg, S., et al. (2006). Fruit and vegetable consumption and risk of coronary heart disease: A meta-analysis of cohort studies. *The Journal of Nutrition*, *136*, 2588–2593.

de Moura, R. S., Miranda, D. Z., Pinto, A. C., et al. (2004). Mechanism of the endothelium-dependent vasodilation and the antihypertensive effect of Brazilian red wine. *Journal of Cardiovascular Pharmacology*, *44*, 302–309.

de Nigris, F., Balestrieri, M. L., Williams-Ignarro, S., et al. (2007). The influence of pomegranate fruit extract in comparison to regular pomegranate juice and seed oil on nitric oxide and arterial function in obese Zucker rats. *Nitric Oxide*, *17*, 50–54.

de Wit, C., & Griffith, T. M. (2010). Connexins and gap junctions in the EDHF phenomenon and conducted vasomotor responses. *Pflugers Archives*, *459*, 897–914.

Del Rio, D., Costa, L. G., Lean, M. E., et al. (2009). Polyphenols and health: What compounds are involved? *Nutrition, Metabolism, and Cardiovascular Diseases*, *20*, 1–6.

Deprez, S., Brezillon, C., Rabot, S., et al. (2000). Polymeric proanthocyanidins are catabolized by human colonic microflora into low-molecular-weight phenolic acids. *The Journal of Nutrition*, *130*, 2733–2738.

Desch, S., Schmidt, J., Kobler, D., et al. (2010). Effect of cocoa products on blood pressure: Systematic review and meta-analysis. *American Journal of Hypertension*, *23*, 97–103.

Di Castelnuovo, A., Rotondo, S., Iacoviello, L., et al. (2002). Meta-analysis of wine and beer consumption in relation to vascular risk. *Circulation*, *105*, 2836–2844.

Ding, E. L., Hutfless, S. M., Ding, X., et al. (2006). Chocolate and prevention of cardiovascular disease: A systematic review. *Nutrition and Metabolism*, *3*, 2.

Djousse, L., Ellison, R. C., McLennan, C. E., et al. (1999). Acute effects of a high-fat meal with and without red wine on endothelial function in healthy subjects. *The American Journal of Cardiology*, *84*, 660–664.

Dora, K. A., McEvoy, L., King, M., et al. (2003). The intensity of agonist-stimulation influences the mechanism for relaxation in rat mesenteric arteries. In P. M. Vanhoutte (Ed.), *EDHF 2002* (pp. 256–260). London, UK: Taylor & Francis.

Duarte, J., Andriambeloson, E., Diebolt, M., et al. (2004). Wine polyphenols stimulate superoxide anion production to promote calcium signaling and endothelial-dependent vasodilatation. *Physiological Research*, *53*, 595–602.

Duarte, J., Jimenez, R., O'Valle, F., et al. (2002). Protective effects of the flavonoid quercetin in chronic nitric oxide deficient rats. *Journal of Hypertension*, *20*, 1843–1854.

Duarte, J., Jimenez, R., Villar, I. C., et al. (2001a). Vasorelaxant effects of the bioflavonoid chrysin in isolated rat aorta. *Planta Medica*, *67*, 567–569.

Duarte, J., Perez-Palencia, R., Vargas, F., et al. (2001b). Antihypertensive effects of the flavonoid quercetin in spontaneously hypertensive rats. *British Journal of Pharmacology*, *133*, 117–124.

Duffy, S. J., Keaney, J. F., Jr., Holbrook, M., et al. (2001). Short- and long-term black tea consumption reverses endothelial dysfunction in patients with coronary artery disease. *Circulation*, *104*, 151–156.

Edirisinghe, I., Burton-Freeman, B., & Tissa, K. C. (2008a). Mechanism of the endothelium-dependent relaxation evoked by a grape seed extract. *Clinical Science*, *114*, 331–337.

Edirisinghe, I., Burton-Freeman, B., Varelis, P., et al. (2008b). Strawberry extract caused endothelium-dependent relaxation through the activation of PI3 kinase/Akt. *Journal of Agricultural and Food Chemistry*, *56*, 9383–9390.

Edwards, G., Dora, K. A., Gardener, M. J., et al. (1998). K+ is an endothelium-derived hyperpolarizing factor in rat arteries. *Nature*, *396*, 269–272.

Edwards, G., Félétou, M., & Weston, A. H. (2010). Endothelium-derived hyperpolarising factors and associated pathways: A synopsis. *Pflugers Archives*, *459*, 863–879.

Emura, K., Yokomizo, A., Toyoshi, T., et al. (2007). Effect of enzymatically modified isoquercitrin in spontaneously hypertensive rats. *Journal of Nutritional Science and Vitaminology*, *53*, 68–74.

Engler, M. B., Engler, M. M., Chen, C. Y., et al. (2004). Flavonoid-rich dark chocolate improves endothelial function and increases plasma epicatechin concentrations in healthy adults. *Journal of the American College of Nutrition*, *23*, 197–204.

Faridi, Z., Njike, V. Y., Dutta, S., et al. (2008). Acute dark chocolate and cocoa ingestion and endothelial function: A randomized controlled crossover trial. *The American Journal of Clinical Nutrition*, *88*, 58–63.

Fisslthaler, B., Popp, R., Kiss, L., et al. (1999). Cytochrome P450 2C is an EDHF synthase in coronary arteries. *Nature*, *401*, 493–497.

Fitzpatrick, D. F., Bing, B., Maggi, D. A., et al. (2002). Vasodilating procyanidins derived from grape seeds. *Annals of the New York Academy of Sciences*, *957*, 78–89.

Fitzpatrick, D. F., Bing, B., & Rohdewald, P. (1998). Endothelium-dependent vascular effects of pycnogenol. *Journal of Cardiovascular Pharmacology*, *32*, 509–515.

Fitzpatrick, D. F., Fleming, R. C., Bing, B., et al. (2000). Isolation and characterization of endothelium-dependent vasorelaxing compounds from grape seeds. *Journal of Agricultural and Food Chemistry*, *48*, 6384–6390.

Fitzpatrick, D. F., Hirschfield, S. L., & Coffey, R. G. (1993). Endothelium-dependent vasorelaxing activity of wine and other grape products. *American Journal of Physiology. Heart and Circulatory Physiology*, *265*, H774–H778.

Fleming, I. (2010). Molecular mechanisms underlying the activation of eNOS. *Pflugers Archives*, *459*, 793–806.

Flesch, M., Schwarz, A., & Bohm, M. (1998). Effects of red and white wine on endothelium-dependent vasorelaxation of rat aorta and human coronary arteries. *The American Journal of Physiology*, *275*, H1183–H1190.

Fleschhut, J., Kratzer, F., Rechkemmer, G., et al. (2006). Stability and biotransformation of various dietary anthocyanins in vitro. *European Journal of Nutrition*, *45*, 7–18.

Galisteo, M., Garcia-Saura, M. F., Jimenez, R., et al. (2004). Effects of chronic quercetin treatment on antioxidant defence system and oxidative status of deoxycorticosterone acetate-salt-hypertensive rats. *Molecular and Cellular Biochemistry*, *259*, 91–99.

Garcia-Saura, M. F., Galisteo, M., Villar, I. C., et al. (2005). Effects of chronic quercetin treatment in experimental renovascular hypertension. *Molecular and Cellular Biochemistry*, *270*, 147–155.

German, J. B., & Walzem, R. L. (2000). The health benefits of wine. *Annual Review of Nutrition*, *20*, 561–593.

Grassi, D., Desideri, G., Necozione, S., et al. (2008). Blood pressure is reduced and insulin sensitivity increased in glucose-intolerant, hypertensive subjects after 15 days of consuming high-polyphenol dark chocolate. *The Journal of Nutrition*, *138*, 1671–1676.

Grassi, D., Mulder, T. P., Draijer, R., et al. (2009). Black tea consumption dose-dependently improves flow-mediated dilation in healthy males. *Journal of Hypertension*, *27*, 774–781.

Griffith, T. M., Chaytor, A. T., & Edwards, D. H. (2004). The obligatory link: Role of gap junctional communication in endothelium-dependent smooth muscle hyperpolarization. *Pharmacological Research*, *49*, 551–564.

Hall, W. L., Formanuik, N. L., Harnpanich, D., et al. (2008). A meal enriched with soy isoflavones increases nitric oxide-mediated vasodilation in healthy postmenopausal women. *The Journal of Nutrition*, *138*, 1288–1292.

Hashemi, M., Kelishadi, R., Hashemipour, M., et al. (2010). Acute and long-term effects of grape and pomegranate juice consumption on vascular reactivity in paediatric metabolic syndrome. *Cardiology in the Young*, *20*, 73–77.

Hashimoto, M., Kim, S., Eto, M., et al. (2001). Effect of acute intake of red wine on flow-mediated vasodilatation of the brachial artery. *The American Journal of Cardiology*, *88*, 1457–1460.

He, F. J., Nowson, C. A., Lucas, M., et al. (2007). Increased consumption of fruit and vegetables is related to a reduced risk of coronary heart disease: Meta-analysis of cohort studies. *Journal of Human Hypertension*, *21*, 717–728.

He, F. J., Nowson, C. A., & MacGregor, G. A. (2006). Fruit and vegetable consumption and stroke: Meta-analysis of cohort studies. *Lancet*, *367*, 320–326.

Hermann, F., Spieker, L. E., Ruschitzka, F., et al. (2006). Dark chocolate improves endothelial and platelet function. *Heart*, *92*, 119–120.

Holt, R. R., Lazarus, S. A., Sullards, M. C., et al. (2002). Procyanidin dimer B2 [epicatechin-(4beta-8)-epicatechin] in human plasma after the consumption of a flavanol-rich cocoa. *The American Journal of Clinical Nutrition*, *76*, 798–804.

Hong, J., Lu, H., Meng, X., et al. (2002). Stability, cellular uptake, biotransformation, and efflux of tea polyphenol (−)-epigallocatechin-3-gallate in HT-29 human colon adenocarcinoma cells. *Cancer Research*, *62*, 7241–7246.

Hooper, L., Kroon, P. A., Rimm, E. B., et al. (2008). Flavonoids, flavonoid-rich foods, and cardiovascular risk: A meta-analysis of randomized controlled trials. *The American Journal of Clinical Nutrition*, *88*, 38–50.

Hou, Z., Sang, S., You, H., et al. (2005). Mechanism of action of (−)-epigallocatechin-3-gallate: Auto-oxidation-dependent inactivation of epidermal growth factor receptor and direct effects on growth inhibition in human esophageal cancer KYSE 150 cells. *Cancer Research*, *65*, 8049–8056.

Huang, Y., Chan, N. W., Lau, C. W., et al. (1999). Involvement of endothelium/nitric oxide in vasorelaxation induced by purified green tea (−)epicatechin. *Biochimica et Biophysica Acta*, *1427*, 322–328.

Jimenez, R., Lopez-Sepulveda, R., Kadmiri, M., et al. (2007). Polyphenols restore endothelial function in DOCA-salt hypertension: Role of endothelin-1 and NADPH oxidase. *Free Radical Biology & Medicine*, *43*, 462–473.

Jin, X., Otonashi-Satoh, Y., Sun, P., et al. (2008). Endothelium-derived hyperpolarizing factor (EDHF) mediates endothelium-dependent vasodilator effects of aqueous extracts from Eucommia ulmoides Oliv. leaves in rat mesenteric resistance arteries. *Acta Medica Okayama*, *62*, 319–325.

Jochmann, N., Lorenz, M., Krosigk, A., et al. (2008). The efficacy of black tea in ameliorating endothelial function is equivalent to that of green tea. *The British Journal of Nutrition*, *99*, 863–868.

Joy, S., Siow, R. C., Rowlands, D. J., et al. (2006). The isoflavone Equol mediates rapid vascular relaxation: Ca2+-independent activation of endothelial nitric-oxide synthase/Hsp90 involving ERK1/2 and Akt phosphorylation in human endothelial cells. *The Journal of Biological Chemistry*, *281*, 27335–27345.

Kamata, K., Kobayashi, T., Matsumoto, T., et al. (2005). Effects of chronic administration of fruit extract (Citrus unshiu Marc) on endothelial dysfunction in streptozotocin-induced diabetic rats. *Biological & Pharmaceutical Bulletin*, *28*, 267–270.

Kane, M. O., Anselm, E., Rattmann, Y. D., et al. (2009). Role of gender and estrogen receptors in the rat aorta endothelium-dependent relaxation to red wine polyphenols. *Vascular Pharmacology*, *51*, 140–146.

Kane, M. O., Etienne-Selloum, N., Madeira, S. V., et al. (2010). Endothelium-derived contracting factors mediate the Ang II-induced endothelial dysfunction in the rat aorta: preventive effect of red wine polyphenols. *Pflugers Archives*, *459*, 671–679.

Karatzi, K., Papamichael, C., Aznaouridis, K., et al. (2004). Constituents of red wine other than alcohol improve endothelial function in patients with coronary artery disease. *Coronary Artery Disease*, *15*, 485–490.

Kim, J. A., Formoso, G., Li, Y., et al. (2007). Epigallocatechin gallate, a green tea polyphenol, mediates NO-dependent vasodilation using signaling pathways in vascular endothelium requiring reactive oxygen species and Fyn. *The Journal of Biological Chemistry*, *282*, 13736–13745.

Koltermann, A., Hartkorn, A., Koch, E., et al. (2007). Ginkgo biloba extract EGb 761 increases endothelial nitric oxide production in vitro and in vivo. *Cellular and Molecular Life Sciences*, *64*, 1715–1722.

Kurita, I., Maeda-Yamamoto, M., Tachibana, H., et al. (2010). Antihypertensive effect of Benifuuki tea containing O-methylated EGCG. *Journal of Agricultural and Food Chemistry*, *58*, 1903–1908.

Kuriyama, S. (2008). The relation between green tea consumption and cardiovascular disease as evidenced by epidemiological studies. *The Journal of Nutrition*, *138*, 1548S–1553S.

Kwak, C. J., Kubo, E., Fujii, K., et al. (2009). Antihypertensive effect of French maritime pine bark extract (Flavangenol): Possible involvement of endothelial nitric oxide-dependent vasorelaxation. *Journal of Hypertension*, *27*, 92–101.

Kwan, C. Y., Zhang, W. B., Sim, S. M., et al. (2004). Vascular effects of Siberian ginseng (*Eleutherococcus senticosus*): Endothelium-dependent NO- and EDHF-mediated relaxation depending on vessel size. *Naunyn-Schmiedeberg's Archives of Pharmacology*, *369*, 473–480.

Law, M. R., & Morris, J. K. (1998). By how much does fruit and vegetable consumption reduce the risk of ischaemic heart disease? *European Journal of Clinical Nutrition*, *52*, 549–556.

Leblais, V., Krisa, S., Valls, J., et al. (2008). Relaxation induced by red wine polyphenolic compounds in rat pulmonary arteries: Lack of inhibition by NO-synthase inhibitor. *Fundamental & Clinical Pharmacology*, *22*, 25–35.

Lee, J. O., Chang, K., Kim, C. Y., et al. (2010). Lysimachia clethroides extract promote vascular relaxation via endothelium-dependent mechanism. *Journal of Cardiovascular Pharmacology*, *55*, 481–488.

Leikert, J. F., Rathel, T. R., Wohlfart, P., et al. (2002). Red wine polyphenols enhance endothelial nitric oxide synthase expression and subsequent nitric oxide release from endothelial cells. *Circulation*, *106*, 1614–1617.

Lekakis, J., Rallidis, L. S., Andreadou, I., et al. (2005). Polyphenolic compounds from red grapes acutely improve endothelial function in patients with coronary heart disease. *European Journal of Cardiovascular Prevention and Rehabilitation*, *12*, 596–600.

Lemos, V. S., Freitas, M. R., Muller, B., et al. (1999). Dioclein, a new nitric oxide- and endothelium-dependent vasodilator flavonoid. *European Journal of Pharmacology*, *386*, 41–46.

Li, H. F., Chen, S. A., & Wu, S. N. (2000). Evidence for the stimulatory effect of resveratrol on Ca(2+) activated K+ current in vascular endothelial cells. *Cardiovascular Research*, *45*, 1035–1045.

Li, Y., Starrett, J. E., Meanwell, N. A., et al. (1997). The discovery of novel openers of Ca2+-dependent large-conductance potassium channels: Pharmacophore search and physiological evaluation of flavonoids. *Bioorganic & Medicinal Chemistry Letters*, *7*, 759–762.

Liu, L., Xu, D. M., & Cheng, Y. Y. (2008). Distinct effects of naringenin and hesperetin on nitric oxide production from endothelial cells. *Journal of Agricultural and Food Chemistry*, *56*, 824–829.

Lopez-Sepulveda, R., Jimenez, R., Romero, M., et al. (2008). Wine polyphenols improve endothelial function in large vessels of female spontaneously hypertensive rats. *Hypertension*, *51*, 1088–1095.

Lorenz, M., Wessler, S., Follmann, E., et al. (2004). A constituent of green tea, epigallocatechin-3-gallate, activates endothelial nitric oxide synthase by a phosphatidylinositol-3-OH-kinase-, cAMP-dependent protein kinase-, and Akt-dependent pathway and leads to endothelial-dependent vasorelaxation. *The Journal of Biological Chemistry*, *279*, 6190–6195.

Luo, X., Ma, M., Chen, B., et al. (2004). Analysis of nine bioactive compounds in Eucommia ulmoides Oliv. and their preparation by HPLC-photodiode array detection and mass spectrometry. *Journal of Liquid Chromatography and Related Technologies*, *27*, 63–81.

Machha, A., Achike, F. I., Mohd, M. A., et al. (2007a). Baicalein impairs vascular tone in normal rat aortas: Role of superoxide anions. *European Journal of Pharmacology*, *565*, 144–150.

Machha, A., Achike, F. I., Mustafa, A. M., et al. (2007b). Quercetin, a flavonoid antioxidant, modulates endothelium-derived nitric oxide bioavailability in diabetic rat aortas. *Nitric Oxide*, *16*, 442–447.

Machha, A., & Mustafa, M. R. (2005). Chronic treatment with flavonoids prevents endothelial dysfunction in spontaneously hypertensive rat aorta. *Journal of Cardiovascular Pharmacology*, *46*, 36–40.

Mackraj, I., Govender, T., & Ramesar, S. (2008). The antihypertensive effects of quercetin in a salt-sensitive model of hypertension. *Journal of Cardiovascular Pharmacology*, *51*, 239–245.

Madeira, S. V., Auger, C., Anselm, E., et al. (2009). eNOS activation induced by a polyphenol-rich grape skin extract in porcine coronary arteries. *Journal of Vascular Research*, *46*, 406–416.

Maher, M. A., Mataczynski, H., Stefaniak, H. M., et al. (2000). Cranberry juice induces nitric oxide-dependent vasodilation in vitro and its infusion transiently reduces blood pressure in anesthetized rats. *Journal of Medicinal Food*, *3*, 141–147.

Manach, C., Williamson, G., Morand, C., et al. (2005). Bioavailability and bioefficacy of polyphenols in humans. I. Review of 97 bioavailability studies. *The American Journal of Clinical Nutrition*, *81*, 230S–242S.

Martin, S., Andriambeloson, E., Takeda, K., et al. (2002). Red wine polyphenols increase calcium in bovine aortic endothelial cells: A basis to elucidate signalling pathways leading to nitric oxide production. *British Journal of Pharmacology*, *135*, 1579–1587.

Matoba, T., Shimokawa, H., Nakashima, M., et al. (2000). Hydrogen peroxide is an endothelium-derived hyperpolarizing factor in mice. *Journal of Clinical Investigation*, *106*, 1521–1530.

Matsui, T., Korematsu, S., Byun, E. B., et al. (2009). Apple procyanidins induced vascular relaxation in isolated rat aorta through NO/cGMP pathway in combination with hyperpolarization by multiple K+ channel activations. *Bioscience, Biotechnology, and Biochemistry*, *73*, 2246–2251.

Mead, A., Atkinson, G., Albin, D., et al. (2006). Dietetic guidelines on food and nutrition in the secondary prevention of cardiovascular disease—Evidence from systematic reviews of randomized controlled trials (second update, January 2006). *Journal of Human Nutrition and Dietetics*, *19*, 401–419.

Mendes, A., Desgranges, C., Cheze, C., et al. (2003). Vasorelaxant effects of grape polyphenols in rat isolated aorta. Possible involvement of a purinergic pathway. *Fundamental & Clinical Pharmacology*, *17*, 673–681.

Mizutani, K., Ikeda, K., Kawai, Y., et al. (1999). Extract of wine phenolics improves aortic biomechanical properties in stroke-prone spontaneously hypertensive rats (SHRSP). *Journal of Nutritional Science and Vitaminology*, *45*, 95–106.

Mukai, Y., & Sato, S. (2009). Polyphenol-containing azuki bean (*Vigna angularis*) extract attenuates blood pressure elevation and modulates nitric oxide synthase and caveolin-1 expressions in rats with hypertension. *Nutrition, Metabolism, and Cardiovascular Diseases*, *19*, 491–497.

Mullen, W., Edwards, C. A., Serafini, M., et al. (2008). Bioavailability of pelargonidin-3-O-glucoside and its metabolites in humans following the ingestion of strawberries with and without cream. *Journal of Agricultural and Food Chemistry*, *56*, 713–719.

Mullen, W., McGinn, J., Lean, M. E., et al. (2002). Ellagitannins, flavonoids, and other phenolics in red raspberries and their contribution to antioxidant capacity and vasorelaxation properties. *Journal of Agricultural and Food Chemistry*, *50*, 5191–5196.

Nagaoka, T., Hein, T. W., Yoshida, A., et al. (2007). Resveratrol, a component of red wine, elicits dilation of isolated porcine retinal arterioles: Role of nitric oxide and potassium channels. *Investigative Ophthalmology & Visual Science*, *48*, 4232–4239.

Nakamura, Y., Matsumoto, H., & Todoki, K. (2002). Endothelium-dependent vasorelaxation induced by black currant concentrate in rat thoracic aorta. *Japanese Journal of Pharmacology*, *89*, 29–35.

Ndiaye, M., Chataigneau, T., Andriantsitohaina, R., et al. (2003). Red wine polyphenols cause endothelium-dependent EDHF-mediated relaxations in porcine coronary arteries via a redox-sensitive mechanism. *Biochemical and Biophysical Research Commununications*, *310*, 371–377.

Ndiaye, M., Chataigneau, T., Chataigneau, M., et al. (2004). Red wine polyphenols induce EDHF-mediated relaxations in porcine coronary arteries through the redox-sensitive activation of the PI3-kinase/Akt pathway. *British Journal of Pharmacology*, *142*, 1131–1136.

Ndiaye, M., Chataigneau, M., Lobysheva, I., Chataigneau, T., et al. (2005). Red wine polyphenol-induced, endothelium-dependent NO-mediated relaxation is due to the redox-sensitive PI3-kinase/Akt-dependent phosphorylation of endothelial NO-synthase in the isolated porcine coronary artery. *The FASEB Journal*, *19*, 455–457.

Negishi, H., Xu, J. W., Ikeda, K., et al. (2004). Black and green tea polyphenols attenuate blood pressure increases in stroke-prone spontaneously hypertensive rats. *The Journal of Nutrition*, *134*, 38–42.

Nguelefack, T. B., Mekhfi, H., Dongmo, A. B., et al. (2009). Hypertensive effects of oral administration of the aqueous extract of *Solanum torvum* fruits in L-NAME treated rats: Evidence from in vivo and in vitro studies. *Journal of Ethnopharmacology*, *124*, 592–599.

Nguelefack-Mbuyo, P. E., Nguelefack, T. B., Dongmo, A. B., et al. (2008). Anti-hypertensive effects of the methanol/methylene chloride stem bark extract of Mammea africana in L-NAME-induced hypertensive rats. *Journal of Ethnopharmacology*, *117*, 446–450.

Nicholson, S. K., Tucker, G. A., & Brameld, J. M. (2008). Effects of dietary polyphenols on gene expression in human vascular endothelial cells. *The Proceedings of the Nutrition Society*, *67*, 42–47.

Nishioka, K., Hidaka, T., Nakamura, S., et al. (2007). Pycnogenol, French maritime pine bark extract, augments endothelium-dependent vasodilation in humans. *Hypertension Research*, *30*, 775–780.

Njike, V. Y., Faridi, Z., Shuval, K., et al. (2009). Effects of sugar-sweetened and sugar-free cocoa on endothelial function in overweight adults. *International Journal of Cardiology*, 10.1016/j.ijcard.2009.12.010, Epub ahead of print.

Novakovic, A., Bukarica, L. G., Kanjuh, V., et al. (2006). Potassium channels-mediated vasorelaxation of rat aorta induced by resveratrol. *Basic & Clinical Pharmacology & Toxicology*, *99*, 360–364.

Padilla, E., Ruiz, E., Redondo, S., et al. (2005). Relationship between vasodilation capacity and phenolic content of Spanish wines. *European Journal of Pharmacology*, *517*, 84–91.

Papamichael, C., Karatzi, K., Karatzis, E., et al. (2006). Combined acute effects of red wine consumption and cigarette smoking on haemodynamics of young smokers. *Journal of Hypertension*, *24*, 1287–1292.

Park, Y. K., Kim, J. S., & Kang, M. H. (2004). Concord grape juice supplementation reduces blood pressure in Korean hypertensive men: Double-blind, placebo controlled intervention trial. *Biofactors*, *22*, 145–147.

Pechanova, O., Bernatova, I., Babal, P., et al. (2004). Red wine polyphenols prevent cardiovascular alterations in L-NAME-induced hypertension. *Journal of Hypertension*, *22*, 1551–1559.

Peng, N., Clark, J. T., Prasain, J., et al. (2005). Antihypertensive and cognitive effects of grape polyphenols in estrogen-depleted, female, spontaneously hypertensive rats. *American Journal of Physiology: Regulatory, Integrative and Comparative Physiology*, *289*, R771–R775.

Penumathsa, S. V., & Maulik, N. (2009). Resveratrol: A promising agent in promoting cardioprotection against coronary heart disease. *Canadian Journal of Physiology and Pharmacology*, *87*, 275–286.

Peters, U., Poole, C., & Arab, L. (2001). Does tea affect cardiovascular disease? A meta-analysis. *American Journal of Epidemiology*, *154*, 495–503.

Potenza, M. A., Marasciulo, F. L., Tarquinio, M., et al. (2007). EGCG, a green tea polyphenol, improves endothelial function and insulin sensitivity, reduces blood pressure, and protects against myocardial I/R injury in SHR. *American Journal of Physiology. Endocrinology and Metabolism*, *292*, E1378–E1387.

Rakici, O., Kiziltepe, U., Coskun, B., et al. (2005). Effects of resveratrol on vascular tone and endothelial function of human saphenous vein and internal mammary artery. *International Journal of Cardiology*, *105*, 209–215.

Renaud, S., & de Lorgeril, M. (1992). Wine, alcohol, platelets, and the French paradox for coronary heart disease. *Lancet*, *339*, 1523–1526.

Rimm, E. B., Williams, P., Fosher, K., et al. (1999). Moderate alcohol intake and lower risk of coronary heart disease: Meta-analysis of effects on lipids and haemostatic factors. *BMJ*, *319*, 1523–1528.

Rivera, L., Moron, R., Sanchez, M., et al. (2008). Quercetin ameliorates metabolic syndrome and improves the inflammatory status in obese Zucker rats. *Obesity*, *16*, 2081–2087.

Rivera, L., Moron, R., Zarzuelo, A., et al. (2009). Long-term resveratrol administration reduces metabolic disturbances and lowers blood pressure in obese Zucker rats. *Biochemical Pharmacology*, *77*, 1053–1063.

Rocha, A. P., Carvalho, L. C., Sousa, M. A., et al. (2007). Endothelium-dependent vasodilator effect of *Euterpe oleracea* Mart. (Açaí) extracts in mesenteric vascular bed of the rat. *Vascular Pharmacology*, *46*, 97–104.

Roghani, M., & Baluchnejadmojarad, T. (2009). Chronic epigallocatechin-gallate improves aortic reactivity of diabetic rats: Underlying mechanisms. *Vascular Pharmacology*, *51*, 84–89.

Roghani, M., Baluchnejadmojarad, T., Vaez-Mahdavi, M. R., et al. (2004). Mechanisms underlying quercetin-induced vasorelaxation in aorta of subchronic diabetic rats: An in vitro study. *Vascular Pharmacology*, *42*, 31–35.

Romano, M. R., & Lograno, M. D. (2009). Epigallocatechin-3-gallate relaxes the isolated bovine ophthalmic artery: Involvement of phosphoinositide 3-kinase-Akt-nitric oxide/cGMP signalling pathway. *European Journal of Pharmacology*, *608*, 48–53.

Romero, M., Jimenez, R., Sanchez, M., et al. (2009). Quercetin inhibits vascular superoxide production induced by endothelin-1: Role of NADPH oxidase, uncoupled eNOS and PKC. *Atherosclerosis*, *202*, 58–67.

Rush, J. W., Quadrilatero, J., Levy, A. S., et al. (2007). Chronic resveratrol enhances endothelium-dependent relaxation but does not alter eNOS levels in aorta of spontaneously hypertensive rats. *Experimental Biology and Medicine*, *232*, 814–822.

Sanchez, M., Galisteo, M., Vera, R., et al. (2006). Quercetin downregulates NADPH oxidase, increases eNOS activity and prevents endothelial dysfunction in spontaneously hypertensive rats. *Journal of Hypertension*, *24*, 75–84.

Sanchez, M., Lodi, F., Vera, R., et al. (2007). Quercetin and isorhamnetin prevent endothelial dysfunction, superoxide production, and overexpression of p47phox induced by angiotensin II in rat aorta. *The Journal of Nutrition*, *137*, 910–915.

Sang, S., Yang, I., Buckley, B., et al. (2007). Autoxidative quinone formation in vitro and metabolite formation in vivo from tea polyphenol (−)-epigallocatechin-3-gallate: Studied by real-time mass spectrometry combined with tandem mass ion mapping. *Free Radical Biology & Medicine*, *43*, 362–371.

Sano, A., Yamakoshi, J., Tokutake, S., et al. (2003). Procyanidin B1 is detected in human serum after intake of proanthocyanidin-rich grape seed extract. *Bioscience, Biotechnology, and Biochemistry*, *67*, 1140–1143.

Sarr, M., Chataigneau, M., Martins, S., et al. (2006). Red wine polyphenols prevent angiotensin II-induced hypertension and endothelial dysfunction in rats: Role of NADPH oxidase. *Cardiovascular Research*, *71*, 794–802.

Sarr, M., Ngom, S., Kane, M. O., et al. (2009). In vitro vasorelaxation mechanisms of bioactive compounds extracted from *Hibiscus sabdariffa* on rat thoracic aorta. *Nutrition and Metabolism*, *6*, 45.

Schroeter, H., Heiss, C., Balzer, J., et al. (2006). (−)-Epicatechin mediates beneficial effects of flavanol-rich cocoa on vascular function in humans. *Proceedings of the National Academy of Sciences of the United States of America*, *103*, 1024–1029.

Schuldt, E. Z., Bet, A. C., Hort, M. A., et al. (2005). An ethyl acetate fraction obtained from a Southern Brazilian red wine relaxes rat mesenteric arterial bed through hyperpolarization and NO-cGMP pathway. *Vascular Pharmacology*, *43*, 62–68.

Serra-Majem, L., Roman, B., Estruch, R., et al. (2006). Scientific evidence of interventions using the Mediterranean diet: A systematic review. *Nutrition Reviews*, *64*, S27–S47.

Seya, K., Furukawa, K., Taniguchi, S., et al. (2003). Endothelium-dependent vasodilatory effect of vitisin C, a novel plant oligostilbene from Vitis plants (Vitaceae), in rabbit aorta. *Clinical Science*, *105*, 73–79.

Shaughnessy, K. S., Boswall, I. A., Scanlan, A. P., et al. (2009). Diets containing blueberry extract lower blood pressure in spontaneously hypertensive stroke-prone rats. *Nutrition Research*, *29*, 130–138.

Shimokawa, H. (2010). Hydrogen peroxide as an endothelium-derived hyperpolarizing factor. *Pflugers Archives*, *459*, 915–922.

Singleton, V. L. (1981). Naturally occurring food toxicants: Phenolic substances of plant origin common in foods. *Advances in Food Research*, *27*, 149–242.

Soares de Moura, R., Costa Viana, F. S., Souza, M. A., et al. (2002). Antihypertensive, vasodilator and antioxidant effects of a vinifera grape skin extract. *Journal of Pharmacy and Pharmacology*, *54*, 1515–1520.

Sofi, F., Cesari, F., Abbate, R., et al. (2008). Adherence to Mediterranean diet and health status: Meta-analysis. *BMJ*, *337*, a1344.

St. Leger, A. S., Cochrane, A. L., & Moore, F. (1979). Ischaemic heart-disease and wine. *Lancet*, *1*, 1294.

Stein, J. H., Keevil, J. G., Wiebe, D. A., et al. (1999). Purple grape juice improves endothelial function and reduces the susceptibility of LDL cholesterol to oxidation in patients with coronary artery disease. *Circulation*, *100*, 1050–1055.

Stoclet, J. C., Kleschyov, A., Andriambeloson, E., et al. (1999). Endothelial NO release caused by red wine polyphenols. *Journal of Physiology and Pharmacology*, *50*, 535–540.

Suzuki, A., Yamamoto, N., Jokura, H., et al. (2006). Chlorogenic acid attenuates hypertension and improves endothelial function in spontaneously hypertensive rats. *Journal of Hypertension*, *24*, 1065–1073.

Taubert, D., Berkels, R., Roesen, R., et al. (2003). Chocolate and blood pressure in elderly individuals with isolated systolic hypertension. *JAMA, the Journal of the American Medical Association*, *290*, 1029–1030.

Tokoudagba, J.-M., Chabert, P., Auger, C., et al. (2010). Recherche de plantes à potentialités antihypertensives dans la biodiversité béninoise. *Ethnopharmacologia*, *44*, 32–41.

Tsang, C., Auger, C., Mullen, W., et al. (2005). The absorption, metabolism and excretion of flavan-3-ols and procyanidins following the ingestion of a grape seed extract by rats. *The British Journal of Nutrition*, *94*, 170–181.

Tsuda, T., Horio, F., & Osawa, T. (1999). Absorption and metabolism of cyanidin 3-O-beta-D-glucoside in rats. *FEBS Letters*, *449*, 179–182.

van't Slot, G., & Humpf, H. U. (2009). Degradation and metabolism of catechin, epigallocatechin-3-gallate (EGCG), and related compounds by the intestinal microbiota in the pig cecum model. *Journal of Agricultural and Food Chemistry*, *57*, 8041–8048.

Vera, R., Galisteo, M., Villar, I. C., et al. (2005). Soy isoflavones improve endothelial function in spontaneously hypertensive rats in an estrogen-independent manner: Role of nitric-oxide synthase, superoxide, and cyclooxygenase metabolites. *The Journal of Pharmacology and Experimental Therapeutics*, *314*, 1300–1309.

Vera, R., Sanchez, M., Galisteo, M., et al. (2007). Chronic administration of genistein improves endothelial dysfunction in spontaneously hypertensive rats: Involvement of eNOS, caveolin and calmodulin expression and NADPH oxidase activity. *Clinical Science*, *112*, 183–191.

Wallerath, T., Li, H., Godtel-Ambrust, U., et al. (2005). A blend of polyphenolic compounds explains the stimulatory effect of red wine on human endothelial NO synthase. *Nitric Oxide*, *12*, 97–104.

Wallerath, T., Poleo, D., Li, H., et al. (2003). Red wine increases the expression of human endothelial nitric oxide synthase: A mechanism that may contribute to its beneficial cardiovascular effects. *Journal of the American College of Cardiology*, *41*, 471–478.

Wattanapitayakul, S. K., Suwatronnakorn, M., Chularojmontri, L., et al. (2007). Kaempferia parviflora ethanolic extract promoted nitric oxide production in human umbilical vein endothelial cells. *Journal of Ethnopharmacology*, *110*, 559–562.

Weiner, C. P., Lizasoain, I., Baylis, S. A., et al. (1994). Induction of calcium-dependent nitric oxide synthases by sex hormones. *Proceedings of the National Academy of Sciences of the United States of America*, *91*, 5212–5216.

Whelan, A. P., Sutherland, W. H., McCormick, M. P., et al. (2004). Effects of white and red wine on endothelial function in subjects with coronary artery disease. *Internal Medicine Journal*, *34*, 224–228.

Widlansky, M. E., Hamburg, N. M., Anter, E., et al. (2007). Acute EGCG supplementation reverses endothelial dysfunction in patients with coronary artery disease. *Journal of the American College of Nutrition*, *26*, 95–102.

Wollny, T., Aiello, L., Di, T. D., et al. (1999). Modulation of haemostatic function and prevention of experimental thrombosis by red wine in rats: A role for increased nitric oxide production. *British Journal of Pharmacology*, *127*, 747–755.

Wu, Y. Z., Li, S. Q., Zu, X. G., et al. (2008). Ginkgo biloba extract improves coronary artery circulation in patients with coronary artery disease: Contribution of plasma nitric oxide and endothelin-1. *Phytotherapy Research*, *22*, 734–739.

Xu, J. W., Ikeda, K., & Yamori, Y. (2004a). Upregulation of endothelial nitric oxide synthase by cyanidin-3-glucoside, a typical anthocyanin pigment. *Hypertension*, *44*, 217–222.

Xu, J. W., Ikeda, K., & Yamori, Y. (2004b). Cyanidin-3-glucoside regulates phosphorylation of endothelial nitric oxide synthase. *FEBS Letters*, *574*, 176–180.

Xu, P. H., Long, Y., Dai, F., et al. (2007). The relaxant effect of curcumin on porcine coronary arterial ring segments. *Vascular Pharmacology*, *47*, 25–30.

Yamamoto, M., Suzuki, A., & Hase, T. (2008). Short-term effects of glucosyl hesperidin and hesperetin on blood pressure and vascular endothelial function in spontaneously hypertensive rats. *Journal of Nutritional Science and Vitaminology*, *54*, 95–98.

Yeboah, J., Crouse, J. R., Hsu, F. C., et al. (2007). Brachial flow-mediated dilation predicts incident cardiovascular events in older adults: The Cardiovascular Health Study. *Circulation*, *115*, 2390–2397.

Zenebe, W., Pechanova, O., & Andriantsitohaina, R. (2003). Red wine polyphenols induce vasorelaxation by increased nitric oxide bioactivity. *Physiological Research*, *52*, 425–432.

Andreas Daiber, Thomas Münzel, and Tommaso Gori

II. Medizinische Klinik, Labor für Molekulare Kardiologie und Abteilung für Kardiologie und Angiologie Universitätsmedizin der Johannes-Gutenberg-Universität, Mainz, Germany

Organic Nitrates and Nitrate Tolerance—State of the Art and Future Developments

Abstract

The hemodynamic and antiischemic effects of nitroglycerin (GTN) are lost upon chronic administration due to the rapid development of nitrate tolerance. The mechanism of this phenomenon has puzzled several generations of scientists, but recent findings have led to novel hypotheses. The formation of reactive oxygen and nitrogen species in the mitochondria and the subsequent inhibition of the nitrate-bioactivating enzyme mitochondrial aldehyde dehydrogenase (ALDH-2) appear to play a central role, at least for

Advances in Pharmacology, Volume 60

1054-3589/10 $35.00
10.1016/S1054-3589(10)60003-8

GTN, that is, bioactivated by ALDH-2. Importantly, these findings provide the opportunity to reconcile the two "traditional" hypotheses of nitrate tolerance, that is, the one postulating a decreased bioactivation and the concurrent one suggesting a role of oxidative stress. Furthermore, recent animal and human experimental studies suggest that the organic nitrates are not a homogeneous group but demonstrate a broad diversity with regard to induction of vascular dysfunction, oxidative stress, and other side effects. In the past, attempts to avoid nitrate-induced side effects have focused on administration schedules that would allow a "nitrate-free interval"; in the future, the role of co-therapies with antioxidant compounds and of activation of endogeneous protective pathways such as the heme oxygenase 1 (HO-1) will need to be explored. However, the development of new nitrates, for example, tolerance-free aminoalkyl nitrates or combination of nitrate groups with established cardiovascular drugs like ACE inhibitors or AT_1-receptor blockers (hybrid molecules) may be of great clinical interest.

I. Introduction

A. Mechanism of Organic Nitrate-Mediated Vasodilation

Organic nitrates belong to the class of the nitrovasodilators and are frequently used in the treatment of stable angina pectoris (Abrams, 1995). The impact of these drugs on cardiac preload and the subsequent reduction in cardiac oxygen consumption are thought to play a central role in the improvement in exercise tolerance associated with the treatment with these drugs. Moreover, organic nitrates are also employed in adjunct (or alternative) of angiotensin converting enzyme inhibitors for the treatment of chronic heart failure (isosorbide dinitrate (ISDN)/hydralazine) (Taylor et al., 2004). Anecdotal evidence points out that also pentaerithrityl tetranitrate (PETN) might be useful in this setting (Gori & Munzel, 2009). As well, the coronary vasodilatory properties of nitrates, and a mild antiaggregant effect, explain their benefit in the setting of acute coronary syndromes such as unstable angina and myocardial infarction. Their principle of action is based on an enzymatic bioactivation process with subsequent formation of nitric oxide (NO) or a related species (RSNO or NO-metal-complex), which results in the activation of soluble guanylyl cyclase (sGC) and increase in the second messenger cGMP, in turn leading to vasodilation (Fig. 1; Katsuki et al., 1977a, 1977b). Although the direct formation of NO from nitrates seems to be an attractive and widely accepted explanation for nitrate-induced vasodilation (Ignarro, 1989), this concept was challenged by two independent studies demonstrating an almost 100-fold discrepancy between glyceryl trinitrate (GTN)-evoked NO formation and vasodilation, whereas for ISMN, there was a direct correlation between these parameters (Kleschyov

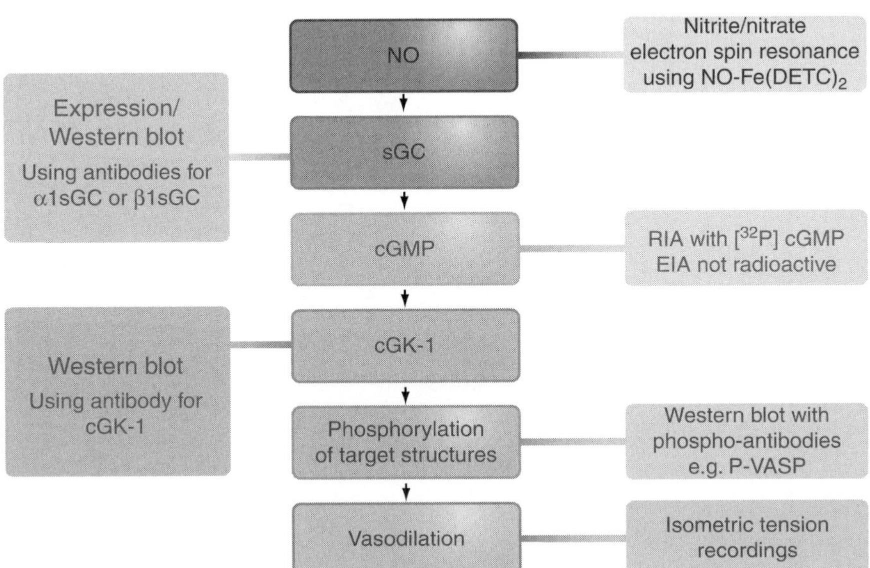

FIGURE 1 Mechanism of organic nitrate-mediated vasodilation via the NO/cGMP cascade. The functionality of this signaling pathway can be determined in each single step: nitric oxide (NO) formation can be assessed by electron paramagnetic resonance (EPR) as a stable paramagnetic iron complex (NO-Fe(DETC)$_2$) or by the nitrite/nitrate ratio; the activity state of soluble guanylyl cyclase (sGC) can be measured by its expression using Western blotting or by cyclic guanosin-3′,5′-monophosphate (cGMP) using radioimmunoassay (RIA) or with a nonradioactive enzymatic immunoassay (EIA); cGMP-dependent kinase (cGK-1) level can be assessed by Western blotting and its activity via phosphorylation of the vasodilator-stimulated phosphoprotein (VASP, one of the substrates of cGK-I) at serine 239 using Western blotting. The most important parameter in this signaling cascade represents the vascular function (or hemodynamic parameters), which is measured by isometric tension recordings using isolated vessel segments in organ chambers.

et al., 2003; Nunez et al., 2005). Thus, it appears that at least GTN-induced vasodilation might be mediated by an NO-related species, rather than NO itself. A detailed discussion on the identity of the vasodilating species, which might be formed by organic nitrates, was provided in previous review articles (Daiber et al., 2008; Munzel et al., 2005).

B. Clinical Trials

While GTN and other nitrates are widely employed in the therapy and prevention of stable angina, the large clinical trials in the setting of myocardial infarction (GISSI-3 for GTN and ISIS-4 for isosorbide-5-mononitrate (ISMN)) conducted 15 years ago revealed only mild effects on mortality and prognosis of coronary artery disease (CAD) patients (Table I). In line with these findings, guidelines suggest that organic nitrates should be used only for

TABLE I Partial List of Clinical Trials on the Therapy with Organic Nitrates

Name of the trial	*Source*	*Patients (with nitrate)*	*Nitrate dose*	*Time interval*	*Placebo random*	*Prognosis*	*Limitations/sub-studies*
GISSI-3	(GISSI-3, 1994)[a]	19,394 (9000)	GTN 10 mg/d	24 h i.v. 6 w t.d.	Yes Yes	n.s. (↑)	GTN + ACE less death
ISIS-4	(ISIS-4, 1995)[b]	58,050 (29,018)	ISMN 30–60 mg/ml	4 weeks p.o.	Yes Yes	n.s. (↑)	None
Nakamura	Nakamura et al. (1999)	~2800 (476)	? ?	? ?	? ?	↓	Retrospective
Ishikawa	Ishikawa et al. (1996)	1002 (621)	ISDN: 40 mg/d GTN: 25 mg/d	18 months p.o. or t.d. t.d.	No Yes	↓	Nitrates also in Ctr
Kanamasa	Kanamasa et al. (2002)	1291 (713)	ISMN: 40–60 mg/d GTN: 25 mg/d ISMN/GTN	p.o. or t.d. 17 months t.d. p.o. and t.d.	No Yes	n.s. (t.d.) ↓ (p.o.) ↓ (comb.)	Nitrates also in Ctr
Taylor A-HefT	Taylor et al. (2004)	1050 (518)	ISDN: 20–120 mg/d Hydralazine: 38–225 mg/d	18 months p.o.	Yes Yes	↑	Only Afro-Americans
GRACE	Ambrosio et al. (2010)	52,693 (10,555)	? ?	? ?	– –	↑	MI damage reduced

i.v., intravenous; t.d., transdermal; p.o., per oral; n.s., not significant; ↑, better; ↓, worse.

[a] "GISSI-3: effects of lisinopril and transdermal glyceryl trinitrate singly and together on 6-week mortality and ventricular function after acute myocardial infarction. Gruppo Italiano per lo Studio della Sopravvivenza nell'infarto Miocardico," 1994.

[b] "ISIS-4: a randomized factorial trial assessing early oral captopril, oral mononitrate, and intravenous magnesium sulfate in 58,050 patients with suspected acute myocardial infarction. ISIS-4 (Fourth International Study of Infarct Survival) Collaborative Group," 1995).

the relief of angina in patients with CAD who are refractory to other therapies (e.g., β-blocker, calcium-channel antagonists, etc.; Fraker et al., 2007). In addition, the introduction of newer antianginal medications such as ranolazine or aliskiren and ivabradine will likely lead to a further reduction of the use of organic nitrates in the next years. While the above large trials did not show a significant impact of short-term therapy with organic nitrates (or metaanalysis), the effects of long-term nitrate therapy remain less well investigated. Nitrates were developed and approved at a time when large-scale, long-term, randomized trials were not required, and the absence of patent rights on these medications does not encourage any industry to finance a large-scale clinical trial. Evidence to date emphasizes the importance of such a trial, as smaller trials have pointed out potential deleterious effects of this therapy: studies in healthy volunteers have demonstrated that nitroglycerin and other nitrates, with the possible exception of PETN, cause endothelial and autonomic dysfunction, and paradoxical coronary vasoconstriction (Caramori et al., 1998; Gori et al., 2002, 2003a; Schulz et al., 2002). In line with these findings, two clinical trials and a metaanalysis showed a trend toward an increase in mortality under nitrate therapy (Table I). Although the limitations of these studies need to be acknowledged (lack of a double-blind design and a placebo control, use of nitrates in formulations different from those currently employed in Western countries), these data definitely suggest that the use of nitrates should not be considered absolutely deprived of potentially important side effects. However, studies have shown that nitrates, at least acutely, have protective properties that mimic those of ischemic preconditioning (for review see Gori & Parker, 2008). These observations suggest that nitrates might have previously unexpected nonhemodynamic properties that might be associated with an increased tolerability toward ischemia and reperfusion damage. Although it needs to be emphasized that these data are very preliminary and large trials will also be necessary to test whether these effects might have clinical implications, a recent *post hoc* analysis of the GRACE trial emphasized that nitrate therapy is associated with a reduction of ST elevation myocardial infarctions and a shift toward clinically more favorable non-ST elevation myocardial infarctions, and with lesser release of markers of cardiac necrosis (Ambrosio et al., 2010).

In the setting of congestive heart failure, the impact of nitrate therapy appears to be more clear. In this setting, the administration of organic nitrates might reduce cardiac work by a reduction in right ventricular filling pressure. The combination of this effect with the predominantly arterial-vasodilator effect of hydralazine, which reduces cardiac afterload, has been associated with an increased cardiac output and improved remodeling in patients with impaired systolic function (Elkayam & Bitar, 2005). While these studies have almost exclusively employed ISDN, recent anecdotic evidence points out that also PETN might offer such benefit (Gori & Munzel,

2009). A dramatic reduction in mortality of Afro-Americans with severe chronic heart failure was observed with the ISDN/hydralazine combination therapy (A-HefT, Table I). Of note, along with its arterial-vasodilator effects, hydralazine has potent antioxidant properties, which might concur to explain the benefit of this combination therapy in congestive heart failure patients.

Thus, despite large clinical use for more than a century and definitive evidence of a benefit in congestive heart failure patients, the impact of organic nitrates on the prognosis of patients with CAD remains under discussion. To the best of our knowledge, the only larg(er) study currently ongoing is the CLEOPATRA trial (Clinical efficacy of Pentalong® in stable angina patients after 12 weeks of routine administration), a multicentric, randomized, double-blind, and placebo-controlled phase-III study enrolling 778 patients with stable *Angina pectoris*. Further studies employing the other nitrates will be needed in the future years. Finally, interest has been shown in the impact of nitrate therapy on (secondary) pulmonary hypertension, and the randomized, placebo-controlled CAESAR study (Clinical efficacy study of Pentalong for pulmonary hypertension in heart failure) will test the the efficacy and safety of PETN in this setting.

II. The Phenomenon of Nitrate Tolerance

A. Clinical Nitrate Tolerance

In 1888, Stewart and coworkers described for the first time the phenomenon of nitrate tolerance as a loss of the hemodynamic effect of GTN (or a requirement for higher dosages to maintain the hemodynamic effect of the drug) in patients undergoing chronic therapy (Stewart, 1888). The development of nitrate tolerance represents a severe limitation for the prolonged therapy of diseases that are by definition chronic (an overview is given in Elkayam, 1991; Abrams, 1988). To date, the best way around this problem has been to recommend that patients receive eccentric (rather than continuous) therapy, that is, administration schedules allowing a >12 h nitrate-free interval (for review, see Parker & Parker, 1998). These strategies do not appear to be devoid of problems: while intermittent administration overcomes the problem of tolerance, it also leaves patients "unprotected" for large time intervals. Paradoxically, indeed, the incidence of ischemia appears to be higher in nitrate-off hours in patients undergoing intermittent therapy (a phenomenon known as rebound ischemia), counterbalancing the beneficial effect of the drug during nitrate-on hours. The mechanisms underlying nitrate tolerance are not fully understood and involve a number of different factors (Fig. 2). For GTN-induced nitrate tolerance, one relevant mechanism may be the impaired bioactivation of GTN itself (Chung & Fung, 1992;

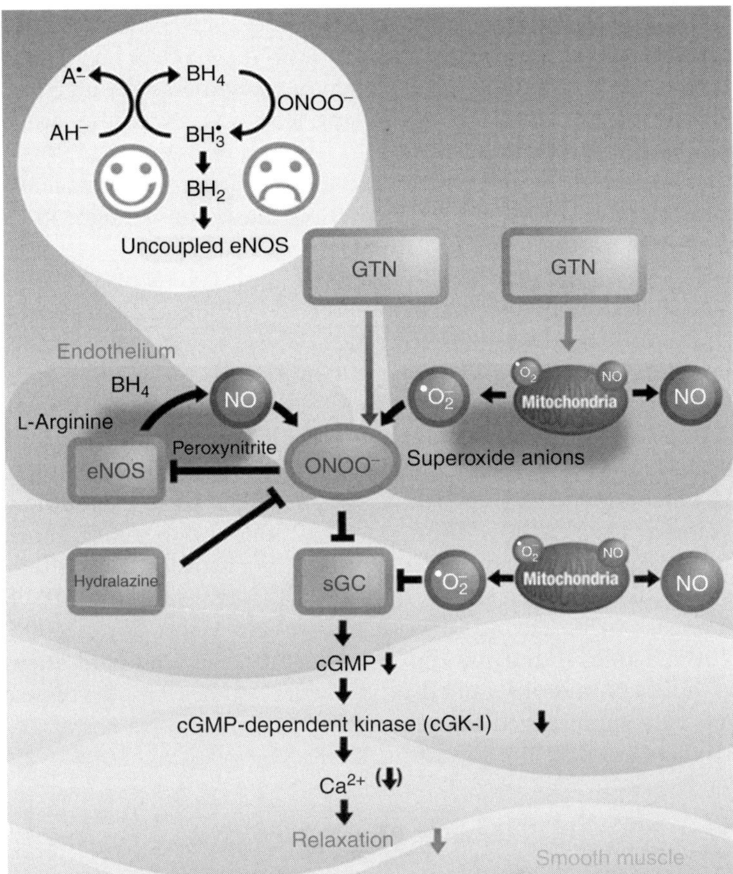

FIGURE 2 Vasodilation by nitroglycerin (GTN). GTN bioactivation to NO or a related species with subsequent activation of soluble guanylyl cyclase (sGC) and smooth muscle relaxation. Mechanisms underlying induction of vascular dysfunction by GTN via induction of mitochondrial and NADPH oxidase-dependent oxidative stress leading to uncoupling of eNOS and desensitization of sGC. Adapted from Gori et al. (2008). With permission of the Wissenschaftliche Verlagsgesellschaft Stuttgart.

Fung, 2004). Other mechanisms involve neurohumoral adaption processes such as the increase in plasma volume, activation of the renin–angiotensin–aldosterone system (RAAS) and increases in plasma levels of vasoconstrictors (e.g., vasopressin and catecholamines; Munzel et al., 1995a, 2005). These mechanisms might result from counterregulatory responses of the organism, overcoming the sustained vasodilation by the organic nitrate and the decrease in cardiac preload by these drugs. A phenomenon, associated to nitrate tolerance is the cross-tolerance toward other nitrovasodilators but also endothelium-dependent (endogenous) vasodilators such as acetylcholine

(=endothelial dysfunction). This was most frequently observed when GTN was administrated chronically *in vivo* (Sage et al., 2000; Schulz et al., 2002) and is only present to a minor extent when *in vitro* tolerance (tachyphylaxis[1]) was induced by short-term bolus challenges with GTN. These latter conditions may be criticized since they are based on supra-pharmacological concentrations of the drug and do not necessarily reflect the clinical situation. The cross-tolerance may, at least in part, be based on changes in the sGC activity (desensitization), the target enzyme of NO, although chronic treatment with GTN rather increases the expression of sGC (Mulsch et al., 2001). But also differences in the activity of the cGMP-degrading phosphodiesterases may contribute to nitrate tolerance (Silver et al., 1998). The processes that contribute to nitrate tolerance and endothelial dysfunction in response to organic nitrates are discussed in detail in previous reviews (Daiber et al., 2008; Munzel et al., 2005). Clinical nitrate tolerance is most pronounced under long-term GTN therapy (Sage et al., 2000; Schulz et al., 2002), whereas ISMN and ISDN (as well as GTN) rather trigger endothelial dysfunction (Sekiya et al., 2005; Thomas et al., 2007) and to some minor extent, depending on the doses, nitrate tolerance (Nordlander & Walter, 1994; Parker et al., 1983; Tauchert et al., 1983; Thadani et al., 1980). Finally, PETN appears to be devoid of both phenomena (Gori et al., 2003a; Jurt et al., 2001; Tables II and III, Fig. 3). In this list, we only paid attention to human studies. Numerous animal experimental studies support these clinical data and were summarized in previous review articles (Daiber et al., 2008, 2009a; Fayers et al., 2003; Johnston, 1998; Munzel et al., 2005).

B. *In Vitro* Nitrate Tolerance (Tachyphylaxis)

Under experimental conditions (with isolated vessels), PETN induces the most pronounced tolerance, followed by GTN and PETriN, whereas the di- and mononitrates induced no or only minor tachyphylaxis (Table III). *In vitro* tolerance (tachyphylaxis) may be quantified by the ratio of half-maximal relaxation concentration (EC_{50}) in tolerant to nontolerant vessel segments, the higher the ratio (EC_{50} [tolerant]/EC_{50} [nontolerant]) the higher the potency of a given organic nitrate (or a nitrovasodilator) to induce *in vitro* tolerance (tachyphylaxis) (Koenig et al., 2007). The order for *in vitro* tolerance induction PETN > GTN = PETriN > ISDN > ISMN was not observed for *in vivo* treatment since other aspects contribute to overall induction of tolerance (e.g., pharmacokinetics, induction of intrinsic protective but also counter-regulatory pathways). *In vitro*, tolerance is mainly based on a overcharge and desensitization of the nitrate-bioactivating system by supra-pharmacological

[1] Nitrate tachyphylaxis may develop *in vivo* and *in vitro* and represents the fast onset of nitrate tolerance in response to repeated or high bolus administration.

TABLE II Representative Clinical–Experimental Studies on the Induction of Nitrate Tolerance and Endothelial Dysfunction

Source	*Patients volunteers*	*Nitrate dose*	*Time interval*	*Placebo random*	*Assessment of endothelial function*	*Tolerance/endothelial dysfunction*
Sage et al. (2000)	30	GTN 10 μg/min	24 h i.v.	Yes Yes	Isom. tension recording with bypass vessels	Yes/n.d.
Jurt et al. (2001)	30	PETN 240 mg/d GTN 0.6 mg/h	24 h p.o. 24 h t.d.	Yes Yes	Forearm plethysmography	PETN no/n.d. GTN yes/n.d.
Schulz et al. (2002)	71	GTN 0.5 μg/kg/min	24–48 h i.v.	Yes Yes	Isom. tension recording with bypass vessels	Yes/yes
Gori et al. (2003a)	28	PETN 240 mg/d GTN 0.6 mg/h	6d p.o. 6d t.d.	Yes Yes	Forearm plethysmography	PETN n.d./no GTN n.d./yes
Sekiya et al. (2005)	54	ISDN 40 mg/d	3 months	Yes Yes	Forearm FMD	n.d./yes
Thomas et al. (2007)	26	ISMN 120 mg/d	7d p.o.	Yes Yes	Forearm plethysmography	n.d./yes
Hink et al. (2007)	50	GTN 2.3 mg/h	24 h i.v.	Yes Yes	Isom. tension recording with bypass vessels	Yes/yes
Schnorbus et al. (2009)	80	PETN 240 mg/d	8 weeks p.o.	Yes Yes	Forearm FMD/NMD	No/no

i.v., intravenous; t.d., transdermal; p.o., per oral; FMD, flow-mediated dilation, NMD, nitroglycerin-mediated dilation; n.d., not determined.

TABLE III pD_2 Values for Different Organic Nitrates in Control and Tolerant Vessels (Koenig et al., 2007)

Vasodilator	*Structure*	*EC_{50} (M) control vessel (nontolerant)*	*EC_{50} (M) tolerant vessel*	*EC_{50} (tolerant)/ EC_{50} (control)*
PETN (pentaerithrityl tetranitrate)	 O_2NO, O_2NO, ONO_2, ONO_2	6.67×10^{-9}	1.02×10^{-7}	15.3
PETriN (pentaerithrityl trinitrate)	 O_2NO, O_2NO, ONO_2, OH	3.55×10^{-8}	4.00×10^{-7}	11.3
PEDN (pentaerithrityl dinitrate)	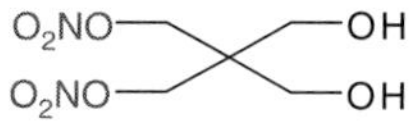 O_2NO, O_2NO, OH, OH	2.34×10^{-6}	3.07×10^{-6}	1.3
PEMN (pentaerithrityl mononitrate)	 O_2NO, HO, OH, OH	5.32×10^{-5}	7.23×10^{-5}	1.4
GTN (glyceryl trinitrate)	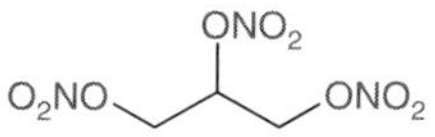 ONO_2, O_2NO, ONO_2	2.21×10^{-8}	2.58×10^{-7}	11.7
ISDN (isosorbide dinitrate)	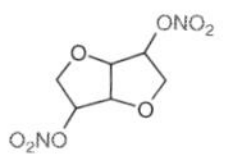 ONO_2, O, O, O_2NO	4.96×10^{-7}	6.20×10^{-7}	1.3
ISMN (isosorbide-5-mononitrate)	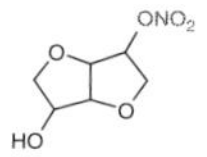 ONO_2, O, O, HO	3.96×10^{-5}	6.67×10^{-5}	1.7
SNAP (*S*-nitroso-*N*-acetyl-DL-penicillamine)	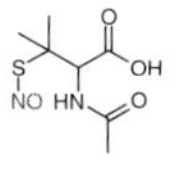 O, S, OH, NO, HN, O	1.37×10^{-8}	1.14×10^{-8}	0.8

EC_{50}, concentration that evokes half-maximal relaxation.

concentrations of the organic nitrate (e.g., desensitization of the ALDH-2). Most surprisingly, chronic PETN *in vivo* therapy demonstrated no tolerance induction at all (Fink & Bassenge, 1997; Wenzel et al., 2007b), whereas GTN, ISDN, and ISMN induced tolerance and endothelial dysfunction (Daiber et al., 2008; Table II). It should be noted that some previous studies have also reported that ISMN does not induce endothelial dysfunction and only moderate tolerance (Kojda et al., 1995; Muller et al., 2003, 2004). One should pay attention to the fact that PETN at high bolus concentrations *in vitro* also causes development of tolerance, which is similar to the one

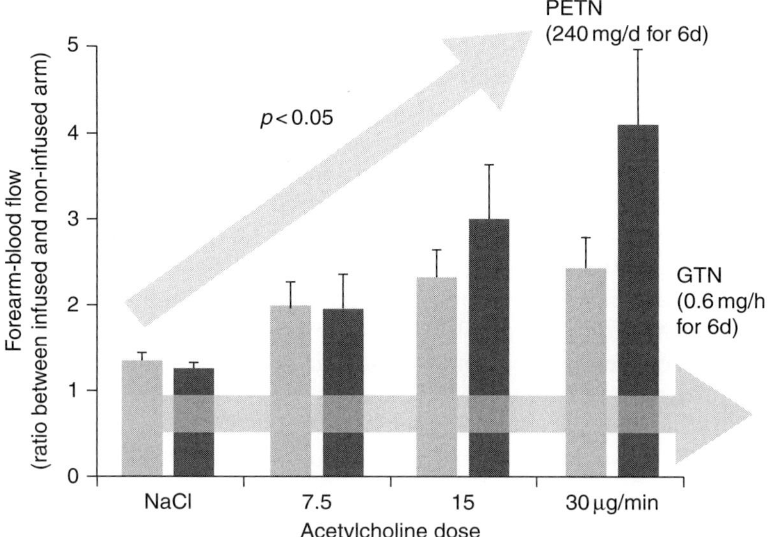

FIGURE 3 Basal forearm-blood flow (FBF) and acetylcholine (ACh)-responses to increasing ACh infusion doses in GTN and PETN treated volunteers. FBF was measured by the forearm plethysmography technique, which is regarded as an indicator of endothelial function and dysfunction. Impaired response of FBF to ACh or even paradoxic vasoconstriction indicates endothelial dysfunction. The ACh-responses in the PETN (dark/right)-treated group are much more pronounced as compared to the GTN (light/left)-treated group. Adapted from Gori et al. (2003a). With permission of Elsevier. (For interpretation of the references to color in this figure legend, the reader is referred to the Web version of this chapter.)

observed with GTN (Daiber et al., 2004c; Koenig et al., 2007). The protective profile of PETN and lack of tolerance induction are mainly based on the activation of protective genes (Oberle et al., 2003; Pautz et al., 2009; Wenzel et al., 2007b) as well as on a slow, controlled uptake of this nitrate in the intestinal duct avoiding an overload of the bioactivating system (Koenig et al., 2007). Clearly, even more complicated is the situation upon acute *in vivo* administration of organic nitrates. In the latter case, mechanisms of *in vivo* and *in vitro* tolerance may be mixed and the pharmacokinetics of the different nitrates play a central role in determining the hemodynamic effects of the drugs and the development of nitrate tolerance (Wenzel et al., 2009).

III. Oxidative Stress

A. The Oxidative Stress Concept and Endothelial Dysfunction

The nitrate tolerance induced by chronic GTN therapy is a complex and multifactorial phenomenon (Fig. 2). The oxidative stress hypothesis, which was postulated by Münzel and coworkers (Munzel et al., 1995c) and was refined during the past years (Daiber et al., 2008), provides an attractive explanation for this phenomenon. In essence, the oxidative stress hypothesis is based on an increased formation of superoxide and peroxynitrite under chronic nitrate therapy (Fig. 2). Superoxide reacts with NO and thereby reduces NO bioavailability. The resulting peroxynitrite leads to uncoupling (loss of activity) of the endothelial NO synthase (eNOS) and an impaired NO/cGMP signaling. Besides this pathway, nitration and inhibition of prostacyclin synthase (suppressing the generation of another important vasodilator) as well as oxidative inactivation of the ALDH-2 may represent other essential key events in the development of nitrate tolerance (Hink et al., 2003; Sydow et al., 2004). The most feasible explanation for the development of endothelial dysfunction (cross-tolerance) is that the formation of reactive oxygen species (ROS) and peroxynitrite leads to uncoupling of eNOS via oxidative depletion of the cofactor tetrahydrobiopterin (BH_4) or a direct oxidative modification resulting in inhibition of the enzyme (Munzel et al., 2005; Schulz et al., 2008). This concept was challenged by a recent report by Mayer and coworkers providing evidence that GTN treatment does not cause BH_4 depletion (neither in cell culture nor in *in vivo*) but rather causes superoxide formation (Schmidt et al., 2010). In the uncoupled state, eNOS does not generate NO but superoxide instead, resulting in a vicious circle of oxidative stress. This may be a possible explanation for GTN-induced cross-tolerance to other vasodilators. Uncoupling of eNOS and impairment of NO/cGMP signaling lead to clinically relevant endothelial dysfunction as observed in healthy volunteers upon chronic therapy with GTN, ISDN, or ISMN (Daiber et al., 2008). The idea of an uncoupled eNOS in GTN-induced nitrate tolerance was based on the observations that in endothelium-denuded tolerant vessels, the potency of GTN was largely improved and superoxide formation in these vessels was significantly reduced in denuded vessels (Munzel et al., 1995c). Pharmacological inhibition of eNOS in tolerant vessels decreased superoxide formation (Munzel et al., 2000) and FBF in GTN-treated volunteers was improved by folic acid (Gori et al., 2001). Finally, BH_4 levels in vessels from tolerant rabbits were significantly decreased (Ikejima et al., 2008).

B. Sources of Oxidative Stress

Several other sources of oxidative stress were discussed to contribute to nitrate tolerance such as NADPH oxidases (Fukatsu et al., 2007; Kurz et al., 1999; Munzel et al., 1995b) and the mitochondrial respiratory chain, which were identified as essential contributors to overall oxidative stress and development of tolerance under chronic and acute nitrate treatment (Daiber et al., 2004c, 2005c; Sydow et al., 2004). The essential role of mitochondrial superoxide formation for the development of nitrate tolerance was demonstrated in a murine genetic model of heterozygous manganese superoxide dismutase (MnSOD, SOD2, the mitochondrial isoform), where the 50%

deficiency in MnSOD significantly increased the susceptibility of the mice for development of nitrate tolerance but also endothelial dysfunction (Daiber et al., 2005c). Further support for mitochondrial oxidative stress as a trigger of nitrate tolerance came from a study using mitochondria-targeted antioxidants to prevent GTN-induced tolerance (Esplugues et al., 2006). Of note, the concept of a role for mitochondrial GTN-driven RONS formation revives the "thiol oxidation" concept of Needleman and coworkers (Jakschik & Needleman, 1973; Needleman et al., 1973). These authors have also observed severe effects of organic nitrates on isolated mitochondria such as swelling and alterations of mitochondrial respiration (Needleman & Hunter, 1966). We would like to add an important note: the redox-sensitive enzyme mitochondrial aldehyde dehydrogenase (ALDH-2) is responsible for the high-affinity bioactivation pathway of GTN (Chen et al., 2002) as well as of PETN (Wenzel et al., 2007c). Since treatment with GTN results in increased mitochondrial oxidative stress one may assume that ALDH-2 will be inactivated in response to sustained GTN therapy (Sydow et al., 2004). These observations provide a new link between nitrate-induced oxidative stress and impaired nitrate bioactivation. Induction of oxidative stress is most pronounced under chronic GTN therapy, and NADPH oxidases and mitochondria are active sources of oxidative stress in the setting of GTN-induced nitrate tolerance (Daiber et al., 2008; Wenzel et al., 2008a). In contrast, ISDN and ISMN treatment do not significantly increase mitochondrial RONS but trigger activation of NADPH (Table IV; Schuhmacher et al., 2010; Thum et al., 2007). However, also this concept was challenged by reports on the lack of effect of diphenylene iodonium (DPI) on nitrate-induced oxidative stress (Ratz et al., 2000b), the lack of effect of inhibiting the endogenous superoxide dismutase with diethyldithiocarbamic acid (DETCA) on impaired GTN potency (Sage et al., 2000) and discrepancies between nitrate tolerance and ascorbate deficiency (Wenzl et al., 2009b; Wolkart et al., 2008). So far, only chronic therapy with the organic nitrate PETN was found to be devoid of endothelial dysfunction, tolerance, and oxidative stress. Tables II and IV describe the results of human studies (animal experimental studies resulted in quite similar findings and were summarized in previous review articles (Daiber et al., 2008, 2009a; Munzel et al., 2005) and a book (Daiber & Münzel, 2006)).

C. Reactive Oxygen and Nitrogen Species

The formation of superoxide in GTN-induced nitrate tolerance was first reported by Münzel and coworkers and attributed to a membrane-bound NADH oxidase (Munzel & Harrison, 1997; Munzel et al., 1995c, 1996b). The involvement of superoxide was evident from observations that liposomal superoxide dismutase improved vascular function in tolerant vessels (Munzel et al., 1995c, 1996b). Moreover, a number of superoxide scavengers were effective in improving the adverse effects of GTN treatment (Daiber et al.,

TABLE IV Representative Clinical–Experimental Studies on the Induction of Oxidative Stress Under Nitrate Therapy

Source	*Patients volunteers*	*Nitrate dose*	*Time interval*	*Placebo random*	*Assessment of oxidative stress*	*Oxidative stress*
Sage et al. (2000)	30	GTN 10 μg/min	24 h i.v.	Yes	Lucigenin ECL	Yes
				Yes		
Jurt et al. (2001)	30	PETN 240 mg/d	24 h p.o.	Yes	Toxic aldehydes/isoprostanes	PETN no
		GTN 0.6 mg/h	24 h t.d.	Yes		GTN yes
Schulz et al. (2002)	71	GTN 0.5 μg/kg/min	24–48 h i.v.	Yes	Dihydroethidine fluorescence	Yes
				Yes		
Keimer et al. (2003)	18	PETN 80 mg/d	5d	Yes	Nitrotyrosine/isoprostanes	PETN no
		ISDN 30 mg/d		Yes	Nitrite/nitrate	ISDN no
Andreassi et al. (2005)	278	Not provided	Not provided	No	DNA damage	Yes
Wenzel et al. (2009)	8 (cross-over)	PETN 80 mg	1× p.o.	No	Serum antioxidants	PETN no
		GTN 0.8 mg	2× s.l.		ALDH-2 inhibition	GTN yes
Schnorbus et al. (2009)	80	PETN 240 mg/d	8 weeks p.o.	Yes	TBARS/MDA	No
				Yes	ALDH-2 inhibition	

i.v., intravenous; t.d., transdermal; p.o., per oral; n.s., not significant.

2008, 2009a; Munzel et al., 2005). Since organic nitrates are thought to produce NO and as a side effect trigger oxidative stress, one of the hypotheses of the "oxidative stress concept" is the formation of peroxynitrite, a potent oxidant and nitrating agent, from NO and superoxide (Beckman & Koppenol, 1996). Peroxynitrous acid (ONOOH) probably undergoes homolytic cleavage thereby generating hydroxyl and nitrogen dioxide radicals and all of these species contribute to oxidative protein modifications, sulfhydryl oxidation, DNA damage, depletion of low-molecular weight antioxidants, lipid peroxidation, and disruption of iron–sulfur clusters (Crow & Beckman, 1995; Daiber & Ullrich, 2002; Koppenol et al., 1992). Protein tyrosine nitration is regarded as a footprint of peroxynitrite *in vivo* (Crow & Beckman, 1995; Daiber & Bachschmid, 2007). Evidence for peroxynitrite formation in the setting of nitrate tolerance (again most pronounced for GTN) comes from increased levels of 3-nitrotyrosine positive proteins in hyperlipidemic and nitrate-treated rabbits (Warnholtz et al., 2002a). In addition, nitrated prostacyclin synthase was identified within the endothelium and subendothelial space in vessels from tolerant rats (Hink et al., 2003) and prostacyclin synthase nitration was previously reported to be a specific marker for peroxynitrite formation *in vivo* (Schmidt et al., 2003; Zou & Bachschmid, 1999). We have previously demonstrated that the antihypertensive drug hydralazine prevents GTN-induced nitrate tolerance (Munzel et al., 1996b) and is a highly effective peroxynitrite scavenger (Daiber et al., 2005a, 2005b), which might also explain the beneficial effects described above in patients with chronic heart failure (A-HefT) (Taylor et al., 2004). More evidence for peroxynitrite formation under chronic nitrate therapy comes from the observation that L-012, a chemiluminescence dye with a very high specificity for peroxynitrite (Daiber et al., 2004a, 2004b), gives remarkable signals in nitrate-treated tissues (Daiber et al., 2004c, 2005c; Wenzel et al., 2008b). Protein tyrosine nitration was also increased in endothelial cell culture upon treatment with GTN, which was blocked by preincubation with uric acid, superoxide dismutase, or chelerythrine (Abou-Mohamed et al., 2004). Increased oxidative stress was also detected in several human studies as summarized in Table IV.

D. Cross-Talk Between Mitochondrial ROS and NADPH Oxidases

The following section provides an attractive explanation for the development of endothelial dysfunction (cross-tolerance to endothelium-dependent vasodilators) under chronic GTN therapy. We have recently reported on the cross-talk between mtROS and cytosolic ROS/RNS in a model of increased mitochondrial oxidative stress (nitroglycerin-induced tolerance). In this system, endothelial dysfunction (sensitive to NADPH oxidases) and vascular dysfunction (sensitive to mitochondria) were dependent on the activation of distinct oxidant sources (Wenzel et al., 2008a). This - cross-talk was blocked by *in vivo* and *ex vivo* administration of the mitochondrial permeability pore inhibitor cyclosporine A (CsA), which selectively improved endothelial dysfunction, whereas nitrite tolerance was not affected. In contrast, the respiratory complex I inhibitor rotenone (Rot) improved endothelial dysfunction and tolerance. Conversely, *in vivo* or *ex vivo* treatment with the K_{ATP} opener diazoxide (Diaz) caused a nitrate tolerance-like phenomenon in control animals, whereas the K_{ATP} inhibitor glibenclamide (Glib) improved tolerance in nitroglycerin-treated animals. Very similar effects of Rot, CsA, Diaz, and Glib have been recently demonstrated by another group in an experimental model of angiotensin-II-induced hypertension (Doughan et al., 2008). A role of K_{ATP} channels for NADPH oxidase-driven activation of mitochondrial ROS formation via changes in the membrane potential was previously proposed (Brandes, 2005). $gp91^{phox-/-}$ and $p47^{phox-/-}$ mice developed tolerance but no endothelial dysfunction in response to nitroglycerin treatment. The findings of this study are summarized and discussed in view of similar observations in different animal models in a recent review article (Daiber, 2010). The mechanism underlying this concept is based on mtROS-driven PKC activation which in turn will activate NADPH oxidases. The NADPH oxidase-dependent cytosolic ROS and RNS formation will then uncouple eNOS, nitrate/inactivate prostacyclin synthase, and desensitize sGC. Previous experimental studies have shown that increased oxidative stress in cellular tissue *per se* is able to activate the oxidase in a positive feedback fashion (Fukui et al., 1997). Thus, nitroglycerin-induced mitochondrial superoxide production may cause a secondary activation of Nox. One may also speculate that via its hypotensive action, nitroglycerin may cause an activation of the renin–angiotensin–aldosterone system (Munzel et al., 1996a), leading to increased circulating levels of angiotensin-II and aldosterone, and therefore, to an activation of the NADPH oxidase. This concept is further corroborated by the demonstration that *in vivo* treatment with an AT_1 receptor blocker was able to prevent the development of nitroglycerin-induced endothelial dysfunction in an animal model of nitrate tolerance (Kurz et al., 1999). It should be noted that some previous studies do not support a beneficial role of AT_1 receptor blockers in the prevention of nitroglycerin-induced nitrate tolerance in human subjects (Longobardi et al., 2004; Milone et al., 1999).

IV. Bioactivation of Organic Nitrates

A. Different Enzymatic Bioactivation Mechanisms

Organic nitrates are metabolized by different systems, which either activate the nitrates to vasodilatory species or decompose them to inactive products. The existence of several synergistic bioactivation pathways can be assumed from observations that inhibitors (and even genetic deletion) for one enzymatic pathway never result in complete suppression of the vasodilatory effects of an organic nitrate but only impairs its potency (e.g., shifts the concentration–relaxation -curve to the right). For GTN, the bioactivation to a vasodilator was reported to be associated with the formation of 1,2-glyceryl dinitrate (1,2-GDN), whereas the formation of 1,3-GDN was rather attributed to degrading pathways (Bennett et al., 1984; Beretta et al., 2008a; Chen et al., 2002; Fujii et al., 1996; Torfgard et al., 1992). Among the metabolizing systems are the xanthine oxidase (XO) (Millar et al., 1998; O'Byrne et al., 2000), glutathione-dependent reductases (GR) (Needleman & Hunter, 1965; Needleman et al., 1969), hemoglobin (Hb) (Bennett et al., 1984, 1986), P450 enzymes (Delaforge et al., 1993; McDonald & Bennett, 1990), or glutathione-*S*-transferases (GST) (Fujii et al., 1996; Lau & Benet, 1990). The bioactivation by mitochondrial aldehyde dehydrogenase (ALDH-2) (Chen et al., 2002) will be discussed in detail below. In the case of GTN, bioactivation yields 1,2- and 1,3-GDN, nitrite, nitrate, and NO. However, it should be noted that most of these enzymes were tested with supra-pharmacological concentrations of GTN (>10 μM), which are of minor clinical relevance (Munzel et al., 2005). Bioactivating pathways produce NO or NO like species such as *S*-nitrosothiols, nitrite at special cellular sites (e.g., mitochondria or red blood cells) as well as the other metabolites described above (Fung et al., 1988; Munzel et al., 2005). Nitrates such as GTN and PETN are bioactivated by at least two different mechanisms and one accounts for the clinically relevant vasodilatory potency at therapeutic concentrations/doses, whereas the other is active at higher, supra-pharmacological concentrations/doses (Fig. 4; Ahlner et al., 1986, 1988; Axelsson et al., 1992; Malta, 1989; Munzel et al., 2005). The bioactivation pathways for ISMN and ISDN (as well as other di- and mononitrate metabolites from GTN and PETN) are less carefully defined but it is assumed that cytochrome P450 enzymes contribute to vasodilation by those less potent nitrates (Fig. 4; Minamiyama et al., 1999, 2004a) although another group reported differences in P450-mediated organic nitrate bioactivation identifying GTN but not ISDN as a substrate (Mulsch et al., 1995). In addition, XO was identified as an ISDN and ISMN-metabolizing enzyme with higher turnover with xanthine instead of NADH as the source of electrons (Doel et al., 2001). However, most of the investigations on nitrate bioactivation by purified enzymes or enriched cellular compartments were performed using high, supra-pharmacological concentrations of the organic nitrates. Therefore, it is strongly recommended to repeat these studies using clinically relevant concentrations of the drugs or even *in vivo* animal or at least cellular models.

B. Low-Molecular Weight Pathways

The low potency pathway leads to formation of measurable amounts of NO in vascular tissues *in vivo* (Mulsch et al., 1995) and *in vitro* (Kleschyov et al., 2003). Therefore, NO is a vasoactive principle of higher concentrations

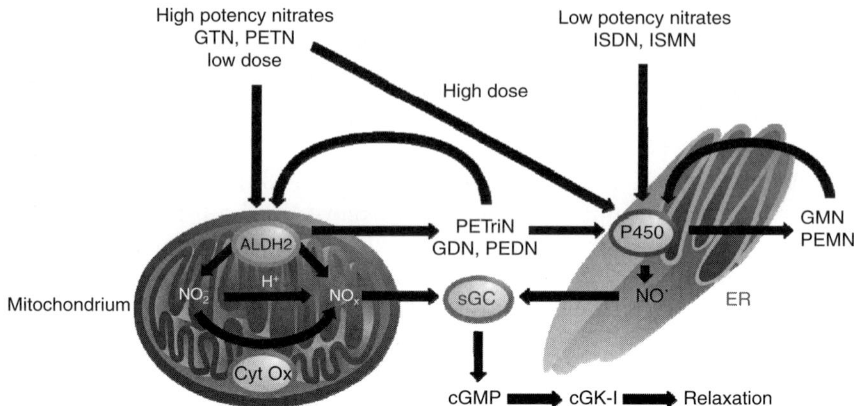

FIGURE 4 High- and low-affinity pathways for nitrate bioactivation. High potency nitrates such as nitroglycerin (GTN), pentaerythrityl tetranitrate (PETN), and pentaerythrityl trinitrate (PETriN) are activated by mitochondrial aldehyde dehydrogenase (ALDH-2), yielding an NO-containing compound. This molecule then activates the soluble guanylyl cyclase, which decreases cytosolic Ca^{2+} by causing extracellular currents and increasing its concentration in intracellular stores such as the sarcoplasmatic reticulum. The right part of the figure describes the bioactivation of low potency nitrates such as isosorbide dinitrate (ISDN), isosorbide-5-mononitrate (ISMN), glyceryl dinitrate (GDN), pentaerithrityl dinitrate (PEDN), and their respective mononitrates GMN and PEMN. These molecules are most probably metabolized by P450 enzyme(s) in the endoplasmic reticulum (ER) directly yielding nitric oxide. The latter mechanism also metabolizes high potency nitrates when they are employed at high concentrations (>1 μM). Adapted from Munzel et al. (2005). With permission of the American Heart Association.

of GTN. Previous studies, which focused on the identification of enzymes and/or low-molecular weight factors which could generate NO from GTN, identified cysteine, *N*-acetyl-cysteine, and thiosalicylic acid (Feelisch et al., 1988; Schroder et al., 1985). Since the nonenzymatic reaction of GTN with thiols requires high concentrations (mM) of these thiols as well as GTN (μM range), this reaction may lack physiological significance (Munzel et al., 2005). Optimized systems observed activation of purified sGC by cysteine (1 mM) and GTN in the low micromolar range, a quite low concentration, but still 100- to 1000-fold less potent than vasodilation of isolated vessels by GTN (Gorren et al., 2005). Another low-molecular weight (but also low affinity) pathway was postulated for the ascorbate system although activation of purified sGC required at least 10 μM GTN in the presence of ascorbate (10 mM) (Kollau et al., 2007). It should be noted that inorganic nitrite was even more effective in this system and evoked more sensitive activation of sGC in the presence of ascorbate. More evidence for a role of ascorbate in GTN bioactivation was provided by two subsequent studies from the same laboratory using ascorbate deficiency in guinea pigs and demonstrating impaired GTN vasodilatory potency in these animals (Wenzl et al., 2009b; Wolkart et al., 2008). The conclusions drawn in these studies were critically discussed in a letter to the Editor by us (Daiber & Gori, 2008): our major concern was that the results could also point toward increased oxidative stress in ascorbate deficient animals and, therefore, increased susceptibility for development of nitrate tolerance (see preceding sections for role of oxidative stress for development of nitrate tolerance). These reports also demonstrated the complications associated with the interpretation of data obtained with low-molecular weight molecules and organic nitrates: *in vitro* activation by these pathways requires high supra-pharmacological concentrations of the organic nitrates. When used *in vivo*, these compounds often significantly improve organic nitrate action or efficiently prevent development of tolerance but since all of them are potent antioxidants, it is unclear whether their action is based on direct bioactivation of the organic nitrates or on secondary antioxidant effects on nitrate-induced RONS formation (see Section V.A for further details).

C. Role of Mitochondrial Aldehyde Dehydrogenase (ALDH-2) for Organic Nitrate Bioactivation

1. Discovery of ALDH-2 as a GTN Reductase: Animal Experimental and Clinical Studies

In 2002, the mitochondrial isoform of aldehyde dehydrogenases (ALDH-2) was identified as a key enzyme in the clinically relevant (high affinity) bioactivation process of GTN and provided new important information for nitrate pharmacology (Chen et al., 2002). In this study, the authors demonstrated that more or less specific ALDH inhibitors impair GTN-induced relaxation of isolated aortic ring segments but also that purified ALDH-2 bioactivates GTN to 1,2-GDN and nitrite in the presence of the dithiol compound dithiothreitol (DTT). It should be noted that there was previous evidence for a role of ALDH in nitrate metabolism based on observations that organic nitrate has disulfiram-like effects (Towell et al., 1985). In subsequent studies, our group and others could demonstrate the role of ALDH-2 in the bioactivation process and development of nitrate tolerance in a more clinical setting using different animal models (Sydow et al., 2004; Zhang et al., 2004). In 2005, involvement of ALDH-2 in the GTN bioactivation process was proven at the molecular level using ALDH-2-deficient (ALDH-$2^{-/-}$) mice that demonstrated impaired relaxation in response to GTN but not SNP or ISDN (Chen et al., 2005). In the same year, there was the first report on a role of ALDH-2 in the bioactivation process in humans (Mackenzie et al., 2005). These authors treated volunteers with the ALDH inhibitor disulfiram, a drug used for treatment of alcoholism (e.g., antabuse) and observed significantly impaired GTN- but not SNP-induced blood flow increases in the forearm. They observed similar effects in East-Asian volunteers with ALDH-2 Glu504Lys mutation polymorphism

(ALDH2*2). Bioactivation of GTN by ALDH-2 is also of high epidemiological interest since a large part of the East-Asian population carries the ALDH2*2 polymorphism and, therefore, demonstrates impaired responsiveness to GTN (Li et al., 2006). According to a recent report, GTN-induced tolerance in humans is also associated with inhibition of ALDH-2 indicating that impaired bioactivation contributes to clinical tolerance (Hink et al., 2007). In 2007, we demonstrated bioactivation of PETN and its trinitrate metabolite PETriN by ALDH-2 using ALDH-2$^{-/-}$ mice (Wenzel et al., 2007c). In contrast, the metabolites of PETN, PEDN, and PEMN as well as the GTN metabolite 1,3-GDN but also ISMN were not bioactivated by ALDH-2. These observations exclude that differences in bioactivation between GTN and ISDN are simply due to structural differences and suggest differences in the reactivity toward thiol groups—the latter parameter determines affinity for bioactivation by ALDH-2.

2. Bioactivation Mechanism of Organic Nitrates by ALDH-2

In a first step, ALDH-2 catalyzes the formation of a hypothetical thionitrate intermediate (Fung, 2004) from the reaction of GTN, PETN, and its trinitrate PETriN with a reactive thiol group at the active site of the enzyme generating the denitrated metabolite (1,2-GDN, PETriN, or PEDN; Fig. 5; Wenzel et al., 2007a, 2007c). The thionitrate stabilizes by spontaneous nucleophilic attack of a second neighboring cysteine thiol group by formation of a disulfide bridge and release of nitrite. Another possibility could be the direct generation of NO by the thionitrate (Fung, 2004) but it is unclear to this date whether this process requires transition metals (Beretta et al., 2008a). The enzymatic activity can be further regulated by reaction of the disulfide bridge with glutathione, which forms a quite stable adduct that can be detected with a specific antibody against GSH. This regulatory process is called glutathionylation and is probably due to a narrow active site of the enzyme prohibiting the attack of a second GSH molecule, which would result in GSSG and the reactivated enzyme. There is also irreversible inhibition of the enzyme, probably via formation of sulfonic acid groups by oxidants such as superoxide or peroxynitrite (Wenzel et al., 2007a), which requires *de novo* synthesis of the ALDH-2. In cells/mitochondria with intact redox-state, the disulfide bridge (or sulfenic acid –SOH upon reaction with peroxynitrite) is reduced by dithiol compounds such as the mitochondrial dihydrolipoic acid or synthetic compounds such as DTT (Chen & Stamler, 2006; Wenzel et al., 2007a).

3. The Role of Inorganic Nitrite in the ALDH-2-Dependent Bioactivation of GTN

The mitochondrial compartment provides different pathways for the reduction of nitrite to NO, an *S*-nitrosothiol or a nitroso-metal-complex:

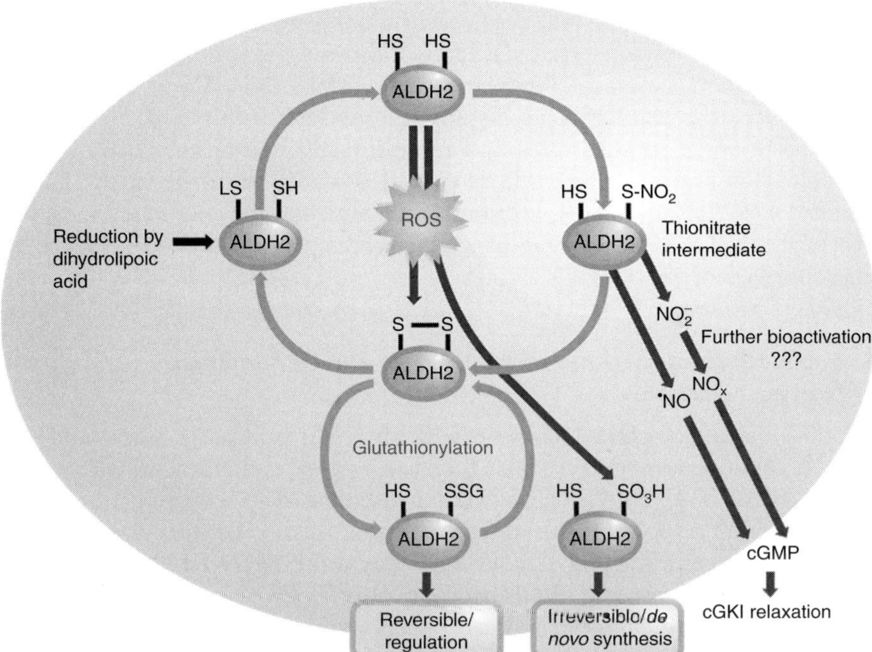

FIGURE 5 Proposed pathway of organic nitrate bioactivation by ALDH-2 in the vasculature. The reductase activity of ALDH-2 converts GTN to nitrite and the denitrated metabolite. In turn, nitrite undergoes either reduction by the respiratory chain or acidic disproportionation in the inner mitochondrial membrane space, to yield an NO-related species. Two neighbored reduced cysteine thiols are essential for this process, which yields nitrite at the expense of the formation of a disulfide group (which temporarily inactivates the enzyme). Restoration of enzymatic activity of ALDH-2 requires oxidation of a dithiol compound (dihydrolipoic acid in the scheme). Obviously, direct oxidation of the ALDH-2 thiols by reactive oxygen and nitrogen species (RONS) such as superoxide and peroxynitrite could also cause formation of a disulfide bridge, inhibiting the enzyme. This mechanism is compatible with evidence of impaired GTN biotransformation as one of the causes of nitrate tolerance. 1,2-GDN, 1,2-glyceryl dinitrate. Adapted from Wenzel et al. (2007a). With permission of the American Society for Biochemistry and Molecular Biology.

acidic disproportionation in the intermembrane space would be one possibility and the reduction by mitochondrial cytochrome *c* oxidase could be another one (see also Fig. 4). Moreover, involvement of nitrite in GTN-induced relaxation would be in good accordance with recent reports on nitrite-mediated protection from ischemic damage (Butler & Feelisch, 2008) and similarities between GTN and nitrite-driven ATP release from erythrocytes (a vasodilatory pathway) (Garcia et al., 2010), providing an attractive explanation for the potent antiischemic properties of GTN. It should be noted that this concept of GTN-induced mitochondrial nitrite formation with subsequent reduction to NO was recently challenged by Mayer and coworkers who demonstrated that sGC activation was triggered by bioactivation of nitrite with mitochondrial cytochrome *c* oxidase and GTN in the presence of mitochondria but these triggers differed markedly with respect to effects of respiratory chain substrates and inhibitors (Kollau et al., 2009). Based on these observations, the authors excluded a role of nitrite in the GTN-induced vasodilation process although there was no direct comparison of the potency of GTN versus nitrite-triggered sGC activation in the presence of mitochondria.

4. Implications of the Role of ALDH-2 for the Development of Nitrate Tolerance

The inactive ALDH-2 enzyme with disulfide bridges is reactivated by suitable dithiol compounds such as the physiological compound dihydrolipoic acid (Wenzel et al., 2007a). However, in case that these reducing equivalents are depleted in response to sustained oxidative stress or by over charge of the system with organic nitrates, ALDH-2 will remain at the level of the inactivated disulfide bridge form and the bioactivation of GTN will become impaired manifesting in nitrate tolerance. This concept would explain why the hemodynamic effect of GTN and other nitrates is reduced in the setting of conditions associated with oxidative stress, such as diabetes and CAD (McVeigh et al., 2002). Even more importantly, organic nitrate-induced oxidative stress may cause irreversible inactivation of the enzyme and different oxidants display distinct potencies for inhibition of the ALDH-2 activity (Table V). This irreversible inactivation process requires *de novo* enzyme synthesis, lasting up to days. In principle, this part of the ALDH-2-based bioactivation is in good accordance with the "thiol oxidation concept" of Needleman and coworkers who observed mitochondrial swelling and impaired oxygen metabolism in organic nitrate-treated mitochondria. These authors postulated that depletion of (mitochondrial) thiol groups is an essential step in the development of nitrate tolerance (Needleman & Hunter, 1966). The ALDH-2-based nitrate bioactivation process (the high-affinity pathway) is highly susceptible to inactivation and the development of nitrate tolerance, *in vivo* as well as *ex vivo* (Daiber et al., 2004c; Koenig et al., 2007). Cotreatment with lipoic acid significantly improved the vascular consequences of GTN-induced nitrate tolerance (Dudek et al., 2008; Wenzel et al., 2007a). It should be noted that lipoic acid is also a potent antioxidant and has beneficial effects in other cardiovascular and neurodegenerative diseases (Bilska & Wlodek, 2005; Lynch, 2001; Wollin & Jones, 2003).

A study from our laboratory showed in vessels obtained from bypass surgery that GTN potency and ALDH activity were decreased in saphenous veins and mammary arteries from *in vivo* GTN-treated patients and the

TABLE V Half-Maximal Inhibition Concentrations of Different Reactive Species and Organic Nitrates for ALDH Activity

	Inhibitor	*IC_{50} values for yeast ALDH (isolated enzyme)*	*IC_{50} values for ALDH-2 (isolated mitochondria)*	*Induction of RONS in mitochondria*
Organic nitrates	GTN	4.44 ± 0.94 μM	~50 μM	High
	PETN	n.d.	~500 μM	Moderate
	PETriN	n.d.	~500 μM	Moderate
	PEDN	n.d.	> 500 μM	Absent
	PEMN	n.d.	> 500 μM	Absent
	ISDN	n.d.	> 500 μM	Moderate
	ISMN	n.d.	> 5000 μM	Absent
Reactive oxygen and nitrogen species	H_2O_2	> 1000 μM	n.d.	n.d.
	ONOO– (bolus)	7.92 ± 4.12 μM	~120 μM	n.d.
	ONOO– (generated by Sin-1)	16.30 ± 4.41 μM	~66 μM	n.d.
	ONOO– (generated by XO/NONOate)	8.48 ± 2.14 μM	n.d.	n.d.
	Superoxide (xanthine oxidase)	17 μM (1 mU/ml)	n.d.	n.d.
	NO (spermine NONOate)	58.98 ± 13.81 μM	n.d.	n.d.

IC_{50}, concentration that evokes half-maximal inhibition; n.d., not determined.

extent of this impairment was quite similar in vessels upon *in vitro* treatment with a pharmacological inhibitor of ALDH-2 (Hink et al., 2007). This observation was supported by a subsequent report performed with human saphenous veins which were tested for ALDH-2 expression and effectiveness of ALDH inhibitors on GTN potency (Huellner et al., 2008). Since we and others have repeatedly shown that ALDH-2 activity was decreased in response to GTN *in vivo* and *in vitro* treatment in animals and humans, we speculated whether this decrease could be used as a reliable marker for nitrate tachyphylaxis (acute overload of the GTN bioactivating system) but also clinical tolerance (Wenzel et al., 2009). For this purpose, we isolated white blood cells (WBCs) from the "buffy coat" and neutrophils (PMNs) by the dextran sedimentation method and size-exclusion centrifugation by Ficoll as previously described (Daiber et al., 2004a). According to our data, WBCs (monocytes/lymphocytes) have higher ALDH-2 activity as compared to PMNs (Daiber et al., 2009b). Therefore, WBCs seemed to be the most reliable marker for ALDH-2 activity in human whole blood and were used for subsequent clinical studies in human volunteers. Our data indicated that WBC ALDH activity is a reliable marker for organic nitrate tachyphylaxis and may be a useful marker for clinical tolerance as well: a single sublingual administration of GTN significantly impaired ALDH-2 activity in human and rat WBC preparations isolated from these subjects (Wenzel et al., 2009). But also chronic GTN infusion resulted in decreased ALDH-2 activity in rat WBC preparations isolated from these animals. There was a highly significant linear correlation of mitochondrial and WBC ALDH-2 activity as well as GTN potency and these parameters decreased with increasing doses of GTN infusion (Fig. 6). A last experiment in the volunteers was designed to test whether lipoic acid co-therapy may improve GTN-induced nitrate tachyphylaxis. Indeed, lipoic acid co-therapy was able to eliminate all negative side effects of GTN administration in humans (such as the decrease in WBC ALDH activity) supporting our previous observation that lipoic acid prevented nitrate tolerance in GTN-treated rats (Wenzel et al., 2007a).

5. Isolated, Purified ALDH-2 and GTN

Recent data from the group of Bernd Mayer demonstrated that isolated, purified human ALDH-2 bioactivates GTN but in the absence of reducing equivalents (e.g., DTT or dihydrolipoic acid) generates RONS causing inactivation of the enzyme (Beretta et al., 2008a, 2008b). According to these results, also cytosolic ALDH-1 in the presence of GTN-evoked activation of purified sGC (measured by cGMP formation) although the potency was 30-fold lower as compared to ALDH-2 (Fig. 7; Beretta et al., 2008a) and therefore is likely to contribute to supra-pharmacological actions of GTN in tissue or cells. GTN even induced some irreversible inactivation of isolated ALDH-2 (Beretta et al., 2008b) as postulated above for RONS-triggered inhibition by sulfonic acid formation at the active site (Wenzel et al., 2007a): when conversion of acetaldehyde was measured by recording the NADH formation and GTN was added, the dehydrogenase activity was significantly impaired. Upon addition of the dithiol, compound DTT the enzymatic acitivity was reactivated, but this rescue could not completely recover the activity—an appreciable portion of the ezyme remained inactive. This means that at least partial irreversible inactivation took place. Already in 2007, we observed partial irreversible inactivation of ALDH-2 in cardiac mitochondria in response to *in vivo* treatment with GTN (Wenzel et al., 2007a). Addition of dihydrolipoic acid only partially restored ALDH-2 activity in these mitochondria. Likewise, treatment of mitochondria with antimycin A, a complex III inhibitor, caused mitochondrial ROS formation triggered partial irreversible inhibition of ALDH-2, which was not completely reversible by addition of dihydrolipoic acid. This concept of dihydrolipoic acid-mediated protection was also confirmed in a human study (Wenzel et al., 2009). Based on these observations, we postulated a redox-regulation

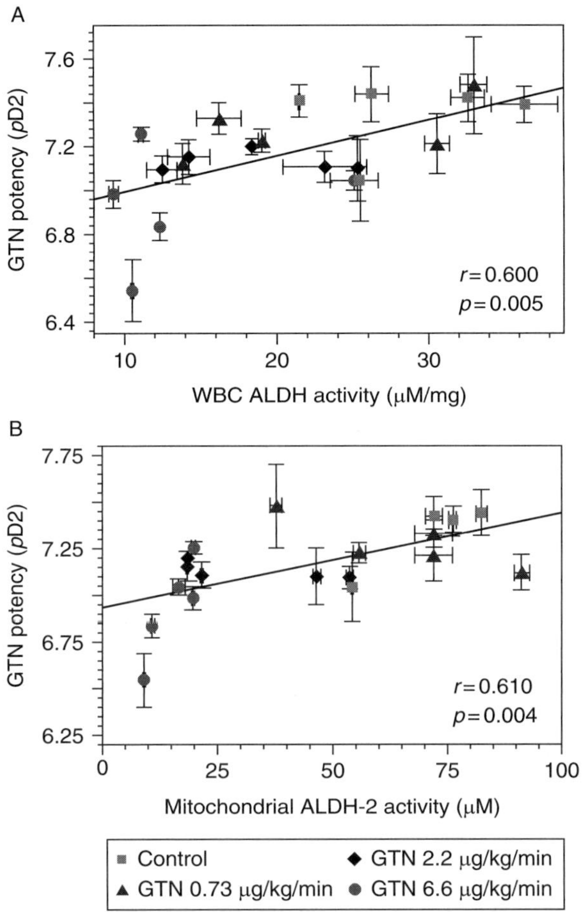

FIGURE 6 Correlations between WBCs ALDH activity, mitochondrial ALDH-2 activity, and GTN potency ($pD_2 = -\log$ of concentration causing half-maximal relaxation). (A) WBC–ALDH activity was plotted for all GTN dose groups (see legend insert) versus the corresponding GTN potency. (B) Mitochondrial ALDH-2 activity was plotted for all GTN dose groups versus the corresponding GTN potency. *r*, correlation coefficient. *p*-Values for linear regressions are 0.005 (A) and 0.004 (B). Adapted from Wenzel et al. (2009). With permission of the American Society for Pharmacology and Experimental Therapeutics.

of ALDH-2 activity by GTN and RONS as well as irreversible inhibition by formation of sulfonic acid by oxidizing species (Daiber & Munzel, 2010b; Daiber et al., 2009b; Wenzel et al., 2007a).

Based on recent reports, the purified ALDH-2 enzyme itself produces RONS that are further increased in the presence of GTN (Wenzl et al., 2009a). As an explanation for this RONS formation, the authors proposed a conformational change taking place upon binding of NAD^+ to the enzyme,

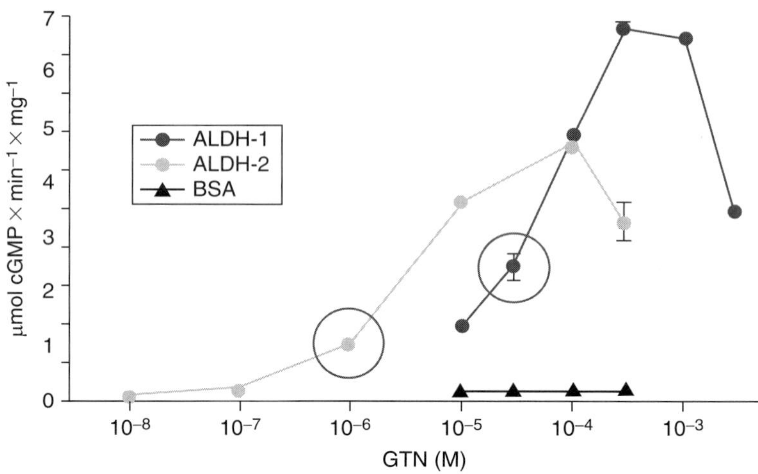

FIGURE 7 The nitroglycerin (GTN) bioactivation by purified, recombinant, human ALDH-1 (50 µg), and ALDH-2 (25 µg). Bioactivation of the nitrate was determined by activation of purified soluble guanylyl cyclase in the same tube and subsequent formation of cGMP. Bovine serum albumin was used as a negative control. Adapted from Beretta et al. (2008a). With permission of the American Society for Biochemistry and Molecular Biology (via rightslink).

since the E268Q mutation reduced the binding affinity of NAD^+ due to the absence of glutamic acid but also suppressed RONS formation by the enzyme (Wenzl et al., 2009a). Interestingly, the nitrate reductase activity of ALDH-2 was significantly affected by E268Q mutation. NO formation from wild-type ALDH-2 and GTN could only be observed in the presence of superoxide dimutase (SOD) providing evidence for involvement of superoxide in the breakdown of NO. In contrast, the E268Q variant displayed significant NO formation in the absence of SOD indicating that superoxide formation in the variant was less pronounced. It should be noted that high GTN concentrations (> 10 µM) were used for the detection of NO. The dehydrogenase and esterase activities are almost absent in the E268Q mutant, whereas basal nitrate reductase activity is similar to the one in wild-type ALDH-2. However, nitrate reductase activity of the wild-type enzyme was sevenfold increased upon addition of NAD^+ (by a structural change) and this increase was completely lost in the E268Q (Wenzl et al., 2009a). Replacement of a cysteine residue at the active site by a serine group (C302S) resulted in complete loss of all three enzymatic activities (dehydrogenase, esterase, and reductase) and a complete suppression of the 1,2-GDN as well as NO formation in the presence of GTN (Wenzl et al., 2009a). The widely distributed East-Asian variant of ALDH-2 (ALDH2*2) with the point mutation E487K displays a similar decrease in NAD^+-binding affinity as compared to E268Q mutant (Larson et al., 2007). Accordingly, the ALDH2*2 variant displayed a significantly reduced dehydrogenase and esterase activity

(Larson et al., 2007) but also impaired GTN reductase activity (Li et al., 2006). Based on these observations, the vasodilatory potency of GTN is blunted in East-Asian individuals with the ALDH2*2 polymorphism as well as in humans under therapy with the ALDH inhibitor disulfiram (Mackenzie et al., 2005). In a recently published work, these findings were supported by data obtained with purified ALDH2*2 variant displaying a reduced dehydrogenase, esterase, and nitrate reductase activity as compared to the wild-type enzyme (ALDH2*1) (Beretta et al., 2010). Accordingly, the ALDH2*2 variant yielded decreased levels of 1,2-GDN and NO in the presence of GTN and finally, evoked less pronounced activation of sGC. Interestingly, the previously described ALDH-2 activator Alda-1 (Chen et al., 2008) induced only minor activation of the dehydrogenase activity of wild-type ALDH-2 but a fourfold increase in ALDH2*2 variant dehydrogenase activity (Beretta et al., 2010). The effect of Alda-1 on esterase activity of the ALDH2*1 enzyme was moderate, but caused a eightfold increase in the ALDH2*2 variant. It was a disappointment to see that Alda-1 did increase neither the GTN bioactivation nor the sGC activity (Beretta et al., 2010) and obviously cannot be clinically employed to increase the vasodilatory potency of GTN in (nitrate-tolerant) patients under chronic treatment or to improve the antiischemic effects of GTN in East-Asian individuals with ALDH2*2 polymorphism.

In a recently completed study from our laboratory, we could demonstrate that purified human ALDH-2 was significantly inactivated by GTN but to a minor extent by PETN (Fig. 8). This finding provides an attractive additional explanation for the tolerance-devoid and beneficial effects of PETN under chronic therapy, which were previously defined to rely on induction of heme oxygenase-1 and ferritin (Oberle et al., 1999, 2003; Wenzel et al., 2007b) as well as on the controlled and slow uptake of the drug upon oral treatment (Koenig et al., 2007; Wenzel et al., 2009). If PETN really prevents inactivation of its bioactivating enzyme ALDH-2, then this could explain the lack of tolerance development. As expected, the RONS formation by ALDH-2 was significantly less pronounced in the presence of PETN as compared to GTN (Fig. 9).

According to these results, ALDH-2 itself could represent a source of RONS in the presence of GTN, as proposed by Mayer and coworkers (Wenzl et al., 2009a). However, this conclusion is at variance with previous data from our group and others indicating that ALDH is rather a RONS scavenger and prevents degradation of NO since knockdown of ALDH-2 in cultured cells increased RONS (Szocs et al., 2007). In a subsequent study, it was demonstrated in ALDH-2 knockout mice that deficiency in ALDH-2 results in increased RONS levels under chronic GTN therapy (Wenzel et al., 2008b). Also acute challenges with GTN did not support a role of ALDH-2 in the GTN-induced RONS formation since those signals were similar in mitochondria from wild-type and ALDH-2 deficient mice (Schuhmacher

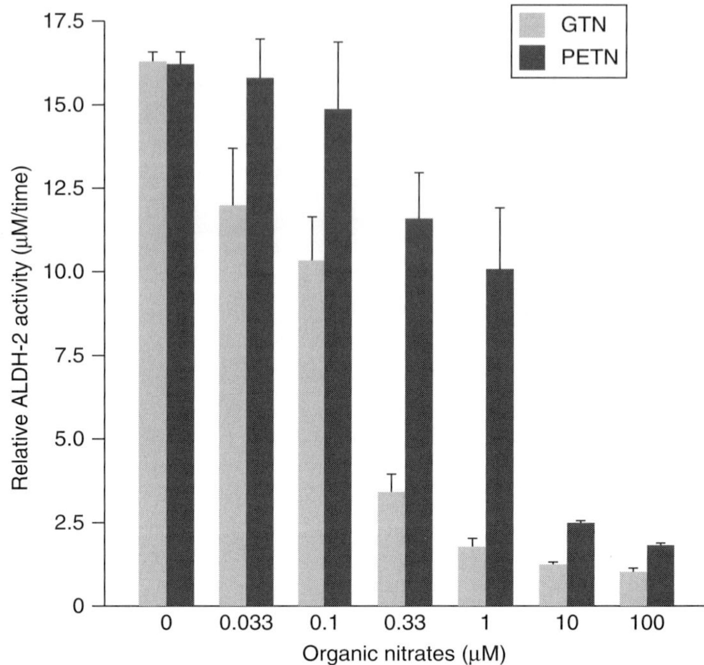

FIGURE 8 Inactivation of purified human ALDH-2 by GTN using an HPLC-based method. The ALDH-2 dehydrogenase activity was determined by quantification of the conversion of 2-hydroxy-3-nitrobenzaldehyde to its benzoic acid product. Inactivation by increasing concentrations of GTN is much more pronounced as compared to PETN.

et al., 2009). A protective role of ALDH-2 is further supported by reports of increased oxidative damage in aged ALDH-$2^{-/-}$ mice (Wenzel et al., 2008c) and doxorubicin-treated ALDH-$2^{-/-}$ mice (Wenzel et al., 2008b) as well as of increased GTN-induced RONS formation in mitochondria treated with pharmacological ALDH inhibitors (Gori et al., 2007). All of these reports speak against an essential role of ALDH-2 in GTN-induced oxidative stress in tolerant tissue. GTN-triggered RONS formation from purified ALDH-2 rather seems to be an artificial side effect as frequently observed with simplified, purified enzymatic systems (e.g., due to lack of essential cofactors or antioxidants). The differential effects of GTN and PETN on ALDH-2 activity are of great scientific and clinical interest and deserve further investigations.

6. ALDH-2 Beyond Nitrate Treatment and Tolerance

In 2008, Chen et al. reported that "activation of aldehyde dehydrogenase-2 reduces ischemic damage to the heart" and vice versa that inhibition of ALDH-

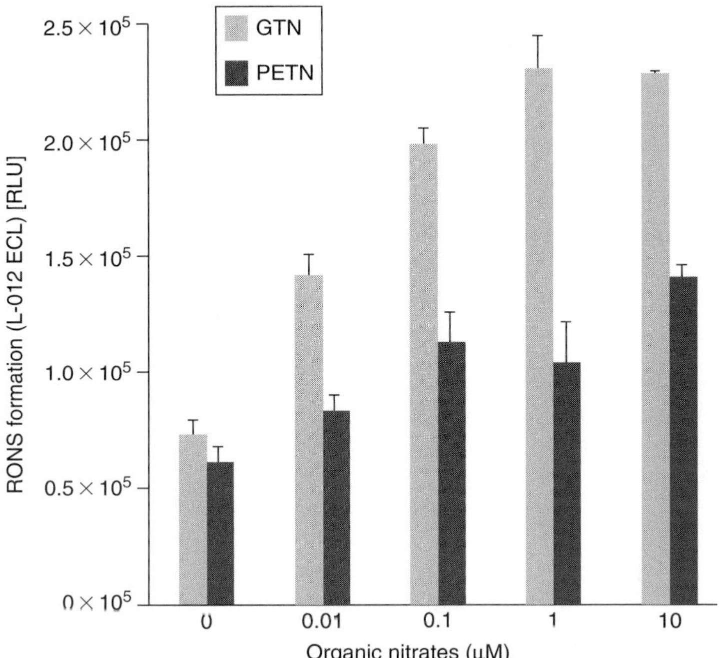

FIGURE 9 RONS production by purified human ALDH-2 in the presence of GTN or PETN. RONS formation was determined by L-012 chemiluminescence. The RONS generation is much less pronounced with PETN as compared to GTN.

2 by nitroglycerin or cyanamide treatment increased infarct area in experimental myocardial infarction (Chen et al., 2008). The cardioprotective role of ALDH-2 is well known from studies in knockout mice and ALDH-2 overexpression (Ren, 2007; Ma et al., 2010). Recently, ALDH-2 deficiency was shown to render mice more susceptible to exogenously triggered oxidative stress. ALDH-$2^{-/-}$ mice were more susceptible to nitroglycerin-, acetaldehyde-, and doxorubicin-induced cardiovascular damage (Wenzel et al., 2008b) but also cultured cells displayed increased oxidative damage in the absence of ALDH-2 (Szocs et al., 2007). Finally, a recent study from our laboratory demonstrated that ALDH-$2^{-/-}$ mice are more susceptible to aging-induced DNA damage, vascular dysfunction, and mitochondrial oxidative stress (Wenzel et al., 2008c). Therefore, ALDH-2 seems to be a marker for oxidative stress but its absence or deficiency in ALDH-2 obviously also increases oxidative stress. Therefore, besides direct oxidative damage by nitroglycerin, ALDH-2 inactivation in the setting of nitrate tolerance may largely contribute to cardiovascular risk. These observations might be compatible with a toxic effect of chronic nitrate therapy: as reported above, based on a retrospective metaanalysis using the databases from two large-scale postinfarction studies, Nakamura et al. presented data that long-term nitrate therapy increases cardiovascular mortality (Nakamura et al., 1999).

In light of the new data presented above, this increased mortality may be secondary to nitrate-mediated inactivation of ALDH-2. Therefore, one may speculate that organic nitrates without inhibitory effect on ALDH-2 activity (e.g., PETN) may have better effects on prognosis in patients. However, future clinical studies have to proof these experimental findings for the clinical situation.

V. Strategies to Overcome Nitrate Tolerance

A. Direct Antioxidants in the Prevention of Nitrate Tolerance

Previously, the nitrate-free interval was the recommended strategy to avoid development of nitrate tolerance although antiischemic protection of patients could be inadequate during the nitrate-free interval and may be even increased due to rebound angina (Freedman et al., 1995; Parker et al., 1995; Schaer et al., 1988). There is also evidence that nitrate-free intervals do not prevent all adverse side effects of nitrate therapy (Munzel et al., 2000). Since oxidative stress obviously plays an important role for the development of nitrate tolerance and endothelial dysfunction (especially under chronic GTN therapy), it did not come as a surprise that numerous groups reported benefical effects of antioxidant co-therapy on nitrate-induced side effects (Fig. 10). Examples are direct antioxidants such as hydralazine (Daiber et al., 2005a, 2005b; Elkayam, 1996; Gogia et al., 1995; Munzel et al., 1996b; Taylor et al., 2004), vitamin C (Bassenge et al., 1998; Daniel & Nawarskas, 2000; Dikalov et al., 1999; Watanabe et al., 1998c), superoxide dismutase (Daiber et al., 2005c; Munzel et al., 1995c, 1996b, 1999), *N*-acetylcysteine (Fung et al., 1989; Ghio et al., 1992; Newman et al., 1990; Packer et al., 1987), ebselen, and uric acid (Hink et al., 2003). It should be noted that several reports also challenged the concept of nitrate tolerance release by hydralazine (Parker et al., 1997), *N*-acetylcysteine (Munzel et al., 1989; Schroder et al., 1988), and SOD (Sage et al., 2000; Wolkart et al., 2008). The antioxidant dihydrolipoic acid probably improves nitrate tolerance by direct antioxidant effects (Moini et al., 2002; Wollin & Jones, 2003), its beneficial effects on ALDH-2 activity (Dudek et al., 2008; Wenzel et al., 2007a) and maybe additional pleiotropic effects (e.g., via cofactor functions in other biological processes; Stoyanovsky et al., 2005; Wollin & Jones, 2003).

B. Indirect Antioxidants in the Prevention of Nitrate Tolerance: Pleiotropic Effects and Synergistic Improvement of Vascular Function

In addition to direct antioxidants, a number of drugs with indirect antioxidant effects were described to efficiently improve GTN-induced nitrate tolerance and endothelial dysfunction (Fig. 10). Among these compounds are statins

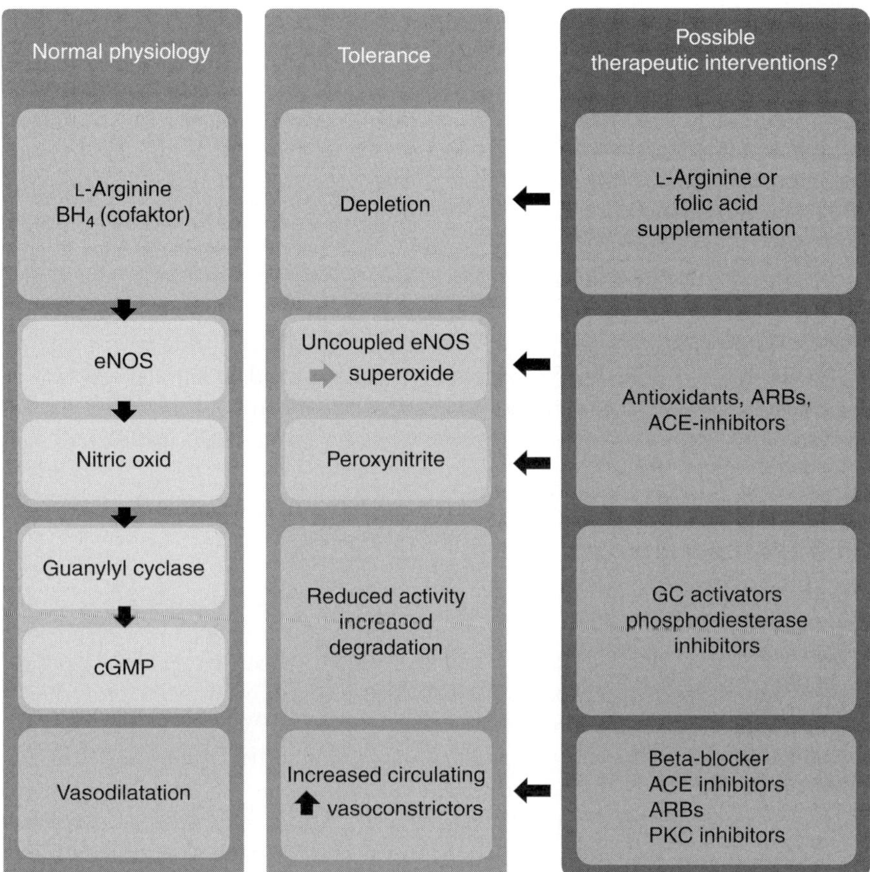

FIGURE 10 Nitrate tolerance and possible therapeutic interventions. The left side indicates constituents of the NO/cGMP signal transduction: functional endothelial nitric oxide synthase (eNOS) with its cofactor tetrahydrobiopterin (BH_4) and substrate L-arginine. Nitrate tolerance causes depletion of BH_4 (by oxidation and decreased *de novo* synthesis) and eventually L-arginine (e.g., degradation to ADMA). Moreover, superoxide and peroxynitrite formation (by NADPH oxidases and mitochondrial sources) as well as eNOS uncoupling (by BH_4 depletion or oxidative modification of eNOS itself) are postulated in tolerant tissue contributing to inactivation of sGC (by superoxide and peroxynitrite) and increased degradation of cGMP by phosphodiesterases. All of these factors as well as humoral counterregulation trigger the increase in circulating vasoconstrictors (activation of RAAS) characterizing the "pseudo-tolerance." Possible co-therapies to suppress the nitrate tolerance compromise induction of eNOS activity by supplementation with BH_4 or its precursor folic acid or the substrate L-arginine (displaying also direct antioxidant effects). Also direct or indirect antioxidant therapies (e.g., ascorbate, hydralazine, AT_1-receptor blockers (ARBs), and ACE inhibitors). sGC activators and PDE inhibitors, to our best knowledge, were not yet tested in the setting of nitrate tolerance. The antioxidant co-therapy normalizes also the levels of circulating vasoconstrictors. The original concept was published by Fayers et al. (2003).

(Fontaine et al., 2003; Inoue et al., 2003; Otto et al., 2005), AT_1-receptor antagonists (Cotter et al., 1998; Hirai et al., 2003; Kurz et al., 1999), ACE inhibitors (Berkenboom et al., 1999; Kurz et al., 1999; Otto et al., 2006; Warnholtz et al., 2002b), PKC inhibitors (Abou-Mohamed et al., 2004; Munzel & Harrison, 1997; Munzel et al., 1995a; Zierhut & Ball, 1996), folic acid (Gori et al., 2001; Gori et al., 2003b), or the β-blocker carvedilol (El-Demerdash, 2006; Fink et al., 1999; Watanabe et al., 1998a). It should be noted that several reports also challenged the concept of nitrate tolerance release by sartans (Longobardi et al., 2004; Milone et al., 1999), NADPH oxidase inhibitor apocynin (Ratz et al., 2000b), or endothelin antagonists (Ratz et al., 2000a). More examples for antioxidant improvement of nitrate tolerance and GTN-induced endothelial dysfunction are summarized in our review article (Munzel et al., 2005) and different strategies to overcome nitrate tolerance are discussed in a previous review article (Schwemmer & Bassenge, 2003).

C. Induction of Protective Genes and Nitrate Tolerance: Is PETN the Better Nitrate?

The important role of oxidative stress for vascular side effects of GTN therapy was also demonstrated in genetically modified mice. The suppression of the activity of the NADPH oxidase isoform Nox2 by genetic deletion of gp91phox and p47phox could suppress the GTN-induced endothelial dysfunction but had no effect on nitrate tolerance (Wenzel et al., 2008a). In contrast, mice with partial deficiency in mitochondrial superoxide dismutase ($MnSOD^{+/-}$) displayed increased susceptibility for the development of nitrate tolerance, endothelial dysfunction, and increased oxidative stress (Daiber et al., 2005c). Interestingly, the nitrate PETN does not induce *in vivo* tolerance and other adverse side effects (Table II). The tolerance-devoid therapy with PETN and its antiatherogenic properties is probably based on its controlled and slow uptake resulting in special pharmacokinetics as well as the distribution of hemodynamic effects on four vasoactive metabolites (Koenig et al., 2007; Wenzel et al., 2009). At least two of these metabolites (PEDN and PEMN) are present at high-plasma levels and do not induce tachyphylaxis. But even more important could be the induction of protective genes by PETN and its trinitrate such as the heme oxygenase-1 and ferritin *in vitro* (Gori et al., 2010; Oberle et al., 1999, 2002, 2003) and *in vivo* (Fig. 11; Daiber & Munzel, 2010a; Schuhmacher et al., 2010; Wenzel et al., 2007b). Via breakdown of porphyrins, HO-1 produces the antioxidant molecule bilirubin (which is formed from biliverdin by biliverdin reductase; Florczyk et al., 2008) and the vasodilator carbon monoxide (CO) (Oberle et al., 2003; Wenzel et al., 2007b). HO-1 in turn stimulates the expression of a second antioxidant protein, ferritin, via the HO-1-dependent release of free iron from endogenous heme sources (Oberle et al., 1999). The induction of extracellular SOD may represent another important intrinsic antioxidant pathway activated by PETN

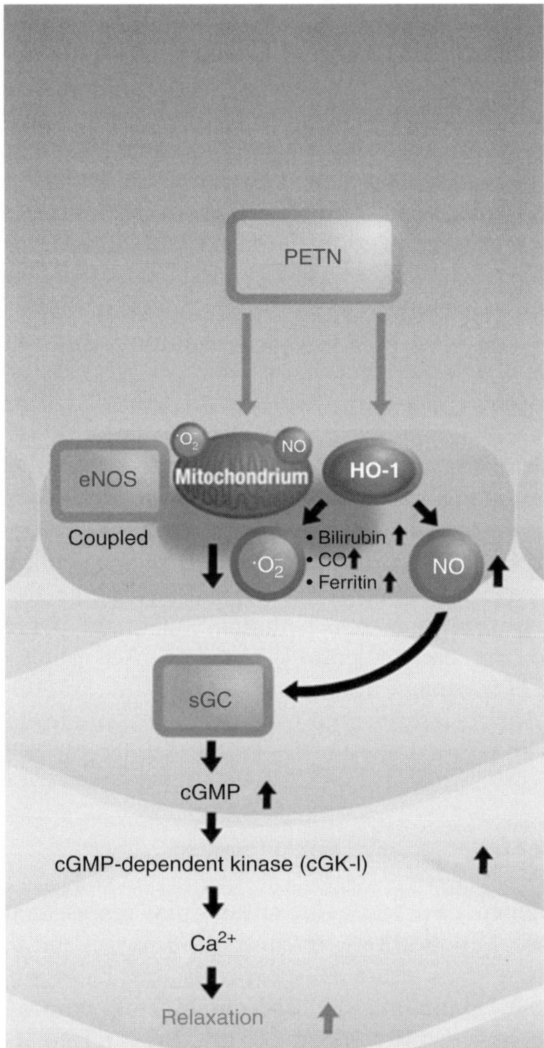

FIGURE 11 Prevention of nitrate tolerance and endothelial dysfunction by PETN. In contrast to GTN, PETN stimulates the expression of the protective protein heme oxygenase-1 (HO-1), and thus triggers formation of the potent antioxidant bilirubin as well as the weak, antiatherosclerotic activator of soluble guanylyl cyclase (sGC), carbon monoxide (CO). As a consequence of HO-1-dependent degradation of metalloporphyrins and increase in free iron, the expression of ferritin is increased providing protection against iron-induced Fenton toxicity. Activation of these intrinsic antioxidative pathways prevents eNOS uncoupling, mitochondrial oxidative stress, inactivation of ALDH-2, thereby maintaining the vasodilatory action of PETN as well as endothelial function. Adapted from Daiber and Munzel (2010a). With permission of Springer (via rightslink).

in vivo treatment (Oppermann et al., 2009) and may explain its beneficial effects in hyperlipidemia (Hacker et al., 2001; Kojda et al., 1995) and arterial hypertension (Schuhmacher et al., 2010). Moreover, a recent study by Kleinert and coworkers suggests that PETN regulates 1215 genes and GTN 532 genes but only 68 genes are regulated by both nitrates (Pautz et al., 2009). A subsequent detailed analysis of gene function revealed that GTN activates a cardiotoxic phenotype, whereas PETN displays a cardioprotective gene induction profile. These data are in good agreement with previous reports on GTN-induced adverse gene regulation (e.g., downregulation of potential GTN bioactivating P450 genes; Minamiyama et al., 2001, 2002, 2004a,2004b, 2006).

These protective properties of PETN manifest in prevention of ALDH-2 inactivation and lack of induction of oxidative stress and nitrate tolerance (Tables II–IV). Moreover, they may explain why PETN mimics ischemic preconditioning (Dragoni et al., 2007). The powerful effect of HO-1 induction on the development of nitrate tolerance was further supported by key observations on the normalization of nitroglycerin-induced nitrate tolerance by hemin (a potent HO-1 inducer) co-therapy and the induction of a tolerance-like phenomenon in PETN-treated rats by apigenin (an HO-1 suppressor) co-treatment (Wenzel et al., 2007b). These results point to a crucial role of this enzyme in the modulation of the degree of tolerance in response to the use of organic nitrates. According to data from our group, PETN treatment induces nitrate tolerance in HO-1$^{+/-}$ mice and low dose nitroglycerin treatment induces severe loss of nitroglycerin potency in these mice with partial deficiency in HO-1 (Schuhmacher et al., 2010).

VI. New Nitrates

The development of new organic nitrates may represent another strategy to find new clinical applications for this class of drugs and to overcome the major limitation of clinical use of organic nitrates. Based on results from a recent study from Lehmann and Daiber, aminoethyl nitrate (AEN) showed an almost similar potency as compared to glyceryl trinitrate (GTN), although being only a mononitrate (Schuhmacher et al., 2009). In contrast to triethanolamine trinitrate (TEAN) and GTN, AEN bioactivation did not depend on ALDH-2 and caused no *in vitro* tolerance. *In vivo* treatment with TEAN and GTN, but not with AEN induced cross-tolerance to acetylcholine-dependent or to GTN-dependent relaxation. Although all nitrates tested induced tolerance, only TEAN and GTN significantly increased mitochondrial oxidative stress *in vitro* and *in vivo*. The results of this study demonstrate that not all high potency nitrates are bioactivated by ALDH-2 and that high potency of a given nitrate is not necessarily associated with induction of oxidative stress or nitrate tolerance. Obviously, there are distinct pathways for bioactivation of organic nitrates with high potency.

AEN may represent a new class of organic nitrates which is devoid of induction of *in vitro* tachyphylaxis as well as oxidative stress and showed an impressively high potency as compared to all tested mononitrates so far but also to di- and trinitrates. Despite these beneficial and pharmacological relevant properties, AEN induced severe *in vivo* tolerance to itself requiring further mechanistic studies to reveal the bioactivation pathway as well as its mechanism of action—the so far known facts are summarized in Fig. 12. This may be of clinical interest since AEN is part of the structure of the potassium channel blocker nicorandil which consists of a fused organic nitrate moiety. Nicorandil is devoid of clinical tolerance (Sekiya et al., 2005). Therefore, future strategies to develop organic nitrate-based vasodilators and hybrid molecules may include the insertion of AEN-like structures.

The development of new nitrate-hybrid molecules, introduction of nitrate functions into established cardiovascular drugs represents another interesting field of nitrate development (Martelli et al., 2006). There are several reports in the literature on nitrate-hybrid molecules: Nicorandil, a potassium channel blocker with nitrate function (Imai et al., 1983; Sakai et al., 1983); 2NTX-99, a thromboxane synthase inhibitor and thromboxane-receptor-antagonist (Buccellati et al., 2006); GT-094, an NSAID with NO function (Hagos et al., 2008); the antifungal drug ketoconazole with a diazen-1-ium-1,2-diolate or an organic nitrate moiety (Konter et al., 2008); NO-donor-tacrine hybrids as hepatoprotective anti-Alzheimer drug candidates (Fang et al., 2008); the NicOx compound nitroaspirin, reviewed in (Gresele & Momi, 2006); NO-releasing celecoxib analogs, inhibitors of inducible cyclooxygenase (Chowdhury et al., 2010); and a vitamin E analog with NO-donor function (Lopez et al., 2007). Probably, we could not list all examples that have been published during the past years but our list provides an overview of the diversity of NO-hybrids. At present, numerous established pharmaceutical enterprises and start-up companies are interested in the development of new hybrid molecules with NO-releasing properties. The idea behind this research is to develop new beneficial functions of established drugs or to improve known principles of action but also to suppress side effects of these drugs.

VII. Conclusion, Outlook, and Clinical Relevance

Organic nitrates should not be considered as a homogenous exchangeable class of vasodilators since they display a considerable diversity (Gori & Daiber, 2009). Especially, GTN induces clinical tolerance, oxidative stress, and endothelial dysfunction, side effects which are shared to more or less extent with the other nitrates, ISMN and ISDN. ISMN is more likely to induce endothelial dysfunction. In contrast, PETN is obviously devoid of these adverse side effects. GTN and PETN, both are bioactivated by the

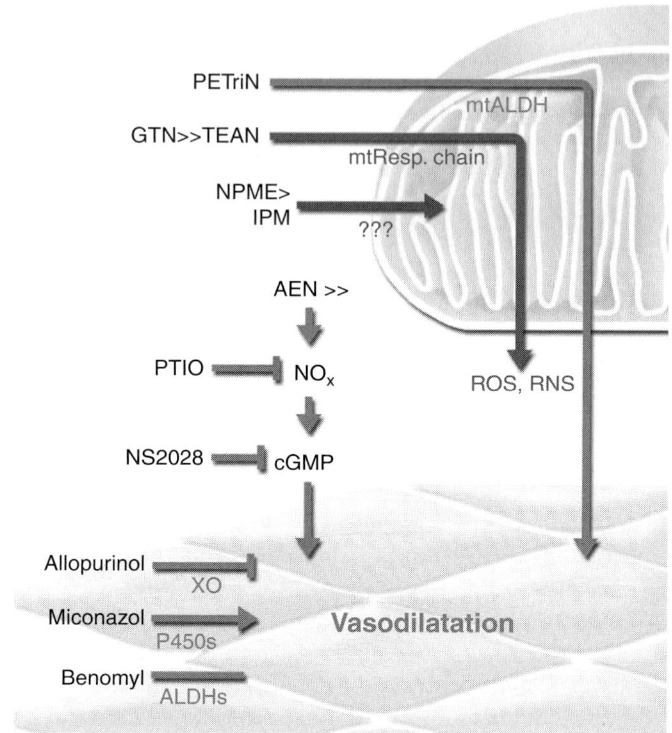

FIGURE 12 Bioactivation, mitochondrial effects, and mechanisms of vasodilatation induced by aminoethyl nitrate (AEN) and other organic nitrates (Schuhmacher et al., 2009). Both the highly potent nitroglycerin (GTN) and the less potent TEAN are bioactivated by mitochondrial aldehyde dehydrogenase (mtALDH) and produce considerable amounts of mitochondrial reactive oxygen and nitrogen species (ROS/RNS, most probably peroxynitrite). In contrast, the trinitrate metabolite of pentaerithrityl tetranitrate, PETriN, also bioactivated by the mtALDH, did not induce mitochondrial oxidative stress. None of the mononitrates were bioactivated by mtALDH. Nevertheless, AEN was almost as potent as GTN and much more potent than methyl-3-nitrooxypropanoate (NPME), which showed significantly higher vasodilator potency than isopropyl nitrate (IPM). None of the mononitrates increased ROS/RNS levels in isolated mitochondria. Whether AEN, NPME, and IPM undergo mitochondrial metabolism remains to be established. AEN-induced vasodilatation was attenuated by the NO scavenger 2-phenyl-4,4,5,5-tetramethylimidazoline-1-oxyl 3-oxide (PTIO) and the inhibitor of soluble guanylate cyclase, NS2028. The lack of effect of miconazol on AEN-induced relaxation makes it unlikely that AEN was metabolized by cytochrome P450 enzymes. Attenuation of AEN-induced relaxation by allopurinol indicated that XO may be involved in bioactivation of AEN. The original concept was published by Bauersachs (2009) and Schuhmacher et al. (2009).

ALDH-2 and GTN but not PETN cause inactivation of this bioactivating pathway. Oxidative stress plays an important role for the development of nitrate tolerance and endothelial dysfunction although the extent and the cellular source of this oxidative stress differ markedly among the ISMN, ISDN, and GTN. In this regard, co-treatment with direct and indirect antioxidants (e.g., ACE inhibitors, statins, hydralazine, or lipoic acid) provides a significant step forward since there is not only good evidence for a release of nitrate-induced side effects by this antioxidant co-therapy but also mechanistic explanation for this beneficial effect: toxic side effects like tolerance, rebound ischemia, and endothelial dysfunction are associated with oxidative stress. Without antioxidant co-therapy, organic nitrates have the potential to evoke clinical-relevant side effects, which could justify the development of new organic nitrates with intrinsic antioxidant properties. Recently, we and Lehmann have characterized a new aminoalkylnitrate, which provided a good compromise between induction of nitrate tolerance and vasodilatory potency. Another strategy is based on the development of hybrid molecules (e.g., combination of NO-releasing function and NSAID in one molecule, see nitroaspirin). There are multiple examples for development of these hybrid molecules, which could provide the basis for a nitrate with intrinsic antioxidant properties and enhancement of therapeutic effects of a given drug.

Acknowledgments

We thank Thilo Weckmüller and Margot Neuser for graphical support. We are indebted to the University Medical Center of the Johannes Gutenberg-University for financial support by MAIFOR and Förderfonds grants (A. D.) as well as Actavis Deutschland GmbH (A. D. and T. G.) for long-lasting support of our research on organic nitrates. Parts of this review article are based on our recent German review article with permission of Wiley-VCH Verlag GmbH (Daiber et al., 2010c).

Conflict of Interest: A. D. and T. G. received honoraries and research support from Actavis Deutschland GmbH (Langenfeld, Germany).

Abbreviations

AEN	aminoethyl nitrate
ALDH-2	mitochondrial aldehyde dehydrogenase
DTT	dithiothreitol
1,2-GDN	1,2-glyceryl dinitrate
GTN	glyceryl trinitrate (nitroglycerin)
HO-1	heme oxygenase 1
ISDN	isosorbide dinitrate
ISMN	isosorbide-5-mononitrate
MnSOD	manganese superoxide dismutase (SOD2, mitochondrial isoform)
PEDN	pentaerithrityl dinitrate
PEMN	pentaerithrityl mononitrate
PETN	pentaerithrityl tetranitrate
PETriN	pentaerithrityl trinitrate

RONS	reactive oxygen and nitrogen species
ROS	reactive oxygen species
sGC	soluble guanylyl cyclase
SOD	superoxide dismutase
SPE/NO	spermine NONOate
TEAN	triethanolamine trinitrate
XO	xanthine oxidase

References

Abou-Mohamed, G., Johnson, J. A., Jin, L., El-Remessy, A. B., Do, K., Kaesemeyer, W. H., et al. (2004). Roles of superoxide, peroxynitrite, and protein kinase C in the development of tolerance to nitroglycerin. *The Journal of Pharmacology and Experimental Therapeutics*, *308*(1), 289–299.

Abrams, J. (1988). A reappraisal of nitrate therapy. *Journal of the American Medical Association*, *259*, 396–401.

Abrams, J. (1995). The role of nitrates in coronary heart disease. *Archives of Internal Medicine*, *155*(4), 357–364.

Ahlner, J., Axelsson, K. L., Ekstram-Ljusegren, M., Friedman, R. L., Grundstrom, N., Karlsson, J. O., et al. (1988). Relaxation of bovine mesenteric artery induced by glyceryl trinitrate is attenuated by pertussis toxin. *Pharmacology & Toxicology*, *62*(3), 155–158.

Ahlner, J., Axelsson, K. L., Ljusegren, M. E., Grundstrom, N., & Andersson, R. G. (1986). Demonstration of a high affinity component of glyceryl trinitrate induced vasodilatation in the bovine mesenteric artery. *Journal of Cyclic Nucleotide and Protein Phosphorylation Research*, *11*(6), 445–456.

Ambrosio, G., Del Pinto, M., Tritto, I., Agnelli, G., Bentivoglio, M., Zuchi, C., et al. (2010). Chronic nitrate therapy is associated with different presentation and evolution of acute coronary syndromes: Insights from 52, 693 patients in the Global Registry of Acute Coronary Events. *European Heart Journal*, *31*(4), 430–438.

Andreassi, M. G., Botto, N., Simi, S., Casella, M., Manfredi, S., Lucarelli, M., Venneri, L., Biagini, A., & Picano, E. (2005). Diabetes and chronic nitrate therapy as co-determinants of somatic DNA damage in patients with coronary artery disease. *Journal of Molecular Medicine*, *83*, 279–286.

Axelsson, K. L., Andersson, C., Ahlner, J., Magnusson, B., & Wikberg, J. E. (1992). Comparative in vitro study of a series of organic nitroesters: Unique biphasic concentration-effect curves for glyceryl trinitrate in isolated bovine arterial smooth muscle and lack of stereoselectivity for some glyceryl trinitrate analogues. *Journal of Cardiovascular Pharmacology*, *19*(6), 953–957.

Bassenge, E., Fink, N., Skatchkov, M., & Fink, B. (1998). Dietary supplement with vitamin C prevents nitrate tolerance. *Journal of Clinical Investigation*, *102*(1), 67–71.

Bauersachs, J. (2009). Aminoethyl nitrate—The novel super nitrate? *British Journal of Pharmacology*, *158*(2), 507–509.

Beckman, J. S., & Koppenol, W. H. (1996). Nitric oxide, superoxide, and peroxynitrite: The good, the bad, and ugly. *The American Journal of Physiology*, *271*(5 Pt 1), C1424–C1437.

Bennett, B. M., Kobus, S. M., Brien, J. F., Nakatsu, K., & Marks, G. S. (1986). Requirement for reduced, unliganded hemoprotein for the hemoglobin- and myoglobin-mediated biotransformation of glyceryl trinitrate. *The Journal of Pharmacology and Experimental Therapeutics*, *237*(2), 629–635.

Bennett, B. M., Nakatsu, K., Brien, J. F., & Marks, G. S. (1984). Biotransformation of glyceryl trinitrate to glyceryl dinitrate by human hemoglobin. *Canadian Journal of Physiology and Pharmacology*, *62*(6), 704–706.

Beretta, M., Gorren, A. C., Wenzl, M. V., Weis, R., Russwurm, M., Koesling, D., et al. (2010). Characterization of the East Asian variant of aldehyde dehydrogenase-2: Bioactivation of nitroglycerin and effects of Alda-1. *The Journal of Biological Chemistry*, *285*(2), 943–952.

Beretta, M., Gruber, K., Kollau, A., Russwurm, M., Koesling, D., Goessler, W., et al. (2008a). Bioactivation of nitroglycerin by purified mitochondrial and cytosolic aldehyde dehydrogenases. *The Journal of Biological Chemistry*, *283*(26), 17873–17880.

Beretta, M., Sottler, A., Schmidt, K., Mayer, B., & Gorren, A. C. (2008b). Partially irreversible inactivation of mitochondrial aldehyde dehydrogenase by nitroglycerin. *The Journal of Biological Chemistry*, *283*(45), 30735–30744.

Berkenboom, G., Fontaine, D., Unger, P., Baldassarre, S., Preumont, N., & Fontaine, J. (1999). Absence of nitrate tolerance after long-term treatment with ramipril: An endothelium-dependent mechanism. *Journal of Cardiovascular Pharmacology*, *34*(4), 547–553.

Bilska, A., & Wlodek, L. (2005). Lipoic acid—The drug of the future? *Pharmacological Reports*, *57*(5), 570–577.

Brandes, R. P. (2005). Triggering mitochondrial radical release: A new function for NADPH oxidases. *Hypertension*, *45*(5), 847–848.

Buccellati, C., Sala, A., Rossoni, G., Capra, V., Rovati, G. E., Di Gennaro, A., et al. (2006). Pharmacological characterization of 2NTX-99 [4-methoxy-N1-(4-trans-nitrooxycyclohexyl)-N3-(3-pyridinylmethyl)-1, 3-benz enedicarboxamide], a potential antiatherothrombotic agent with antithromboxane and nitric oxide donor activity in platelet and vascular preparations. *The Journal of Pharmacology and Experimental Therapeutics*, *317*(2), 830–837.

Butler, A. R., & Feelisch, M. (2008). Therapeutic uses of inorganic nitrite and nitrate: From the past to the future. *Circulation*, *117*(16), 2151–2159.

Caramori, P. R., Adelman, A. G., Azevedo, E. R., Newton, G. E., Parker, A. B., & Parker, J. D. (1998). Therapy with nitroglycerin increases coronary vasoconstriction in response to acetylcholine. *Journal of the American College of Cardiology*, *32*(7), 1969–1974.

Chen, C. H., Budas, G. R., Churchill, E. N., Disatnik, M. H., Hurley, T. D., & Mochly-Rosen, D. (2008). Activation of aldehyde dehydrogenase-2 reduces ischemic damage to the heart. *Science*, *321*(5895), 1493–1495.

Chen, Z., Foster, M. W., Zhang, J., Mao, L., Rockman, H. A., Kawamoto, T., et al. (2005). An essential role for mitochondrial aldehyde dehydrogenase in nitroglycerin bioactivation. *Proceedings of the National Academy of Sciences of the United States of America*, *102* (34), 12159–12164.

Chen, Z., & Stamler, J. S. (2006). Bioactivation of nitroglycerin by the mitochondrial aldehyde dehydrogenase. *Trends in Cardiovascular Medicine*, *16*(8), 259–265.

Chen, Z., Zhang, J., & Stamler, J. S. (2002). Identification of the enzymatic mechanism of nitroglycerin bioactivation. *Proceedings of the National Academy of Sciences of the United States of America*, *99*(12), 8306–8311.

Chowdhury, M. A., Abdellatif, K. R., Dong, Y., Yu, G., Huang, Z., Rahman, M., et al. (2010). Celecoxib analogs possessing a *N*-(4-nitrooxybutyl)piperidin-4-yl or *N*-(4-nitrooxybutyl)-1, 2, 3, 6-tetrahydropyridin-4-yl nitric oxide donor moiety: Synthesis, biological evaluation and nitric oxide release studies. *Bioorganic & Medicinal Chemistry Letters*, *20*(4), 1324–1329.

Chung, S. J., & Fung, H. L. (1992). A common enzyme may be responsible for the conversion of organic nitrates to nitric oxide in vascular microsomes. *Biochemical and Biophysical Research Communications*, *185*(3), 932–937.

Cotter, G., Metzkor-Cotter, E., Kaluski, E., Blatt, A., Litinsky, I., Baumohl, Y., et al. (1998). Usefulness of losartan, captopril, and furosemide in preventing nitrate tolerance and improving control of unstable angina pectoris. *The American Journal of Cardiology*, *82* (9), 1024–1029.

Crow, J. P., & Beckman, J. S. (1995). Reaction between nitric oxide, superoxide, and peroxynitrite: Footprints of peroxynitrite in vivo. [Book]. *Advances in Pharmacology*, *35*, 17–43.

Daiber, A. (2010). Redox signaling (cross-talk) from and to mitochondria involves mitochondrial pores and reactive oxygen species. *Biochim Biophys Acta*, *1797*, 897–906.

Daiber, A., Gori, T., & Münzel, T. (2010c). Mechanismen und klinischer Stellenwert der Nitrattoleranz. Diversität und Nebenwirkungen organischer Nitrate. *Pharmazie in Unserer Zeit*, *39*, 375–384.

Daiber, A., August, M., Baldus, S., Wendt, M., Oelze, M., Sydow, K., et al. (2004a). Measurement of NAD(P)H oxidase-derived superoxide with the luminol analogue L-012. *Free Radical Biology & Medicine*, *36*(1), 101–111.

Daiber, A., & Bachschmid, M. (2007). Enzyme inhibition by peroxynitrite-mediated tyrosine nitration and thiol oxidation. *Current Enzyme Inhibition*, *3*, 103–117.

Daiber, A., & Gori, T. (2008). Vascular tolerance to nitroglycerin in ascorbate deficiency—Results are in favor of an important role of oxidative stress in nitrate tolerance. *Cardiovascular Research*, *79*, 722–723.

Daiber, A., Mulsch, A., Hink, U., Mollnau, H., Warnholtz, A., Oelze, M., et al. (2005a). The oxidative stress concept of nitrate tolerance and the antioxidant properties of hydralazine. *The American Journal of Cardiology*, *96*(7B), 25i–36i.

Daiber, A., & Munzel, T. (2010a). Characterization of the antioxidant properties of pentaerithrityl tetranitrate (PETN)-induction of the intrinsic antioxidative system heme oxygenase-1 (HO-1). *Methods in Molecular Biology*, *594*, 311–326.

Daiber, A., & Munzel, T. (2010b). Nitrate reductase activity of mitochondrial aldehyde dehydrogenase (ALDH-2) as a redox sensor for cardiovascular oxidative stress. *Methods in Molecular Biology*, *594*, 43–55.

Daiber, A., & Münzel, T. (2006). *Oxidativer Stress.* " Steinkopff Verlag Darmstadt, Darmstadt, Germany: Redoxregulation und NO-Bioverfügbarkeit—Experimentelle und klinische Aspekte.

Daiber, A., Oelze, M., August, M., Wendt, M., Sydow, K., Wieboldt, H., et al. (2004b). Detection of superoxide and peroxynitrite in model systems and mitochondria by the luminol analogue L-012. *Free Radical Research*, *38*(3), 259–269.

Daiber, A., Oelze, M., Coldewey, M., Bachschmid, M., Wenzel, P., Sydow, K., et al. (2004c). Oxidative stress and mitochondrial aldehyde dehydrogenase activity: A comparison of pentaerythritol tetranitrate with other organic nitrates. *Molecular Pharmacology*, *66*(6), 1372–1382.

Daiber, A., Oelze, M., Coldewey, M., Kaiser, K., Huth, C., Schildknecht, S., et al. (2005b). Hydralazine is a powerful inhibitor of peroxynitrite formation as a possible explanation for its beneficial effects on prognosis in patients with congestive heart failure. *Biochemical and Biophysical Research Communications*, *338*(4), 1865–1874.

Daiber, A., Oelze, M., Sulyok, S., Coldewey, M., Schulz, E., Treiber, N., et al. (2005c). Heterozygous deficiency of manganese superoxide dismutase in mice (mn-sod+/−): A novel approach to assess the role of oxidative stress for the development of nitrate tolerance. *Molecular Pharmacology*, *68*(3), 579–588.

Daiber, A., Oelze, M., Wenzel, P., Wickramanayake, J. M., Schuhmacher, S., Jansen, T., et al. (2009a). Nitrate tolerance as a model of vascular dysfunction: Roles for mitochondrial aldehyde dehydrogenase and mitochondrial oxidative stress. *Pharmacological Reports*, *61*(1), 33–48.

Daiber, A., & Ullrich, V. (2002). Radikalchemie im organismus: stickstoffmonoxid, superoxid und peroxynitrit. *Chemie in Unserer Zeit*, *36*(6), 366–375.

Daiber, A., Wenzel, P., Oelze, M., & Munzel, T. (2008). New insights into bioactivation of organic nitrates, nitrate tolerance and cross-tolerance. *Clinical Research in Cardiology*, *97* (1), 12–20.

Daiber, A., Wenzel, P., Oelze, M., Schuhmacher, S., Jansen, T., & Munzel, T. (2009b). Mitochondrial aldehyde dehydrogenase (ALDH-2)—Maker of and marker for nitrate tolerance in response to nitroglycerin treatment. *Chemico-Biological Interactions*, *178* (1–3), 40–47.

Daniel, T. A., & Nawarskas, J. J. (2000). Vitamin C in the prevention of nitrate tolerance. *The Annals of Pharmacotherapy*, *34*(10), 1193–1197.

Delaforge, M., Servent, D., Wirsta, P., Ducrocq, C., Mansuy, D., & Lenfant, M. (1993). Particular ability of cytochrome P-450 CYP3A to reduce glyceryl trinitrate in rat liver microsomes: Subsequent formation of nitric oxide. *Chemico-Biological Interactions*, *86*(2), 103–117.

Dikalov, S., Fink, B., Skatchkov, M., & Bassenge, E. (1999). Comparison of glyceryl trinitrate-induced with pentaerythrityl tetranitrate-induced in vivo formation of superoxide radicals: Effect of vitamin C. *Free Radical Biology & Medicine*, *27*(1–2), 170–176.

Doel, J. J., Godber, B. L., Eisenthal, R., & Harrison, R. (2001). Reduction of organic nitrates catalysed by xanthine oxidoreductase under anaerobic conditions. *Biochimica et Biophysica Acta*, *1527*(1–2), 81–87.

Doughan, A. K., Harrison, D. G., & Dikalov, S. I. (2008). Molecular mechanisms of angiotensin II-mediated mitochondrial dysfunction: Linking mitochondrial oxidative damage and vascular endothelial dysfunction. *Circulation Research*, *102*(4), 488–496.

Dragoni, S., Gori, T., Lisi, M., Di Stolfo, G., Pautz, A., Kleinert, H., et al. (2007). Pentaerythrityl tetranitrate and nitroglycerin, but not isosorbide mononitrate, prevent endothelial dysfunction induced by ischemia and reperfusion. *Arteriosclerosis, Thrombosis, and Vascular Biology*, *27*(9), 1955–1959.

Dudek, M., Bednarski, M., Bilska, A., Iciek, M., Sokolowska-Jezewicz, M., Filipek, B., et al. (2008). The role of lipoic acid in prevention of nitroglycerin tolerance. *European Journal of Pharmacology*, *591*(1–3), 203–210.

El-Demerdash, E. (2006). Evidences for prevention of nitroglycerin tolerance by carvedilol. *Pharmacological Research*, *53*, 380–385.

Elkayam, U. (1991). Tolerance to organic nitrates: Evidence, mechanisms, clinical relevance, and strategies for prevention. *Annals of Internal Medicine*, *114*, 667–677.

Elkayam, U. (1996). Prevention of nitrate tolerance with concomitant administration of hydralazine. *The Canadian Journal of Cardiology*, *12*(Suppl. C), 17C–21C.

Elkayam, U., & Bitar, F. (2005). Effects of nitrates and hydralazine in heart failure: Clinical evidence before the african american heart failure trial. *The American Journal of Cardiology*, *96*(7B), 37i–43i.

Esplugues, J. V., Rocha, M., Nunez, C., Bosca, I., Ibiza, S., Herance, J. R., et al. (2006). Complex I dysfunction and tolerance to nitroglycerin: An approach based on mitochondrial-targeted antioxidants. *Circulation Research*, *99*(10), 1067–1075.

Fang, L., Appenroth, D., Decker, M., Kiehntopf, M., Roegler, C., Deufel, T., et al. (2008). Synthesis and biological evaluation of NO-donor-tacrine hybrids as hepatoprotective anti-Alzheimer drug candidates. *Journal of Medicinal Chemistry*, *51*(4), 713–716.

Fayers, K. E., Cummings, M. H., Shaw, K. M., & Laight, D. W. (2003). Nitrate tolerance and the links with endothelial dysfunction and oxidative stress. *British Journal of Clinical Pharmacology*, *56*(6), 620–628.

Feelisch, M., Noack, E., & Schroder, H. (1988). Explanation of the discrepancy between the degree of organic nitrate decomposition, nitrite formation and guanylate cyclase stimulation. *European Heart Journal*, *9*(Suppl. A), 57–62.

Fink, B., & Bassenge, E. (1997). Unexpected, tolerance-devoid vasomotor and platelet actions of pentaerythrityl tetranitrate. *Journal of Cardiovascular Pharmacology*, *30*(6), 831–836.

Fink, B., Schwemmer, M., Fink, N., & Bassenge, E. (1999). Tolerance to nitrates with enhanced radical formation suppressed by carvedilol. *Journal of Cardiovascular Pharmacology*, *34* (6), 800–805.

Florczyk, U. M., Jozkowicz, A., & Dulak, J. (2008). Biliverdin reductase: New features of an old enzyme and its potential therapeutic significance. *Pharmacological Reports*, *60*(1), 38–48.

Fontaine, D., Otto, A., Fontaine, J., & Berkenboom, G. (2003). Prevention of nitrate tolerance by long-term treatment with statins. *Cardiovascular Drugs and Therapy*, *17*(2), 123–128.

Fraker, T. D., Jr., Fihn, S. D., Gibbons, R. J., Abrams, J., Chatterjee, K., Daley, J., et al. (2007). 2007 chronic angina focused update of the ACC/AHA 2002 Guidelines for the management of patients with chronic stable angina: A report of the American College of Cardiology/American Heart Association Task Force on Practice Guidelines Writing Group to develop the focused update of the 2002 Guidelines for the management of patients with chronic stable angina. *Circulation*, *116*(23), 2762–2772.

Freedman, S. B., Daxini, B. V., Noyce, D., & Kelly, D. T. (1995). Intermittent transdermal nitrates do not improve ischemia in patients taking beta-blockers or calcium antagonists: Potential role of rebound ischemia during the nitrate-free period. *Journal of the American College of Cardiology*, *25*(2), 349–355.

Fujii, T., Adachi, T., Usami, Y., Tatematsu, M., & Hirano, K. (1996). Variable glyceryl dinitrate formation as a function of glutathione *S*-transferase. *Biological & Pharmaceutical Bulletin*, *19*(8), 1093–1096.

Fukatsu, A., Hayashi, T., Miyazaki-Akita, A., Matsui-Hirai, H., Furutate, Y., Ishitsuka, A., et al. (2007). Possible usefulness of apocynin, an NADPH oxidase inhibitor, for nitrate tolerance: Prevention of NO donor-induced endothelial cell abnormalities. *American Journal of Physiology. Heart and Circulatory Physiology*, *293*(1), H790–H797.

Fukui, T., Ishizaka, N., Rajagopalan, S., Laursen, J. B., Capers, Q. T., Taylor, W. R., et al. (1997). p22phox mRNA expression and NADPH oxidase activity are increased in aortas from hypertensive rats. *Circulation Research*, *80*(1), 45–51.

Fung, H. L. (2004). Biochemical mechanism of nitroglycerin action and tolerance: Is this old mystery solved? *Annual Review of Pharmacology and Toxicology*, *44*, 67–85.

Fung, H. L., Chong, S., Kowaluk, E., Hough, K., & Kakemi, M. (1988). Mechanisms for the pharmacologic interaction of organic nitrates with thiols. Existence of an extracellular pathway for the reversal of nitrate vascular tolerance by *N*-acetylcysteine. *The Journal of Pharmacology and Experimental Therapeutics*, *245*(2), 524–530.

Fung, H. L., Chung, S. J., Chong, S., Hough, K., Kakami, M., & Kowaluk, E. (1989). Cellular mechanisms of nitrate action. *Zeitschrift für Kardiologie*, *78*(Suppl. 2), 14–17, discussion 64-17.

Garcia, J. I., Seabra, A. B., Kennedy, R., & English, A. M. (2010). Nitrite and nitroglycerin induce rapid release of the vasodilator ATP from erythrocytes: Relevance to the chemical physiology of local vasodilation. *Journal of Inorganic Biochemistry*, *104* (3), 289–296.

Ghio, S., de Servi, S., Perotti, R., Eleuteri, E., Montemartini, C., & Specchia, G. (1992). Different susceptibility to the development of nitroglycerin tolerance in the arterial and venous circulation in humans. Effects of *N*-acetylcysteine administration. *Circulation*, *86* (3), 798–802.

Gogia, H., Mehra, A., Parikh, S., Raman, M., Ajit-Uppal, J., Johnson, J. V., et al. (1995). Prevention of tolerance to hemodynamic effects of nitrates with concomitant use of hydralazine in patients with chronic heart failure. *Journal of the American College of Cardiology*, *26*(7), 1575–1580.

Gori, T., Al-Hesayen, A., Jolliffe, C., & Parker, J. D. (2003a). Comparison of the effects of pentaerythritol tetranitrate and nitroglycerin on endothelium-dependent vasorelaxation in male volunteers. *The American Journal of Cardiology*, *91*(11), 1392–1394.

Gori, T., Burstein, J. M., Ahmed, S., Miner, S. E., Al-Hesayen, A., Kelly, S., et al. (2001). Folic acid prevents nitroglycerin-induced nitric oxide synthase dysfunction and nitrate tolerance: A human in vivo study. *Circulation*, *104*(10), 1119–1123.

Gori, T., & Daiber, A. (2009). Non-hemodynamic effects of organic nitrates and the distinctive characteristics of pentaerithrityl tetranitrate. *American Journal of Cardiovascular Drugs*, *9*(1), 7–15.

Gori, T., Daiber, A., Di Stolfo, G., Sicuro, S., Dragoni, S., Lisi, M., et al. (2007). Nitroglycerine causes mitochondrial reactive oxygen species production: In vitro mechanistic insights. *The Canadian Journal of Cardiology*, *23*(12), 990–992.

Gori, T., Dragoni, S., Di Stolfo, G., Sicuro, S., Liuni, A., Luca, M. C., et al. (2010). Tolerance to nitroglycerin-induced preconditioning of the endothelium: A human in vivo study. *American Journal of Physiology. Heart and Circulatory Physiology*, *298*(2), H340–H345.

Gori, T., Floras, J. S., & Parker, J. D. (2002). Effects of nitroglycerin treatment on baroreflex sensitivity and short-term heart rate variability in humans. *Journal of the American College of Cardiology*, *40*(11), 2000–2005.

Gori, T., & Munzel, T. (2009). Symptomatic and hemodynamic benefit of pentaerythrityl tetranitrate and hydralazine in a case of congestive heart failure. *Clinical Research in Cardiology*, *98*(10), 677–679.

Gori, T., Munzel, T., & Daiber, A. (2008). Nicht-hämodynamische Wirkungen organischer Nitrate und unterschiedliche Charakteristika von Pentaerithrityltetranitrat. *Arzneimittelterapie*, *26*, 290–298.

Gori, T., & Parker, J. D. (2008). Nitrate-induced toxicity and preconditioning: A rationale for reconsidering the use of these drugs. *Journal of the American College of Cardiology*, *52*(4), 251–254.

Gori, T., Saunders, L., Ahmed, S., & Parker, J. D. (2003b). Effect of folic acid on nitrate tolerance in healthy volunteers: Differences between arterial and venous circulation. *Journal of Cardiovascular Pharmacology*, *41*(2), 185–190.

Gorren, A. C., Russwurm, M., Kollau, A., Koesling, D., Schmidt, K., & Mayer, B. (2005). Effects of nitroglycerin/L-cysteine on soluble guanylate cyclase: Evidence for an activation/inactivation equilibrium controlled by nitric oxide binding and heme oxidation. *The Biochemical Journal*, *390*(Pt 2), 625–631.

Gresele, P., & Momi, S. (2006). Pharmacologic profile and therapeutic potential of NCX 4016, a nitric oxide-releasing aspirin, for cardiovascular disorders. *Cardiovascular Drug Reviews*, *24*(2), 148–168.

Hacker, A., Muller, S., Meyer, W., & Kojda, G. (2001). The nitric oxide donor pentaerythritol tetranitrate can preserve endothelial function in established atherosclerosis. *British Journal of Pharmacology*, *132*(8), 1707–1714.

Hagos, G. K., Abdul-Hay, S. O., Sohn, J., Edirisinghe, P. D., Chandrasena, R. E., Wang, Z., et al. (2008). Anti-inflammatory, antiproliferative, and cytoprotective activity of NO chimera nitrates of use in cancer chemoprevention. *Molecular Pharmacology*, *74*(5), 1381–1391.

Hink, U., Daiber, A., Kayhan, N., Trischler, J., Kraatz, C., Oelze, M., et al. (2007). Oxidative inhibition of the mitochondrial aldehyde dehydrogenase promotes nitroglycerin tolerance in human blood vessels. *Journal of the American College of Cardiology*, *50*(23), 2226–2232.

Hink, U., Oelze, M., Kolb, P., Bachschmid, M., Zou, M. H., Daiber, A., et al. (2003). Role for peroxynitrite in the inhibition of prostacyclin synthase in nitrate tolerance. *Journal of the American College of Cardiology*, *42*(10), 1826–1834.

Hirai, N., Kawano, H., Yasue, H., Shimomura, H., Miyamoto, S., Soejima, H., et al. (2003). Attenuation of nitrate tolerance and oxidative stress by an angiotensin II receptor blocker in patients with coronary spastic angina. *Circulation*, *108*(12), 1446–1450.

Huellner, M. W., Schrepfer, S., Weyand, M., Weiner, H., Wimplinger, I., Eschenhagen, T., et al. (2008). Inhibition of aldehyde dehydrogenase type 2 attenuates vasodilatory action of nitroglycerin in human veins. *The FASEB Journal*, *22*(7), 2561–2568.

Ignarro, L. J. (1989). Endothelium-derived nitric oxide: Pharmacology and relationship to the actions of organic nitrate esters. *Pharmaceutical Research*, *6*(8), 651–659.

Ikejima, H., Imanishi, T., Tsujioka, H., Kuroi, A., Muragaki, Y., Mochizuki, S., Goto, M., Yoshida, K., & Akasaka, T. (2008). Effect of pioglitazone on nitroglycerin-induced impairment of nitric oxide bioavailability by a catheter-type nitric oxide sensor. *Circulation Journal*, *72*, 998–1002.

Imai, S., Ushijima, T., Nakazawa, M., Nabata, H., & Sakai, K. (1983). Mechanism of relaxant effects of nicorandil on the dog coronary artery. *Archives Internationales de Pharmacodynamie et de Thérapie*, *265*(2), 274–282.

Inoue, T., Takayanagi, K., Hayashi, T., & Morooka, S. (2003). Fluvastatin attenuates nitrate tolerance in patients with ischemic heart disease complicating hypercholesterolemia. *International Journal of Cardiology*, *90*(2–3), 181–188.

Ishikawa, K., Kanamasa, K., Ogawa, I., Takenaka, T., Naito, T., Kamata, N., Yamamoto, T., Nakai, S., Hama, J., Oyaizu, M., Kimura, A., Yamamoto, K., Aso, N., Arai, M., Yabushita, H., & Katori, Y. (1996). Long-term nitrate treatment increases cardiac events in patients with healed myocardial infarction. Secondary Prevention Group. *Japanese Circulation Journal*, *60*, 779–788.

Jakschik, B., & Needleman, P. (1973). Sulfhydryl reactivity of organic nitrates: Biochemical basis for inhibition of glyceraldehyde-P dehydrogenase and monoamine oxidase. *Biochemical and Biophysical Research Communications*, *53*(2), 539–544.

Johnston, G. D. (1998). Ulster says 'NO'; explosion, resistance and tolerance. Nitric oxide and the actions of organic nitrates. *The Ulster Medical Journal*, *67*(2), 79–90.

Jurt, U., Gori, T., Ravandi, A., Babaei, S., Zeman, P., & Parker, J. D. (2001). Differential effects of pentaerythritol tetranitrate and nitroglycerin on the development of tolerance and evidence of lipid peroxidation: A human in vivo study. *Journal of the American College of Cardiology*, *38*(3), 854–859.

Kanamasa, K., Hayashi, T., Kimura, A., Ikeda, A., & Ishikawa, K. (2002). Long-term, continuous treatment with both oral and transdermal nitrates increases cardiac events in healed myocardial infarction patients. *Angiology*, *53*, 399–408.

Katsuki, S., Arnold, W., Mittal, C., & Murad, F. (1977a). Stimulation of guanylate cyclase by sodium nitroprusside, nitroglycerin and nitric oxide in various tissue preparations and comparison to the effects of sodium azide and hydroxylamine. *Journal of Cyclic Nucleotide Research*, *3*(1), 23–35.

Katsuki, S., Arnold, W. P., & Murad, F. (1977b). Effects of sodium nitroprusside, nitroglycerin, and sodium azide on levels of cyclic nucleotides and mechanical activity of various tissues. *Journal of Cyclic Nucleotide Research*, *3*(4), 239–247.

Keimer, R., Stutzer, F. K., Tsikas, D., Troost, R., Gutzki, F. M., & Frölich, J. C. (2003). Lack of oxidative stress during sustained therapy with isosorbide dinitrate and pentaerythrityl tetranitrate in healthy humans: a randomized, double-blind crossover study. *Journal Cardiovascular Pharmacology*, *41*, 284–292.

Kleschyov, A. L., Oelze, M., Daiber, A., Huang, Y., Mollnau, H., Schulz, E., et al. (2003). Does nitric oxide mediate the vasodilator activity of nitroglycerin? *Circulation Research*, *93*(9), e104–e112.

Koenig, A., Lange, K., Konter, J., Daiber, A., Stalleicken, D., Glusa, E., et al. (2007). Potency and in vitro tolerance of organic nitrates: Partially denitrated metabolites contribute to the tolerance-devoid activity of pentaerythrityl tetranitrate. *Journal of Cardiovascular Pharmacology*, *50*(1), 68–74.

Kojda, G., Stein, D., Kottenberg, E., Schnaith, E. M., & Noack, E. (1995). In vivo effects of pentaerythrityl-tetranitrate and isosorbide-5-mononitrate on the development of atherosclerosis and endothelial dysfunction in cholesterol-fed rabbits. *Journal of Cardiovascular Pharmacology*, *25*(5), 763–773.

Kollau, A., Beretta, M., Gorren, A. C., Russwurm, M., Koesling, D., Schmidt, K., et al. (2007). Bioactivation of nitroglycerin by ascorbate. *Molecular Pharmacology*, *72*(1), 191–196.

Kollau, A., Beretta, M., Russwurm, M., Koesling, D., Keung, W. M., Schmidt, K., et al. (2009). Mitochondrial nitrite reduction coupled to soluble guanylate cyclase activation: Lack of evidence for a role in the bioactivation of nitroglycerin. *Nitric Oxide*, *20*(1), 53–60.

Konter, J., Mollmann, U., & Lehmann, J. (2008). NO-donors. Part 17: Synthesis and antimicrobial activity of novel ketoconazole-NO-donor hybrid compounds. *Bioorganic & Medicinal Chemistry*, *16*(17), 8294–8300.

Koppenol, W. H., Moreno, J. J., Pryor, W. A., Ischiropoulos, H., & Beckman, J. S. (1992). Peroxynitrite, a cloaked oxidant formed by nitric oxide and superoxide. *Chemical Research in Toxicology*, *5*(6), 834–842.

Kurz, S., Hink, U., Nickenig, G., Borthayre, A. B., Harrison, D. G., & Munzel, T. (1999). Evidence for a causal role of the renin-angiotensin system in nitrate tolerance. *Circulation*, *99*(24), 3181–3187.

Larson, H. N., Zhou, J., Chen, Z., Stamler, J. S., Weiner, H., & Hurley, T. D. (2007). Structural and functional consequences of coenzyme binding to the inactive asian variant of mitochondrial aldehyde dehydrogenase: Roles of residues 475 and 487. *The Journal of Biological Chemistry*, *282*(17), 12940–12950.

Lau, D. T., & Benet, L. Z. (1990). Nitroglycerin metabolism in subcellular fractions of rabbit liver. Dose dependency of glyceryl dinitrate formation and possible involvement of multiple isozymes of glutathione *S*-transferases. *Drug Metabolism and Disposition*, *18*(3), 292–297.

Li, Y., Zhang, D., Jin, W., Shao, C., Yan, P., Xu, C., et al. (2006). Mitochondrial aldehyde dehydrogenase-2 (ALDH2) Glu504Lys polymorphism contributes to the variation in efficacy of sublingual nitroglycerin. *Journal of Clinical Investigation*, *116*(2), 506–511.

Longobardi, G., Ferrara, N., Leosco, D., Abete, P., Furgi, G., Cacciatore, F., et al. (2004). Angiotensin II-receptor antagonist losartan does not prevent nitroglycerin tolerance in patients with coronary artery disease. *Cardiovascular Drugs and Therapy*, *18*(5), 363–370.

Lopez, G. V., Blanco, F., Hernandez, P., Ferreira, A., Piro, O. E., Batthyany, C., et al. (2007). Second generation of alpha-tocopherol analogs-nitric oxide donors: Synthesis, physicochemical, and biological characterization. *Bioorganic & Medicinal Chemistry*, *15*(18), 6262–6272.

Lynch, M. A. (2001). Lipoic acid confers protection against oxidative injury in non-neuronal and neuronal tissue. *Nutritional Neuroscience*, *4*(6), 419–438.

Ma, H., Guo, R., Yu, L., Zhang, Y., & Ren, J. (2010). Aldehyde dehydrogenase 2 (ALDH2) rescues myocardial ischaemia/reperfusion injury: role of autophagy paradox and toxic aldehyde. *European Heart Journal*, in press.

Mackenzie, I. S., Maki-Petaja, K. M., McEniery, C. M., Bao, Y. P., Wallace, S. M., Cheriyan, J., et al. (2005). Aldehyde dehydrogenase 2 plays a role in the bioactivation of nitroglycerin in humans. *Arteriosclerosis, Thrombosis, and Vascular Biology*, *25*(9), 1891–1895.

Malta, E. (1989). Biphasic relaxant curves to glyceryl trinitrate in rat aortic rings. Evidence for two mechanisms of action. *Naunyn-Schmiedeberg's Archives of Pharmacology*, *339*(1–2), 236–243.

Martelli, A., Rapposelli, S., & Calderone, V. (2006). NO-releasing hybrids of cardiovascular drugs. *Current Medicinal Chemistry*, *13*(6), 609–625.

McDonald, B. J., & Bennett, B. M. (1990). Cytochrome P-450 mediated biotransformation of organic nitrates. *Canadian Journal of Physiology and Pharmacology*, *68*(12), 1552–1557.

McVeigh, G. E., Morgan, D. R., Allen, P., Trimble, M., Hamilton, P., Dixon, L. J., et al. (2002). Early vascular abnormalities and de novo nitrate tolerance in diabetes mellitus. *Diabetes, Obesity & Metabolism*, *4*(5), 336–341.

Millar, T. M., Stevens, C. R., Benjamin, N., Eisenthal, R., Harrison, R., & Blake, D. R. (1998). Xanthine oxidoreductase catalyses the reduction of nitrates and nitrite to nitric oxide under hypoxic conditions. *FEBS Letters*, *427*(2), 225–228.

Milone, S. D., Azevedo, E. R., Forster, C., & Parker, J. D. (1999). The angiotensin II-receptor antagonist losartan does not prevent hemodynamic or vascular tolerance to nitroglycerin. *Journal of Cardiovascular Pharmacology*, *34*(5), 645–650.

Minamiyama, Y., Imaoka, S., Takemura, S., Okada, S., Inoue, M., & Funae, Y. (2001). Escape from tolerance of organic nitrate by induction of cytochrome P450. *Free Radical Biology & Medicine*, *31*(11), 1498–1508.

Minamiyama, Y., Takemura, S., Akiyama, T., Imaoka, S., Inoue, M., Funae, Y., et al. (1999). Isoforms of cytochrome P450 on organic nitrate-derived nitric oxide release in human heart vessels. *FEBS Letters*, *452*(3), 165–169.

Minamiyama, Y., Takemura, S., Hai, S., Suehiro, S., & Okada, S. (2006). Vitamin E deficiency accelerates nitrate tolerance via a decrease in cardiac P450 expression and increased oxidative stress. *Free Radical Biology & Medicine*, *40*(5), 808–816.

Minamiyama, Y., Takemura, S., Nishino, Y., & Okada, S. (2002). Organic nitrate tolerance is induced by degradation of some cytochrome P450 isoforms. *Redox Report*, *7*(5), 339–342.

Minamiyama, Y., Takemura, S., Yamasaki, K., Hai, S., Hirohashi, K., Funae, Y., et al. (2004a). Continuous administration of organic nitrate decreases hepatic cytochrome P450. *The Journal of Pharmacology and Experimental Therapeutics*, *308*(2), 729–735.

Minamiyama, Y., Takemura, S., Yamasaki, K., Hai, S., Hirohashi, K., & Okada, S. (2004b). Continuous treatment with organic nitrate affects hepatic cytochrome P450. *Redox Report*, *9*(6), 360–364.

Moini, H., Packer, L., & Saris, N. E. (2002). Antioxidant and prooxidant activities of alpha-lipoic acid and dihydrolipoic acid. *Toxicology and Applied Pharmacology*, *182*(1), 84–90.

Muller, S., Konig, I., Meyer, W., & Kojda, G. (2004). Inhibition of vascular oxidative stress in hypercholesterolemia by eccentric isosorbide mononitrate. *Journal of the American College of Cardiology*, *44*(3), 624–631.

Muller, S., Laber, U., Mullenheim, J., Meyer, W., & Kojda, G. (2003). Preserved endothelial function after long-term eccentric isosorbide mononitrate despite moderate nitrate tolerance. *Journal of the American College of Cardiology*, *41*(11), 1994–2000.

Mulsch, A., Bara, A., Mordvintcev, P., Vanin, A., & Busse, R. (1995). Specificity of different organic nitrates to elicit NO formation in rabbit vascular tissues and organs in vivo. *British Journal of Pharmacology*, *116*(6), 2743–2749.

Mulsch, A., Oelze, M., Kloss, S., Mollnau, H., Topfer, A., Smolenski, A., et al. (2001). Effects of in vivo nitroglycerin treatment on activity and expression of the guanylyl cyclase and cGMP-dependent protein kinase and their downstream target vasodilator-stimulated phosphoprotein in aorta. *Circulation*, *103*(17), 2188–2194.

Munzel, T., Daiber, A., & Mulsch, A. (2005). Explaining the phenomenon of nitrate tolerance. *Circulation Research*, *97*(7), 618–628.

Munzel, T., Giaid, A., Kurz, S., Stewart, D. J., & Harrison, D. G. (1995a). Evidence for a role of endothelin 1 and protein kinase C in nitroglycerin tolerance. *Proceedings of the National Academy of Science*, *92*(11), 5244–5248.

Munzel, T., & Harrison, D. G. (1997). Evidence for a role of oxygen-derived free radicals and protein kinase C in nitrate tolerance. *Journal of Molecular Medicine*, *75*(11–12), 891–900.

Munzel, T., Heitzer, T., Kurz, S., Harrison, D. G., Luhman, C., Pape, L., et al. (1996a). Dissociation of coronary vascular tolerance and neurohormonal adjustments during long-term nitroglycerin therapy in patients with stable coronary artery disease. *Journal of the American College of Cardiology*, *27*(2), 297–303.

Munzel, T., Hink, U., Yigit, H., Macharzina, R., Harrison, D. G., & Mulsch, A. (1999). Role of superoxide dismutase in in vivo and in vitro nitrate tolerance. *British Journal of Pharmacology*, *127*, 1224–1230.

Munzel, T., Holtz, J., Mulsch, A., Stewart, D. J., & Bassenge, E. (1989). Nitrate tolerance in epicardial arteries or in the venous system is not reversed by *N*-acetylcysteine in vivo, but tolerance-independent interactions exist. *Circulation*, *79*(1), 188–197.

Munzel, T., Kurz, S., Rajagopalan, S., Tarpey, M., Freeman, B., & Harrison, D. G. (1995b). Identification of the membrane bound NADH oxidase as the major source of superoxide anion in nitrate tolerance. *Endothelium, 3*(suppl.), s14 (abstract).

Munzel, T., Kurz, S., Rajagopalan, S., Thoenes, M., Berrington, W. R., Thompson, J. A., et al. (1996b). Hydralazine prevents nitroglycerin tolerance by inhibiting activation of a membrane-bound NADH oxidase. A new action for an old drug. *Journal of Clinical Investigation, 98*(6), 1465–1470.

Munzel, T., Mollnau, H., Hartmann, M., Geiger, C., Oelze, M., Warnholtz, A., et al. (2000). Effects of a nitrate-free interval on tolerance, vasoconstrictor sensitivity and vascular superoxide production. *Journal of the American College of Cardiology, 36*(2), 628–634.

Munzel, T., Sayegh, H., Freeman, B. A., Tarpey, M. M., & Harrison, D. G. (1995c). Evidence for enhanced vascular superoxide anion production in nitrate tolerance. A novel mechanism underlying tolerance and cross-tolerance. *Journal of Clinical Investigation, 95*(1), 187–194.

Nakamura, Y., Moss, A. J., Brown, M. W., Kinoshita, M., & Kawai, C. (1999). Long-term nitrate use may be deleterious in ischemic heart disease: A study using the databases from two large-scale postinfarction studies. Multicenter Myocardial Ischemia Research Group. *American Heart Journal, 138*(3 Pt 1), 577–585.

Needleman, P., Blehm, D. J., & Rotskoff, K. S. (1969). Relationship between glutathione-dependent dentration and the vasodilator effectiveness of organic nitrates. *The Journal of Pharmacology and Experimental Therapeutics, 165*(2), 286–288.

Needleman, P., & Hunter, F. E. Jr., (1965). The transformation of glyceryl trinitrate and other nitrates by glutathione-organic nitrate reductase. *Molecular Pharmacology, 1*(1), 77–86.

Needleman, P., & Hunter, F. E. Jr., (1966). Effects of organic nitrates on mitochondrial respiration and swelling: Possible correlations with the mechanism of pharmacologic action. *Molecular Pharmacology, 2*(2), 134–143.

Needleman, P., Jakschik, B., & Johnson, E. M. Jr., (1973). Sulfhydryl requirement for relaxation of vascular smooth muscle. *The Journal of Pharmacology and Experimental Therapeutics, 187*(2), 324–331.

Newman, C. M., Warren, J. B., Taylor, G. W., Boobis, A. R., & Davies, D. S. (1990). Rapid tolerance to the hypotensive effects of glyceryl trinitrate in the rat: Prevention by *N*-acetyl-L- but not *N*-acetyl-D-cysteine. *British Journal of Pharmacology, 99*(4), 825–829.

Nordlander, R., & Walter, M. (1994). Once- versus twice-daily administration of controlled-release isosorbide-5-mononitrate 60 mg in the treatment of stable angina pectoris. A randomized, double-blind, cross-over study. The Swedish Multicentre Group. *European Heart Journal, 15*(1), 108–113.

Nunez, C., Victor, V. M., Tur, R., Alvarez-Barrientos, A., Moncada, S., Esplugues, J. V., et al. (2005). Discrepancies between nitroglycerin and NO-releasing drugs on mitochondrial oxygen consumption, vasoactivity, and the release of NO. *Circulation Research, 97*(10), 1063–1069.

Oberle, S., Abate, A., Grosser, N., Hemmerle, A., Vreman, H. J., Dennery, P. A., et al. (2003). Endothelial protection by pentaerithrityl trinitrate: Bilirubin and carbon monoxide as possible mediators. *Experimental Biology and Medicine (Maywood, NJ), 228*(5), 529–534.

Oberle, S., Abate, A., Grosser, N., Vreman, H. J., Dennery, P. A., Schneider, H. T., et al. (2002). Heme oxygenase-1 induction may explain the antioxidant profile of pentaerythrityl trinitrate. *Biochemical and Biophysical Research Communications, 290*(5), 1539–1544.

Oberle, S., Schwartz, P., Abate, A., & Schroder, H. (1999). The antioxidant defense protein ferritin is a novel and specific target for pentaerithrityl tetranitrate in endothelial cells. *Biochemical and Biophysical Research Communications, 261*(1), 28–34.

O'Byrne, S., Shirodaria, C., Millar, T., Stevens, C., Blake, D., & Benjamin, N. (2000). Inhibition of platelet aggregation with glyceryl trinitrate and xanthine oxidoreductase. *The Journal of Pharmacology and Experimental Therapeutics*, *292*(1), 326–330.

Oppermann, M., Balz, V., Adams, V., Dao, V. T., Bas, M., Suvorava, T., et al. (2009). Pharmacological induction of vascular extracellular superoxide dismutase expression in vivo. *Journal of Cellular and Molecular Medicine*, *13*(7), 1271–1278.

Otto, A., Fontaine, J., & Berkenboom, G. (2006). Ramipril treatment protects against nitrate-induced oxidative stress in eNOS−/− mice: An implication of the NADPH oxidase pathway. *Journal of Cardiovascular Pharmacology*, *48*(1), 842–849.

Otto, A., Fontaine, D., Fontaine, J., & Berkenboom, G. (2005). Rosuvastatin treatment protects against nitrate-induced oxidative stress. *Journal of Cardiovascular Pharmacology*, *46*(2), 177–184.

Packer, M., Lee, W. H., Kessler, P. D., Gottlieb, S. S., Medina, N., & Yushak, M. (1987). Prevention and reversal of nitrate tolerance in patients with congestive heart failure. *The New England Journal of Medicine*, *317*(13), 799–804.

Parker, J. O., Fung, H. L., Ruggirello, D., & Stone, J. A. (1983). Tolerance to isosorbide dinitrate: Rate of development and reversal. *Circulation*, *68*, 1074–1080.

Parker, J. D., & Parker, J. O. (1998). Nitrate therapy for stable angina pectoris. *The New England Journal of Medicine*, *338*(8), 520–531.

Parker, J. D., Parker, A. B., Farrell, B., & Parker, J. O. (1995). Intermittent transdermal nitroglycerin therapy. Decreased anginal threshold during the nitrate-free interval. *Circulation*, *91*(4), 973–978.

Parker, J. D., Parker, A. B., Farrell, B., & Parker, J. O. (1997). The effect of hydralazine on the development of tolerance to continuous nitroglycerin. *The Journal of Pharmacology and Experimental Therapeutics*, *280*(2), 866–875.

Pautz, A., Rauschkolb, P., Schmidt, N., Art, J., Oelze, M., Wenzel, P., et al. (2009). Effects of nitroglycerin or pentaerithrityl tetranitrate treatment on the gene expression in rat hearts: Evidence for cardiotoxic and cardioprotective effects. *Physiological Genomics*, *38*(2), 176–185.

Ratz, J. D., Fraser, A. B., Rees-Milton, K. J., Adams, M. A., & Bennett, B. M. (2000a). Endothelin receptor antagonism does not prevent the development of in vivo glyceryl trinitrate tolerance in the rat. *The Journal of Pharmacology and Experimental Therapeutics*, *295*(2), 578–585.

Ratz, J. D., McGuire, J. J., Anderson, D. J., & Bennett, B. M. (2000b). Effects of the flavoprotein inhibitor, diphenyleneiodonium sulfate, on ex vivo organic nitrate tolerance in the rat. *The Journal of Pharmacology and Experimental Therapeutics*, *293*(2), 569–577.

Ren, J. (2007). Acetaldehyde and alcoholic cardiomyopathy: Lessons from the ADH and ALDH2 transgenic models. *Novartis Foundation Symposium*, *285*, 69–76, discussion 76–69, 198–199.

Sage, P. R., de la Lande, I. S., Stafford, I., Bennett, C. L., Phillipov, G., Stubberfield, J., et al. (2000). Nitroglycerin tolerance in human vessels: Evidence for impaired nitroglycerin bioconversion. *Circulation*, *102*(23), 2810–2815.

Sakai, K., Akima, M., Shiraki, Y., & Hoshino, E. (1983). Effects of a new antianginal agent, nicorandil, on the cardiovascular system of the miniature pig: With special reference to coronary vasospasm. *The Journal of Pharmacology and Experimental Therapeutics*, *227*(1), 220–228.

Schaer, D. H., Buff, L. A., & Katz, R. J. (1988). Sustained antianginal efficacy of transdermal nitroglycerin patches using an overnight 10-hour nitrate-free interval. *The American Journal of Cardiology*, *61*(1), 46–50.

Schmidt, K., Rehn, M., Stessel, H., Wolkart, G., & Mayer, B. (2010). Evidence against tetrahydrobiopterin depletion of vascular tissue exposed to nitric oxide/superoxide or nitroglycerin. *Free Radical Biology & Medicine*, *48*(1), 145–152.

Schmidt, P., Youhnovski, N., Daiber, A., Balan, A., Arsic, M., Bachschmid, M., et al. (2003). Specific nitration at tyrosine 430 revealed by high resolution mass spectrometry as basis for redox regulation of bovine prostacyclin synthase. *The Journal of Biological Chemistry*, *278*(15), 12813–12819.

Schnorbus, B., Schiewe, R., Ostad, M. A., Medler, C., Wachtlin, D., Wenzel, P., Daiber, A., Münzel, T., Warnholtz, A (2009). Effects of pentaerythritol tetranitrate on endothelial function in coronary artery disease: results of the PENTA study. *Clinical Research Cardiology*, *99*, 115–124.

Schroder, H., Leitman, D. C., Bennett, B. M., Waldman, S. A., & Murad, F. (1988). Glyceryl trinitrate-induced desensitization of guanylate cyclase in cultured rat lung fibroblasts. *The Journal of Pharmacology and Experimental Therapeutics*, *245*(2), 413–418.

Schroder, H., Noack, E., & Muller, R. (1985). Evidence for a correlation between nitric oxide formation by cleavage of organic nitrates and activation of guanylate cyclase. *Journal of Molecular and Cellular Cardiology*, *17*(9), 931–934.

Schuhmacher, S., Schulz, E., Oelze, M., Konig, A., Roegler, C., Lange, K., et al. (2009). A new class of organic nitrates: Investigations on bioactivation, tolerance and cross-tolerance phenomena. *British Journal of Pharmacology*, *158*(2), 510–520.

Schuhmacher, S., Wenzel, P., Schulz, E., Oelze, M., Mang, C., Kamuf, J., et al. (2010). Pentaerythritol tetranitrate improves angiotensin II-induced vascular dysfunction via induction of heme oxygenase-1. *Hypertension*, *55*(4), 897–904.

Schulz, E., Jansen, T., Wenzel, P., Daiber, A., & Munzel, T. (2008). Nitric oxide, tetrahydrobiopterin, oxidative stress, and endothelial dysfunction in hypertension. *Antioxidants Redox Signaling*, *10*(6), 1115–1126.

Schulz, E., Tsilimingas, N., Rinze, R., Reiter, B., Wendt, M., Oelze, M., et al. (2002). Functional and biochemical analysis of endothelial (dys)function and NO/cGMP signaling in human blood vessels with and without nitroglycerin pretreatment. *Circulation*, *105*(10), 1170–1175.

Schwemmer, M., & Bassenge, E. (2003). New approaches to overcome tolerance to nitrates. *Cardiovascular Drugs and Therapy*, *17*(2), 159–173.

Sekiya, M., Sato, M., Funada, J., Ohtani, T., Akutsu, H., & Watanabe, K. (2005). Effects of the long-term administration of nicorandil on vascular endothelial function and the progression of arteriosclerosis. *Journal of Cardiovascular Pharmacology*, *46*(1), 63–67.

Silver, P. J., Pagani, E. D., Dundore, R. L., de Garavilla, L., Bode, D. C., & Bacon, E. R. (1998). Cardiovascular activity of WIN 65579, a novel inhibitor of cyclic GMP phosphodiesterase 5. *European Journal of Pharmacology*, *349*(2–3), 263–268.

Stewart, D. D. (1888). Remarkable tolerance to nitroglycerin. *Philadelphia Polyclinic*, *6*, 43.

Stoyanovsky, D. A., Tyurina, Y. Y., Tyurin, V. A., Anand, D., Mandavia, D. N., Gius, D., et al. (2005). Thioredoxin and lipoic acid catalyze the denitrosation of low molecular weight and protein *S*-nitrosothiols. *Journal of the American Chemical Society*, *127*(45), 15815–15823.

Sydow, K., Daiber, A., Oelze, M., Chen, Z., August, M., Wendt, M., et al. (2004). Central role of mitochondrial aldehyde dehydrogenase and reactive oxygen species in nitroglycerin tolerance and cross-tolerance. *Journal of Clinical Investigation*, *113*(3), 482–489.

Szocs, K., Lassegue, B., Wenzel, P., Wendt, M., Daiber, A., Oelze, M., et al. (2007). Increased superoxide production in nitrate tolerance is associated with NAD(P)H oxidase and aldehyde dehydrogenase 2 downregulation. *Journal of Molecular and Cellular Cardiology*, *42*(6), 1111–1118.

Tauchert, M., Jansen, W., Osterspey, A., Fuchs, M., Hombach, V., & Hilger, H. H. (1983). Dose dependence of tolerance during treatment with mononitrates. *Zeitschrift für Kardiologie*, *72*(Suppl. 3), 218–228.

Taylor, A. L., Ziesche, S., Yancy, C., Carson, P., D'Agostino, R., Jr., Ferdinand, K., et al. (2004). Combination of isosorbide dinitrate and hydralazine in blacks with heart failure. *The New England Journal of Medicine*, *351*(20), 2049–2057.

Thadani, U., Manyari, D., Parker, O., & Fung, H. L. (1980). Tolerance to the circulatory effects of oral isosorbide dinitrate. *Circulation, 61*, 526–535.

Thomas, G. R., DiFabio, J. M., Gori, T., & Parker, J. D. (2007). Once daily therapy with isosorbide-5-mononitrate causes endothelial dysfunction in humans: Evidence of a free-radical-mediated mechanism. *Journal of the American College of Cardiology, 49*(12), 1289–1295.

Thum, T., Fraccarollo, D., Thum, S., Schultheiss, M., Daiber, A., Wenzel, P., et al. (2007). Differential effects of organic nitrates on endothelial progenitor cells are determined by oxidative stress. *Arteriosclerosis, Thrombosis, and Vascular Biology, 27*, 748–754.

Torfgard, K. E., Ahlner, J., Axelsson, K. L., Norlander, B., & Bertler, A. (1992). Tissue disposition of glyceryl trinitrate, 1, 2-glyceryl dinitrate, and 1, 3-glyceryl dinitrate in tolerant and nontolerant rats. *Drug Metabolism and Disposition, 20*(4), 553–558.

Towell, J., Garthwaite, T., & Wang, R. (1985). Erythrocyte aldehyde dehydrogenase and disulfiram-like side effects of hypoglycemics and antianginals. *Alcoholism, Clinical and Experimental Research, 9*(5), 438–442.

Warnholtz, A., Mollnau, H., Heitzer, T., Kontush, A., Moller-Bertram, T., Lavall, D., et al. (2002a). Adverse effects of nitroglycerin treatment on endothelial function, vascular nitrotyrosine levels and cGMP-dependent protein kinase activity in hyperlipidemic Watanabe rabbits. *Journal of the American College of Cardiology, 40*(7), 1356–1363.

Warnholtz, A., Tsilimingas, N., Wendt, M., & Munzel, T. (2002b). Mechanisms underlying nitrate-induced endothelial dysfunction: Insight from experimental and clinical studies. *Heart Failure Reviews, 7*(4), 335–345.

Watanabe, H., Kakihana, M., Ohtsuka, S., & Sugishita, Y. (1998a). Preventive effects of carvedilol on nitrate tolerance—A randomized, double-blind, placebo-controlled comparative study between carvedilol and arotinolol. *Journal of the American College of Cardiology, 32*(5), 1201–1206.

Watanabe, H., Kakihana, M., Ohtsuka, S., & Sugishita, Y. (1998c). Randomized, double-blind, placebo-controlled study of the preventive effect of supplemental oral vitamin C on attenuation of development of nitrate tolerance. *Journal of the American College of Cardiology, 31*(6), 1323–1329.

Wenzel, P., Hink, U., Oelze, M., Schuppan, S., Schaeuble, K., Schildknecht, S., et al. (2007a). Role of reduced lipoic acid in the redox regulation of mitochondrial aldehyde dehydrogenase (ALDH-2) activity. Implications for mitochondrial oxidative stress and nitrate tolerance. *The Journal of Biological Chemistry, 282*(1), 792–799.

Wenzel, P., Hink, U., Oelze, M., Seeling, A., Isse, T., Bruns, K., et al. (2007c). Number of nitrate groups determines reactivity and potency of organic nitrates: A proof of concept study in ALDH-$2^{-/-}$ mice. *British Journal of Pharmacology, 150*, 526–533.

Wenzel, P., Mollnau, H., Oelze, M., Schulz, E., Wickramanayake, J. M., Muller, J., et al. (2008a). First evidence for a crosstalk between mitochondrial and NADPH oxidase-derived reactive oxygen species in nitroglycerin-triggered vascular dysfunction. *Antioxidants Redox Signaling, 10*(8), 1435–1447.

Wenzel, P., Muller, J., Zurmeyer, S., Schuhmacher, S., Schulz, E., Oelze, M., et al. (2008b). ALDH-2 deficiency increases cardiovascular oxidative stress—Evidence for indirect antioxidative properties. *Biochemical and Biophysical Research Communications, 367* (1), 137–143.

Wenzel, P., Oelze, M., Coldewey, M., Hortmann, M., Seeling, A., Hink, U., et al. (2007b). Heme oxygenase-1: A novel key player in the development of tolerance in response to organic nitrates. *Arteriosclerosis, Thrombosis, and Vascular Biology, 27*(8), 1729–1735.

Wenzel, P., Schuhmacher, S., Kienhofer, J., Muller, J., Hortmann, M., Oelze, M., et al. (2008c). Manganese superoxide dismutase and aldehyde dehydrogenase deficiency increase mitochondrial oxidative stress and aggravate age-dependent vascular dysfunction. *Cardiovascular Research, 80*(2), 280–289.

Wenzel, P., Schulz, E., Gori, T., Ostad, M. A., Mathner, F., Schildknecht, S., et al. (2009). Monitoring white blood cell mitochondrial aldehyde dehydrogenase activity: Implications for nitrate therapy in humans. *The Journal of Pharmacology and Experimental Therapeutics*, *330*(1), 63–71.

Wenzl, M. V., Beretta, M., Gorren, A. C., Zeller, A., Baral, P. K., Gruber, K., et al. (2009a). Role of the general base Glu-268 in nitroglycerin bioactivation and superoxide formation by aldehyde dehydrogenase-2. *The Journal of Biological Chemistry*, *284*(30), 19878–19886.

Wenzl, M. V., Wolkart, G., Stessel, H., Beretta, M., Schmidt, K., & Mayer, B. (2009b). Different effects of ascorbate deprivation and classical vascular nitrate tolerance on aldehyde dehydrogenase-catalysed bioactivation of nitroglycerin. *British Journal of Pharmacology*, *156*(8), 1248–1255.

Wolkart, G., Wenzl, M. V., Beretta, M., Stessel, H., Schmidt, K., & Mayer, B. (2008). Vascular tolerance to nitroglycerin in ascorbate deficiency. *Cardiovascular Research*, *79*(2), 304–312.

Wollin, S. D., & Jones, P. J. (2003). Alpha-lipoic acid and cardiovascular disease. *The Journal of Nutrition*, *133*(11), 3327–3330.

Zhang, J., Chen, Z., Cobb, F. R., & Stamler, J. S. (2004). Role of mitochondrial aldehyde dehydrogenase in nitroglycerin-induced vasodilation of coronary and systemic vessels: An intact canine model. *Circulation*, *110*(6), 750–755.

Zierhut, W., & Ball, H. A. (1996). Prevention of vascular nitroglycerin tolerance by inhibition of protein kinase C. *British Journal of Pharmacology*, *119*(1), 3–5.

Zou, M. H., & Bachschmid, M. (1999). Hypoxia-reoxygenation triggers coronary vasospasm in isolated bovine coronary arteries via tyrosine nitration of prostacyclin synthase. *The Journal of Experimental Medicine*, *190*(1), 135–139.

Aimin Xu*,†,‡, Yu Wang†,‡, Karen S.L. Lam*,†, and Paul M. Vanhoutte†,‡

*Department of Medicine, University of Hong Kong, Hong Kong, China

†Research Centre of Heart, Brain, Hormone and Healthy Aging, University of Hong Kong, Hong Kong, China

‡Department of Pharmacology and Pharmacy, University of Hong Kong, Hong Kong, China

Vascular Actions of Adipokines: Molecular Mechanisms and Therapeutic Implications

Abstract

Adipose tissue is a critical regulator of vascular function, which until recently had been virtually ignored. Almost all blood vessels are surrounded by perivascular adipose tissue, which is actively involved in the maintenance of vascular homeostasis by producing "vasocrine" signals such as adipokines. Adiponectin and adipocyte fatty acid binding protein (A-FABP), both of which are major adipokines predominantly produced in adipose tissue, have recently been shown to be pivotal modulators of vascular function. Adiponectin has

Advances in Pharmacology, Volume 60 1054-3589/10 $35.00

10.1016/S1054-3589(10)60002-6

multiple beneficial effects on cardiovascular health. It prevents obesity-induced endothelial dysfunction by inducing nitric oxide production, suppressing endothelial cell activation, inhibiting reactive oxygen species and apoptosis, and promoting endothelial cell repair. By contrast, A-FABP plays a detrimental role in vascular dysfunction and atherosclerosis, mainly by acting as a lipid sensor to transmit toxic lipids-induced vascular inflammation through induction of endoplasmic reticulum stress. Decreased production of adiponectin and/or elevated expression of A-FABP are important contributors to the pathogenesis of obesity-induced endothelial dysfunction and cardiovascular disease. This chapter highlights recent advances in both clinical investigations and animal studies promoting the understanding of the roles of adiponectin and A-FABP in the modulation of vascular function, and discusses the possibilities of using these two adipokines as therapeutic targets to design new drugs for preventing vascular disease associated with obesity and diabetes.

I. Introduction

As a consequence of overnutrition and sedentary lifestyle, the prevalence of overweight and obesity has increased dramatically to an epidemic level worldwide. The World Health Organization estimates that globally there are over 1.1 billion overweight adults, 300 million of whom are clinically obese, defined as having a body mass index of (BMI) $\geq$ 30 kg/m^2. Obesity constitutes the greatest threat to global human health, as it increases the risk of a cluster of most common and severe diseases. In particular, obesity is closely associated with increased cardiovascular morbidity and mortality, including hypertension, stroke, and ischemic heart disease. A report by the Prospective Studies Collaboration (PSC), with data from almost 900,000 participants, 57 prospective studies, and 4 continents, shows that the median survival rate is reduced by 8–10 years for those morbidly obese subjects with BMI at 40–45 kg/m^2 compared to those with normal BMI, primarily due to the increased death from vascular disease (Whitlock et al., 2009).

Although the precise mechanism that links obesity with cardiovascular disease remains poorly understood, a growing body of evidence suggests that dysfunction of adipose tissue (fat) in obesity plays a central role. Besides its role as a depot for energy storage, adipose tissue secretes a large number of bioactive molecules (termed as adipokines or adipocytokines) actively involved in energy metabolism, insulin sensitivity, and vascular homeostasis (Tilg & Moschen, 2006). Most of the adipokines can modulate vascular function through their direct actions on the endothelium or by their indirect effects on inflammation. In obesity, enlarged adipose tissue is infiltrated with activated macrophages, leading to augmented production of various proinflammatory adipokines, such as tumor necrosis factor (TNF)α, interleukin (IL)6, monocyte chemoattractant protein (MCP)-1, resistin, leptin, serum amyloid A3, lipocalin-2, and plasminogen activator inhibitor (PAI)-1 (Shoelson et al., 2007). However, the production of adiponectin, which possesses multiple beneficial effects on the endothelial cells, is markedly decreased. Aberrant production of these adipokines has now been recognized as a key mediator of obesity-induced vascular disorders. This chapter highlights recent advances in the understanding of the role of major adipokines in regulating vascular function and explores the potential of these adipokines to be therapeutically modulated as a future treatment approach for obesity-related cardiovascular disease.

II. Cross-Talk Between Adipose Tissue and Blood Vessels

A complex interplay exists between the adipose tissue and the vasculature. On hand, angiogenesis contributes to adipogenesis, expansion and remodeling of adipose tissue through multiple mechanisms, in order to supply nutrients, growth factors, and circulating stem cells, and to facilitate the infiltration of monocytes and neutrophils (Cao, 2010). On the other hand, virtually all blood vessels are surrounded by a layer of perivascular adipose tissue (PVAT), which exerts profound effects on vascular functions (Rajsheker et al., 2010). A unique feature of PVAT is that adipocytes are not separated from the blood vessel wall by a fascial layer, but encroach into the adventitial region, thus allowing factors secreted from PVAT readily access into blood vessels (Chatterjee et al., 2009).

PVAT modulates vascular function through several mechanisms, including inflammation, modulation of vascular tone, and proliferation of smooth muscle cells (Rajsheker et al., 2010.). In humans, the infiltration of inflammatory cells in the PVAT surrounding atherosclerotic aortae is markedly increased compared to those from healthy individuals, and this change is associated with an elevated expression of inflammatory genes (Mazurek et al., 2003). In animal studies, balloon injury of porcine coronary arteries induces inflammation extending into the adventitia and the PVAT (Okamoto et al., 2001). These findings suggest that PVAT is highly inflamed in the setting of both vascular injury and atherosclerosis.

The effects of PVAT on vascular reactivity have been extensively investigated in the past decade. Since the first report in 1991, it has been widely accepted that healthy PVAT possesses anticontractile activity, by releasing adipocyte-derived relaxing factors (ADRF), which activate potassium channels in vascular smooth muscle cells and thereby inhibit vasoconstriction to various agonists including phenylephrine, serotonin, angiotensin II, and U46619 (a thromboxane A2 mimic) (Gao, 2007). In addition, PVAT also enhances endothelium-dependent vasodilatation by increasing the release of nitric oxide. Although ADRF was initially identified in conduit vessels, a similar PVAT-dependent vasodilator response has also been detected in small mesenteric arteries of the rat (Verlohren et al., 2004). Notably, the anticontractile activity of mesenteric adipose tissue is diminished in spontaneously hypertensive rats, suggesting that impaired functions of PVAT may contribute to hypertension in this animal model (Galvez et al., 2006). Consistent with these animal findings, a study in humans demonstrates that PVAT from healthy individuals exerts anticontractile activity on small arteries by increasing nitric oxide bioavailability (Greenstein et al., 2009). However, this anticontractile activity of PVAT is lost in obese subjects, primarily due to an increase in inflammation and oxidative stress.

In addition to releasing ADRF, PVAT also promotes vasoconstriction by producing vasoconstrictors (namely adipose tissue-derived constricting factor, ADCF) (Gao, 2007). The ADCF-dependent vasoconstriction is mediated by reactive oxygen species (ROS) and is blocked by inhibitors of NADPH oxidase. Taken together, these findings suggest that PVAT plays a dual regulatory role in modulating vascular tone, attenuating vasoconstriction to agonists by ADRF and promoting vasoconstriction by ADCF. In obesity, dysfunction of adipose tissue and local inflammation results in impaired production of ADCF and elevated release of ADCF, thereby leading to endothelial dysfunction and hypertension (Gao, 2007). Nevertheless, the identity of these factors has yet to be established.

III. Modulation of Vascular Function by Adipokines

Since the discovery of leptin in 1994, the role of adipose tissue as an endocrine organ has been extensively studied in animal models as well as in humans (Tilg & Moschen, 2006). It is now well established that adipokines released from adipose tissue are the key component of the "adipo-vascular axis," actively participating in the maintenance of vascular homeostasis by either their direct actions on the vasculature, or their indirect effects on insulin sensitivity and inflammation (Goldstein & Scalia, 2007). The majority of adipokines such as leptin, TNFα, interleukin-1β, PAI-1, and adipocyte fatty acid binding protein (A-FABP) possess proinflammatory properties and exert detrimental effects on vascular function. Only few adipokines, especially adiponectin, possess anti-inflammatory and vasculo-protective properties. Discordant production of adipokines in adipose tissue has been proposed as a key mediator that links obesity to vascular dysfunction. Among several dozens of adipokines identified so far, adiponectin and A-FABP have attracted attention due to their potential as therapeutic targets for treating obesity-related vascular disease.

A. Adiponectin: As a Vasculo-protective Adipokine

1. Structural Features of Adiponectin

Human adiponectin is composed of 247 amino acid residues, including the N-terminal hyper-variable region, followed by a conserved collagen-like domain comprising 22 Gly-X-Y repeats and a COOH-terminal C1q-like globular domain (Wang et al., 2008). In the circulation, adiponectin is present predominantly as three distinct oligomeric complexes (Waki et al., 2003; Xu et al., 2005). The monomeric form of adiponectin has never been detected under native conditions. The basic building block of adiponectin is a tightly associated homotrimer, which is formed *via* hydrophobic interactions within its globular domains. Two trimers self-associate to form a disulfide-linked hexamer, which further assembles into a bouquet-like higher molecular weight (HMW) multimeric complex that consists of 12–18 protomers (Tsao et al., 2003). The assembly of hexameric and HMW forms of adiponectin depends on the formation of a disulfide bond mediated by an NH_2-terminal conserved cysteine residue within the hyper-variable region. Mutation of this cysteine to either alanine or serine leads exclusively to trimer formation (Pajvani et al., 2003; Tsao et al., 2003).

Different oligomers of adiponectin possess distinct biological activities. The trimeric adiponectin is the most potent form involved in its insulin-sensitizing actions in skeletal muscle (Tsao et al., 2003). However, the HMW oligomeric complex of the adipokine is the major bioactive form responsible for inhibition of hepatic glucose production (Pajvani et al., 2003; Wang et al., 2006b) and protection of endothelial cells from apoptosis (Kobayashi et al., 2004). In humans, the oligomeric complex distribution, but not the absolute amount of total adiponectin, determines insulin sensitivity (Trujillo & Scherer, 2005).

Adiponectin is extensively modified at the posttranslational level during its secretion from adipocytes (Wang et al., 2008). Several highly conserved lysine and proline residues within the collagen-like domain are hydroxylated. Hydroxylysine residues at positions 68, 71, 80, and 104 are further glycosylated by a glucosyl α(1-2)galactosyl group (Wang et al., 2004). These posttranslational modifications appear to be important for the formation of the HMW oligomeric complex of adiponectin. Depletion of hydroxylation and glycosylation by mutation of the four conserved lysines to arginines abrogate the intracellular assembly of the HMW oligomers *in vitro* as well as *in vivo* (Wang et al., 2006b).

Another type of posttranslational modification on adiponectin is succination, a process where *S*-(2-succinyl)cysteine is formed by the nonenzymatic reaction of fumarate with cysteine residues (Frizzell et al., 2009). Adiponectin is succinated on Cys36, thereby blocking its oligomerization through inhibition of disulfide bond formation. Notably, the extent of succination on adiponectin is elevated in diabetes, suggesting that this modification may contribute to impaired adiponectin secretion in obesity-related disorders.

Adiponectin is also modified by a α-2,8-linked disialic acid moiety (Neu5Aca2 $\rightarrow$ 8Neu5Aca2-3Gal) (Sato et al., 2001). This modification occurs on O-linked glycans situated on Thr residues within the hypervariable region (Richards et al., 2010). Loss of sialylation has no effect on adiponectin oligomerization, secretion, or activity, but causes its rapid clearance from the circulation, suggesting a role for this modification in determining the half-life of circulating adiponectin.

2. Multiple Functions of Adiponectin

The role of adiponectin as an important metabolic regulator has been extensively studied in the past decade (Berg et al., 2001; Fruebis et al., 2001; Yamauchi et al., 2001). It is now widely accepted that adiponectin is an insulin-sensitizing adipokine with multiple beneficial effects on glucose and lipid homeostasis, through activation of AMP-activated protein kinase (AMPK). In addition to its role as an insulin sensitizer, adiponectin can protect against almost all the major obesity-related disorders, including hypertension (Ohashi et al., 2006), atherosclerosis (Okamoto et al., 2002), nonalcoholic fatty liver disease (NAFLD), steatohepatitis (NASH) (Xu et al., 2003), coronary heart disease (Shibata et al., 2005), airway inflammation (Shore et al., 2006), and several types of cancers (Wang et al., 2006a, 2007).

Adiponectin exerts its pleiotropic beneficial effects through its direct actions on multiple target tissues. In skeletal muscles, adiponectin reduces lipid accumulation by enhancing fatty acid β-oxidation and increases glucose uptake (Fruebis et al., 2001; Yamauchi et al., 2001). In the liver, adiponectin inhibits both gluconeogenesis and lipogenesis, resulting in alleviation of hyperglycemia and hepatic steatosis (Berg et al., 2001; Wang et al., 2002; Xu et al., 2003, 2004). In addition, adiponectin ameliorates NASH and liver fibrosis by inhibiting the activation of both Kupffer cells and hepatic stellate cells (Ding et al., 2005; Thakur et al., 2006). The protective effects of adiponectin against myocardial infarction are attributed to its inhibition of local TNFα production and cardiomyocyte apoptosis induced by ischemia/reperfusion injury (Shibata et al., 2005).

3. Vascular Actions of Adiponectin

The beneficial effects of adiponectin on the vasculature have been documented in both clinical investigations and animal studies (Zhu et al., 2008). At the cellular level, adiponectin protects against vascular dysfunction through its actions in almost all the major types of cells in the vasculature, including endothelial cells, smooth muscle cells, platelets, leucocytes, and macrophages.

a. Hypoadiponectinemia and Vascular Diseases in Humans Unlike most other adipokines, circulating levels of adiponectin are decreased in obese individuals and patients with type 2 diabetes and cardiovascular disease (Zhu et al., 2008). Low level of serum adiponectin (hypoadiponectinemia) is a significant predictor of endothelial dysfunction in both peripheral and coronary arteries, independent of the index of insulin resistance, adiposity, and hyperlipidemia (Tan et al., 2004; Torigoe et al., 2007). In addition, hypoadiponectinemia is an independent risk factor for hypertension (Chow et al.,

2007; Ryo et al., 2004). Patients with essential hypertension have significantly lower concentrations of plasma adiponectin compared to normotensive healthy subjects, even after adjustment for confounding factors by multiple regression analysis (Mallamaci et al., 2002). However, in a study of 981 70-year-old Caucasians, serum levels of adiponectin were found to be positively associated with several measures of vascular function, including plaque grayscale median (indicating lower fat content in the plaques) and carotid artery distensibility (indicating higher wall elasticity), and this was independent of potential confounders (Gustafsson et al., 2010).

The inverse correlation between circulating adiponectin levels and carotid intima-media thickness (IMT, a marker of subclinical atherosclerosis) has been reported in both healthy subjects and diabetic patients (Zhu et al., 2008). Hypoadiponectinemia is also a predictor for coronary artery disease, acute coronary syndrome, myocardial infarction, and ischemic cerebrovascular disease, independently of classical cardiovascular risk factors (Szmitko et al., 2007). In another nested case–control study in 18,225 male participants without a previous history of cardiovascular disease, high plasma levels of adiponectin were associated with a significantly reduced risk of myocardial infarction over a follow-up period of 6 years (Pischon et al., 2004), independently of hypertension, diabetes, or inflammation. Taken together, these clinical data suggest an etiological role of adiponectin deficiency in the development of various vascular diseases in humans.

b. Vasculo-Protective Effects of Adiponectin in Animals In line with the aforementioned clinical observations, both loss-of-functional and gain-of-functional studies in various animal models have consistently demonstrated that adiponectin has protective effects against endothelial dysfunction (Shimabukuro et al., 2003), atherosclerosis (Lam & Xu, 2005), hypertension (Ohashi et al., 2006), retinopathy (Higuchi et al., 2009), and cerebral ischemic injury (Nishimura et al., 2008). Adiponectin knockout mice displayed a significantly increased neointimal hyperplasia after carotid injury (Matsuda et al., 2002), an impaired endothelium-dependent vasodilatation (Ouchi et al., 2003b), and elevated systemic blood pressure (Shimabukuro et al., 2003), and pulmonary hypertension (Summer et al., 2009), as well as an increased susceptibility to myocardial infarction (Shibata et al., 2005), and ischemic brain injury (Nishimura et al., 2008). By contrast, elevation of circulating adiponectin by either genetic or pharmacological approaches led to a marked alleviation of atherosclerotic lesions in apoE $^{-/-}$ mice (Okamoto et al., 2002) and in a rabbit model with spontaneous atherosclerosis (Li et al., 2007), and also caused a significant improvement in endothelial dysfunction, hypertension, retinopathy, heart failure, and diabetic cardiomyopathy (Higuchi et al., 2009; Zhu et al., 2008).

c. Pleiotropic Actions of Adiponectin on the Endothelium Endothelium plays a central role in maintaining vascular homeostasis by secreting a large number of bioactive molecules involved in the modulation of vascular reactivity. Endothelial dysfunction is the earliest change in several types of vascular diseases such as hypertension and atherosclerosis (Félétou & Vanhoutte, 2006; Vanhoutte, 1997; Vanhoutte et al., 2009; Zhu et al., 2008). Recent data obtained from both animal and cell culture studies demonstrate that the vasculoprotective effects of adiponectin can be attributed to its multiple actions on endothelial cells (Fig. 1), as summarized below.

i. Stimulation of Nitric Oxide (NO) Production Adiponectin increases NO production in several types of cultured endothelial cells as well as in rodent models, through activation of eNOS (Chen et al., 2003; Cheng et al., 2009; Hattori et al., 2003). It induces eNOS phosphorylation at Ser^{1177} and also promotes the association between eNOS and heat shock protein 90 (HSP90), a complex required for the maximal activation of the enzyme.

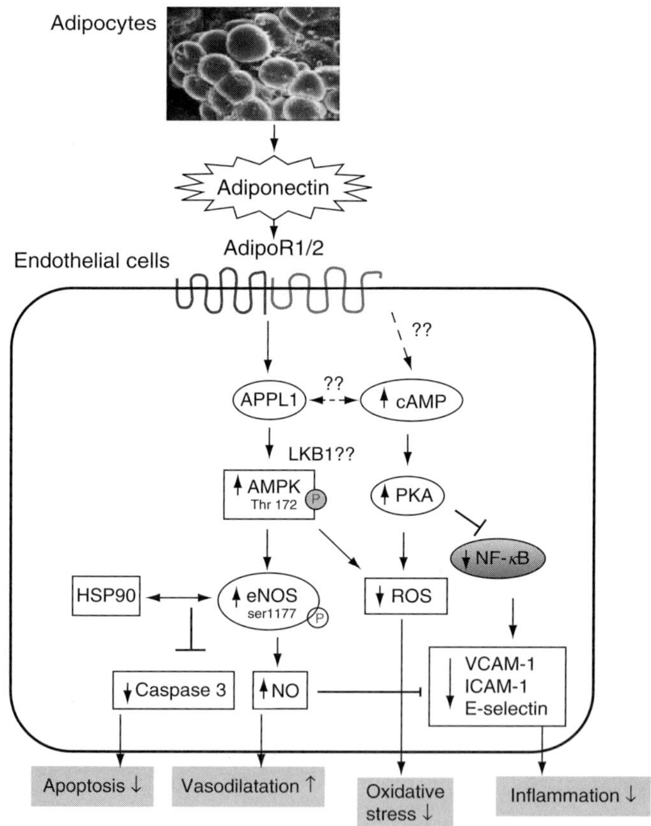

FIGURE 1 The signaling pathways underlying the multiple actions of adiponectin in endothelial cells.

Adiponectin-mediated activation of eNOS follows that of AMPK (Cheng et al., 2007). It facilitates the interaction between its two receptors (adipoR1 and adipoR2) with APPL1, a multiple domain adaptor protein that transmits adiponectin signaling from the receptor to AMPK (Cheng et al., 2007). Suppression of APPL1 expression by RNAi attenuates adiponectin-induced phosphorylation of AMPK at Thr^{172} and eNOS at Ser^{1177}, and inhibits the complex formation between eNOS and HSP90, resulting in a marked attenuation of NO production in endothelial cells. Overexpression of a constitutively active form of AMPK alone is sufficient to stimulate eNOS activation and NO production, even when APPL1 expression is suppressed, suggesting that AMPK acts downstream of APPL1 and is directly responsible for both eNOS phosphorylation and its interaction with HSP90. However, the detailed signaling events that link APPL1 with AMPK activation remain to be defined. APPL1 may promote the translocation of LKB1 from the nucleus to the cytosol, thereby increasing the accessibility of LKB1 to its downstream kinase AMPK for further activation (Zhou et al., 2009).

ii. Suppression of Endothelial Cell Inflammation Adiponectin exerts its anti-inflammatory activities on endothelial cells by inhibiting both TNF-α and resistin-induced expression of adhesion molecules and interleukin 8, which in turn results in the attenuation of monocyte attachment to endothelial cells (Kobashi et al., 2005). In both apoE$^{-/-}$ mice and a rabbit model of atherosclerosis, ementation with adiponectin decreases the expression of adhesion molecules in the aorta (Li et al., 2007; Okamoto et al., 2002). The anti-inflammatory effects of adiponectin on endothelial cells are possibly mediated by protein kinase A (PKA)-dependent suppression of nuclear factor-kappa B (NF-κB) activation (Ouchi et al., 2000).

iii. Inhibition of Oxidative Stress Adiponectin, by suppressing NAD (P)H oxidase, inhibits the production of ROS in endothelial cells induced by both high glucose and oxidized-low-density lipoprotein (LDL) (Motoshima et al., 2004; Ouedraogo et al., 2006). Consistent with these *in vitro* findings, clinical studies report a negative association between plasma levels of adiponectin and markers of oxidative stress (e.g., urinary 8-epi-prostaglandin-F2α) (Zhu et al., 2008).

The suppressive effect of adiponectin on endothelial ROS accumulation appears to be mediated by PKA, but not AMPK. Indeed, pretreatment of endothelial cells with the cAMP-dependent PKA inhibitor H-89, but not the AMPK inhibitor compound C, abrogated the suppression by adiponectin of high glucose-induced ROS production (Ouedraogo et al., 2006). Furthermore, activation of cAMP signaling with forskolin or dibutyryl-cAMP mimicked the effects of adiponectin in decreasing ROS production, while activation of AMPK by its chemical activator AICAR had little effect.

iv. Suppression of the Leukocyte–Endothelium Interaction Adiponectin inhibits the leukocyte–endothelium interaction, an important step in the pathogenesis of both macrovascular and microvascular disease. In the microcirculation of adiponectin knockout mice, the rolling and adhesion of leukocytes is markedly elevated compared to wild-type littermates (Ouedraogo et al., 2007). The changes in the adiponectin knockout mice are accompanied by a significantly reduced NO level and an elevated expression of E-selectin and VCAM-1 in the endothelium. On the other hand, systemic administration of recombinant adiponectin to adiponectin knockout mice restores endothelial NO to a physiological level and suppresses the expression of the adhesion molecules, leading to decreased leukocyte–endothelium interactions (Ouedraogo et al., 2007). The suppressive effects of adiponectin on leukocyte adhesion and adhesion molecule expression are abolished by the eNOS inhibitor N_{ω}-nitro-L-arginine methyl ester (L-NAME), suggesting that the eNOS/NO signaling cascade is indispensible for this activity.

v. Antiapoptotic Activities Adiponectin inhibits apoptosis and caspase-3 activity in human umbilical vein endothelial cells (HUVEC), by activation of the AMPK signaling pathway (Kobayashi et al., 2004). In addition, globular adiponectin prevents angiotensin II-induced apoptosis of bovine endothelial cells by restoring eNOS–HSP90 interaction and eNOS activation (Lin et al., 2004).

d. Adiponectin Actions on Endothelial Progenitor Cells Endothelial progenitor cells (EPCs) are now recognized as an important contributor to endothelial repair following vascular damage (Kawamoto & Losordo, 2008). In response to stimuli such as tissue ischemia, EPCs can be mobilized into the bloodstream, and then home or migrate toward the area of vascular damage, where they adhere, proliferate, and differentiate into mature endothelium, thereby leading to reendothelialization and neovascularization. Deceased numbers and/or impaired function of EPCs are causally associated with endothelial dysfunction and cardiovascular disease (Fadini et al., 2007).

Both clinical and animal studies suggest that adiponectin promotes endothelial repair by increasing the functionality of EPCs. In humans, plasma levels of adiponectin correlate positively with the number of circulating EPCs (Matsuo et al., 2007). In adiponectin knockout mice, the mobilization of EPCs from the bone marrow into the bloodstream is abrogated in response to hindlimb ischemia (Shibata et al., 2008), and this change is associated with impaired angiogenesis and vascular repair in the injured area (Ouchi et al., 2003a). By contrast, replenishment with recombinant adiponectin increases EPC mobilization and promotes endothelial repair after vessel injury (Shibata et al., 2008). *In vitro* studies demonstrate that adiponectin enhances proliferation and migration of EPCs isolated from human peripheral blood

(Shibata et al., 2008), and also promotes the differentiation of EPCs into tube-like structures (Nakamura et al., 2009; Werner et al., 2007). In addition, adiponectin reverses diabetes-induced impairment in vascular recruitment of EPCs through the AMPK signaling cascade (Sambuceti et al., 2009).

e. Adiponectin as a Potential Candidate of ADRF Adiponectin might be one of the long sought-after ADRFs that mediate the anticontractile activity of PVAT. In humans and rodents, the secretion of adiponectin from obese PVAT is significantly decreased compared to that from lean individuals (Chatterjee et al., 2009), and this change is associated with a compromised anticontractile activity of PVAT (Greenstein et al., 2009). Healthy adipose tissue around human small arteries secretes factors that influence vasodilatation by increasing nitric oxide bioavailability (Greenstein et al., 2009). However, this effect is abolished by incubation with a soluble adiponectin type 1 receptor-blocking fragment (Greenstein et al., 2009), suggesting that in human subcutaneous adipose tissue, adiponectin is the predominant mediator of the vasodilatation. This conclusion is further reinforced by similar findings in the rat, in which adiponectin, if released adjacent to a small artery by perivascular adipocytes, induces vasodilatation (Greenstein et al., 2009). In contrast to these findings, another study on Sprague–Dawley rats and adiponectin knockout mice concludes that adiponectin acts as a humoral vasodilator, but not as the ADRF responsible for the anticontractile activity of PVAT (Fesus et al., 2007).

f. Effects of Adiponectin on Smooth Muscle Cells Adiponectin inhibits both proliferation and migration of smooth muscle cells induced by several atherogenic growth factors, including heparin-binding epidermal growth factor-like growth factor, platelet-derived growth factor (PDGF)-BB, and basic fibroblast growth factor (Wang et al., 2005; Zhu et al., 2008). These inhibitory effects of adiponectin are primarily attributed to its oligomerization-dependent interaction with the atherogenic growth factors, subsequently leading to the blockade of their binding to the respective cell membrane receptors (Wang et al., 2005). Adiponectin-deficient mice exhibit enhanced proliferation of vascular smooth muscle cells and increased neointimal thickening after mechanical injury (Kubota et al., 2002). By contrast, adenovirus-mediated expression of adiponectin in these mice attenuated the extent of neointimal proliferation (Okamoto et al., 2002).

B. A-FABP: An Inflammatory Mediator Linking Obesity and Vascular Disease

A-FABP, also termed aP2 and FABP4, is one of the most abundant proteins in mature adipocytes, accounting for approximately 6% of the total cellular protein (Makowski & Hotamisligil, 2004). In addition,

A-FABP is expressed in macrophages (Kazemi et al., 2005; Pelton et al., 1999), a major source of proinflammatory cytokines in obese adipose tissue. The endogenous ligands of A-FABP includes oleic acid, retinoic acid, arachidonic acid, and 15-deoxy-$\Delta^{12,14}$-prostaglandin J_2 (Hoo et al., 2008). Upon association with its ligands, A-FABP can translocate from the cytosol to the nucleus, where it delivers the ligands to the nuclear receptor peroxisome proliferator-activated receptor γ (PPARγ), thereby enhancing the transcriptional activity of the receptor (Furuhashi & Hotamisligil, 2008). A-FABP appears to be an important player in lipolysis, as both basal (Coe et al., 1999) and hormone-stimulated lipolysis in response to β-adrenergic activation (Scheja et al., 1999) is impaired in A-FABP knockout mice. The stimulatory effect of A-FABP on lipolysis is mediated by its physical interaction with hormone-sensitive lipase (HSL). A-FABP knockout mice are partially protected from insulin resistance induced by dietary and genetic obesity, suggesting that this lipid chaperone is also involved in regulating insulin sensitivity (Furuhashi & Hotamisligil, 2008).

In addition to its role in lipid metabolism and insulin sensitivity, both clinical investigations and animal studies suggest that A-FABP is a central player in mediating obesity-related vascular disease, primarily by inducing insulin resistance and potentiating lipids-induced inflammation (Hoo et al., 2008).

1. Circulating A-FABP as a Risk Factor of Cardiovascular Disease in Humans

Although A-FABP was initially identified as a cytoplasmic protein, this protein is also released from adipose tissue into the bloodstream in humans (Xu et al., 2006). A-FABP concentrations in human plasma range from 10 to 50 ng/mL, a level that is much higher than that of several other major adipokines secreted from adipose tissue, including leptin, TNF-α, and IL6. Plasma levels of A-FABP correlate positively with measures of adiposity (BMI, waist–hip ratio, waist circumference, and fat percentage), suggesting that adipose tissue is the predominant contributor of A-FABP in the circulation (Xu et al., 2006). An elevated circulating level of A-FABP in obese individuals has also been reported in several different ethnic groups (Hsu et al., 2010; Stejskal & Karpisek, 2006). However, weight loss by gastric banding surgery reduces circulating levels of A-FABP in obese subjects (Haider et al., 2007). Epidemiological studies demonstrate a close association between serum levels of A-FABP and a cluster of obesity-related cardiometabolic risk factors, endothelial dysfunction, and macrovascular complications of diabetes.

First, circulating levels of A-FABP are closely correlated with several key features of the metabolic syndrome, including adverse lipid profiles (increased serum triglyceride and LDL-cholesterol, and decreased HDL-cholesterol), hyperglycemia, and hypertension, independently of sex, age, and adiposity (Hsu et al., 2010 ; Mohlig et al., 2007; Stejskal & Karpisek, 2006; Xu et al.,

2007). A 5-year prospective study including 495 nondiabetic adults demonstrates that individuals with higher A-FABP levels at baseline have a progressively worse cardiometabolic risk profile and an increasing risk of metabolic syndrome (Xu et al., 2007). The baseline A-FABP levels predict the development of the metabolic syndrome during the 5-year follow-up, independently of adiposity, insulin resistance, and other classical risk factors (Xu et al., 2007). In addition, circulating levels of A-FABP are positively associated with the pathogenesis of nonalcoholic fatty liver disease, which is now recognized as the hepatic manifestation of metabolic syndrome (Milner et al., 2009).

Second, the circulating level of A-FABP is a strong predictor of diabetes in a 10-year follow-up study, independently of the traditional risk factors including obesity, insulin resistance, or glycemic indices (Tso et al., 2007).

Third, circulating levels of A-FABP are independently associated with measures of endothelial dysfunction (Xiao et al., 2010), coronary atherosclerotic burden (Miyoshi et al., 2010), and various types of cardiovascular disease (Rhee et al., 2009; Yeung et al., 2007). In addition, serum levels of A-FABP augment as the numbers of stenotic vessel increase from normal to three-vessel disease, suggesting an etiological role of A-FABP in coronary artery disease (Rhee et al., 2009). Another cross-sectional study including 237 diabetic patients demonstrates that serum A-FABP is independently associated with diabetic nephropathy staging, and is markedly elevated in patients with macrovascular complications of the disease (Yeung et al., 2009).

Finally, the serum level of A-FABP is positively correlated with those of lipocalin-2 and hsCRP, two inflammation markers related to atherosclerosis (Xu et al., 2007). By contrast, A-FABP levels correlate inversely with those of adiponectin (Xu et al., 2007). Taken in conjunction, these clinical observations suggest that A-FABP is a proinflammatory factor that links obesity with vascular disease.

2. *A-FABP as a Mediator of Atherosclerosis in Animals*

Consistent with the clinical observations, animal studies also support an etiological role of A-FABP in vascular disease (Furuhashi & Hotamisligil, 2008). Targeted disruption of the A-FABP gene causes a marked reduction of atherosclerotic lesions along the whole aorta in apoE$^{-/-}$ mice whether fed standard chow (Makowski et al., 2001) or a hypercholesterolemic Western diet (Boord et al., 2002). When mice are challenged with a high-fat atherogenic Western diet for 1 year, the survival rate of apoE$^{-/-}$ mice null for A-FABP is 67% higher than those of apoE$^{-/-}$ control mice, primarily due to the increased stability of the atherosclerotic plaques (Boord et al., 2004). Likewise, pharmacological inhibition of the action of A-FABP also renders a signifcant protection against atherosclerotic plaque formation in apoE$^{-/-}$ mice (Erbay et al., 2009; Furuhashi et al., 2007).

The proatherogenic role of A-FABP is mediated by its direct actions on macrophages, independently of lipid metabolism and insulin sensitivity

(Hoo et al., 2008). This conclusion is supported by the finding that bone-marrow transplantation of A-FABP$^{-/-}$ macrophages into apoE$^{-/-}$ mice reduces the size of atherosclerotic lesions to a level comparable to that observed in apoE$^{-/-}$ mice with total A-FABP deficiency, suggesting that the macrophage-specific actions of A-FABP are the predominant contributor to atherosclerotic plaque formation (Makowski et al., 2001). A-FABP expression in macrophages is induced by several atherogenic and proinflammatory factors, such as oxidized LDL (Fu et al., 2000), saturated free fatty acids and Toll-like receptor activators (Kazemi et al., 2005), and is suppressed by the cholesterol-lowering statins (Llaverias et al., 2004). Adenovirus-mediated overexpression of A-FABP in human macrophages can induce foam cell formation by increasing intracellular cholesterol ester accumulation (Fu et al., 2002). By contrast, depletion of A-FABP expression in macrophages prevents oxidized LDL-induced foam cell formation by increasing cholesterol efflux, and also inhibits IκB kinase/NF-κB activity, resulting in suppression of inflammatory functions including reduced expression of both cyclooxygenase-2 and inducible nitric oxide synthase as well as impaired production of proinflammatory cytokines (Makowski et al., 2001, 2005). Incubation of macrophages with a chemical inhibitor of A-FABP decreases the production of inflammatory cytokines in a way similar to that observed in A-FABP-deficient macrophage cells (Furuhashi et al., 2007).

3. A Positive Feedback Regulation Between A-FABP and JNK

C-Jun N-terminal kinases (JNK) is a central mediator of obesity-related pathologies, including insulin resistance, type 2 diabetes, and vascular dysfunction (Hirosumi et al., 2002; Sabio et al., 2008). In apoE$^{-/-}$ mice, JNK is selectively activated in the atherosclerotic lesion area (Ricci et al., 2004). Genetic ablation of JNK2 (one of the three JNK isoforms) protects apoE$^{-/-}$ mice from developing high cholesterol diet-induced atherosclerosis, and also prevents hypercholesterolemia-induced endothelial dysfunction and oxidative stress in the aorta (Osto et al., 2008).

A-FABP may potentiate vascular inflammation by forming a positive feedback loop with JNK and activator protein-1 (AP-1) (Hui et al., 2010). Indeed, in response to proinflammatory stimuli, activated JNK in macrophages increases A-FABP expression by inducing the phosphorylation of c-Jun, which in turn binds to a highly conserved AP-1 *cis*-element within the A-FABP gene promoter and enhances the gene transcription. Vice versa, elevated A-FABP potentiates JNK phosphorylation and subsequent activation of the AP-1 complex, leading to elevated production of proinflammatory cytokines (Hui et al., 2010). Interestingly, pharmacological inhibition of A-FABP not only reduces JNK phosphorylation and AP-1 activity but also decreases A-FABP expression by suppressing transcriptional activation of the gene promoter, suggesting the existence of an autoregulatory mechanism that tightly controls the feedback loop between A-FABP, JNK, and its downstream target c-Jun.

The reciprocal regulation between A-FABP and JNK is also supported by animal studies demonstrating that genetic or pharmacological inhibition of A-FABP suppresses JNK activity in adipose tissue of obese mice as well as in atherosclerotic lesion areas of apoE$^{-/-}$ mice (Erbay et al., 2009; Furuhashi et al., 2007). By contrast, genetic inhibition of JNK reduces A-FABP expression in obese adipose tissue.

4. A-FABP as a Lipid Sensor to Induce Cellular Stress

Endoplasmic reticulum (ER) stress, initiated by protein overload or malfolding, has been proposed as a central player in mediating obesity-related inflammation, insulin resistance, and vascular disorders by activation of JNK (Hotamisligil, 2010). In both humans and mice, ER stress is present in obese adipose tissue as well as in macrophages of atherosclerotic plaques (Ozcan et al., 2004). Alleviation of ER stress by either genetic or chemical approaches prevents obesity-induced inflammation and insulin resistance, and also ameliorates atherosclerosis in apoE$^{-/-}$ mice (Erbay et al., 2009; Ozcan et al., 2006).

A-FABP is an obligatory mediator coupling toxic lipids to ER stress and inflammation in macrophages *in vitro* and *in vivo* (Erbay et al., 2009). In macrophages, toxic lipids (such as palmitate) induce A-FABP expression and concurrently mitigate ER stress, leading to JNK activation. In apoE$^{-/-}$ mice, both ER stress and A-FABP expression coexist in macrophages of the atherosclerotic lesion areas (Erbay et al., 2009). Genetic depletion of A-FABP or chemical inhibition of this lipid chaperone leads to alleviation of ER stress and attenuation of JNK activation, thereby reducing atherosclerosis. Similarly, attenuation of ER stress using the chemical chaperone 4-phenylbutyric acid (PBA) also prevents toxic lipids-induced inflammation in macrophages and reduces atherosclerosis in apoE$^{-/-}$ mice. Further analysis using a lipidomics strategy led to the discovery of steroyl CoA desaturase-1 (SCD1) as an intermediate component that couples A-FABP to ER stress. SCD1 activity converts toxic saturated lipids to bioactive monounsaturated lipid moieties, leading to alleviation of lipid-induced ER stress (Erbay et al., 2009). A-FABP suppresses SCD1 expression by inhibiting the nuclear receptor LXR-α, thereby leading to accumulation of saturated toxic lipids and ER stress. Taken in conjunction, these findings uncover a novel signaling cascade from A-FABP to ER stress, inflammation, and finally atherosclerosis (Fig. 2).

IV. Adiponectin and A-FABP as Therapeutic Targets for Vascular Diseases

Because decreased adiponectin production and enhanced A-FABP expression in adipose tissue are causally associated with obesity-related inflammation and vascular dysfunction, both adipokines represent promising

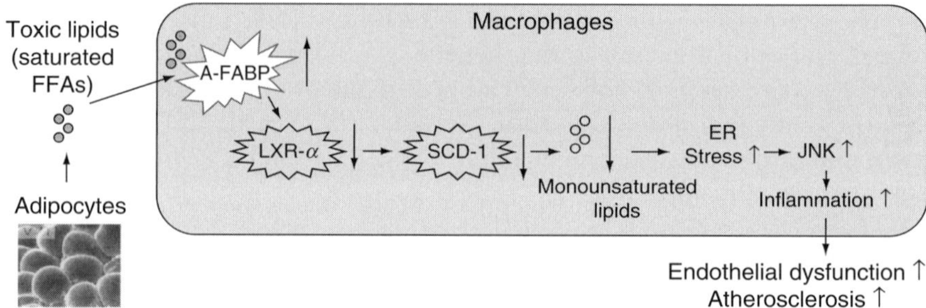

FIGURE 2 A-FABP acts as a lipid sensor mediating toxic lipids-induced inflammation through ER stress in macrophages. In obesity, toxic lipids (such as palmitate) released from adipose tissue bind to A-FABP and also induce A-FABP expression, which in turn suppresses SCD1 expression via inhibition of LXR-α, consequently leading to impaired production of monounsaturated lipids and ER stress. ER stress induces JNK activation and inflammation in macrophages, resulting in vascular dysfunction and atherosclerosis.

therapeutic targets for designing novel drugs to treat cardiovascular disease associated with obesity and diabetes.

A. Elevation of Adiponectin Production

Many currently available therapies for cardiovascular diseases, such as life style modifications, caloric restriction, pharmacological and dietary interventions, increase plasma levels of adiponectin in rodents and/or humans (Zhu et al., 2008). In particular, the PPARγ agonists thiazolidinediones (TZDs), a class of antidiabetic drugs that also possess vasculoprotective and anti-inflammatory properties, increase adiponectin production in both humans and rodents (Combs et al., 2001; Iwaki et al., 2003; Yamauchi et al., 2001). In diabetic patients, TZDs-mediated increases in adiponectin, especially its HMW oligomeric complexes, correlate well with the improvement in insulin sensitivity (Pajvani et al., 2004). Several beneficial effects of TZDs, such as insulin sensitization (Kubota et al., 2006), cardioprotection following myocardial infarction and antioxidative stress (Tao et al., 2010), are abrogated in adiponectin knockout mice, suggesting an indispensible role of adiponectin in this process.

Renin-angiotensin system blocking drugs, including angiotensin-converting-enzyme inhibitors (ACEIs) and angiotensin II receptor blockers (ARBs), also increase plasma adiponectin levels in humans (Clasen et al., 2005; Furuhashi et al., 2003; Koh et al., 2004, 2005). Losartan alone or in combination with simvastatin significantly leads to a significant elevation of plasma levels of adiponectin in hypertensive patients. In addition, several other drugs with either antidiabetic and/or vasculoprotective properties, including glimepiride (a glucose-lowering agent), nebivolol (a β-adrenergic blocker), and rimonabant (a cannabinoid CB1 receptor antagonist), also increase the plasma adiponectin concentration in humans (Zhu et al., 2008). Furthermore, increased plasma levels of adiponectin are observed in humans or rodents receiving dietary fish oils, which possess beneficial effects on cardiovascular health (Zhu et al., 2008). However, whether the beneficial effects of these agents on cardiovascular disease are mediated by adiponectin remains to be demonstrated.

There is growing interest in using adiponectin as a biomarker to screen new compounds with antidiabetic and vasculoprotective potentials. Using this strategy, two structurally related natural compounds (astragaloside II and isoastragaloside I) were identified from the medicinal herb *Radix Astragali* that possesses such an activity (Xu et al., 2009). Astragaloside II and isoastragaloside I selectively increase adiponectin secretion in primary adipocytes, and raise circulating concentrations of adiponectin in mice with dietary or genetic obesity. Furthermore, chronic treatment with these two compounds improves obesity-related insulin resistance and metabolic disorders through induction of adiponectin (Xu et al., 2009). Whether these two natural compounds possess cardiovascular benefits remains to be determined.

B. Chemical Inhibitors of A-FABP

In light of the etiological role of A-FABP in the pathogenesis of obesity-related insulin resistance, metabolic syndrome, and vascular dysfunction, pharmacological agents that inhibit A-FABP function may represent a new class of therapeutic drugs for prevention of these diseases. Indeed, a number of A-FABP inhibitors have already been identified, including carbazole- and indole-based inhibitors, benzylamino-6-(trifluoromethyl) pyrimidin-4(1H) inhibitors, and a biphenyl azole inhibitor (also known as BMS309403, developed by Bristol-Myers Squibb) (Furuhashi & Hotamisligil, 2008).

Among these A-FABP inhibitors, BMS309403 has been proven to possess multiple therapeutic effects in rodent models (Erbay et al., 2009; Furuhashi et al., 2007). This small-molecule compound interacts with the fatty acid binding pocket within the interior of A-FABP to inhibit binding of endogenous fatty acids (Sulsky et al., 2007). The compound is orally active, potent, and highly selective to A-FABP over other isoforms of FABP. In macrophages, treatment with BMS309403 prevents toxic lipids induced ER stress, JNK activation, production of proinflammatory cytokines as well as reduces foam cell formation (Erbay et al., 2009; Hui et al., 2010). In animal models, oral administration of BMS309403 improved insulin sensitivity and glucose tolerance associated with both dietary and genetic obesity (Furuhashi et al., 2007). Furthermore, BMS309403 markedly reduced the extent of atherosclerotic lesion in apoE$^{-/-}$ mice (Furuhashi et al., 2007), and also reversed the impairment in endothelial NO production and vasodilatation (Xu A. and Vanhoutte P. M., unpublished observation). These beneficial effects of BMS309403 were accompanied by inhibition of JNK activity. However, whether the A-FABP inhibitors are effective in humans remains to be determined.

V. Conclusion

Both clinical data and animal studies consistently demonstrate that adiponectin and A-FABP, which are the two most abundant adipokines in adipose tissue, exert opposite effects on vascular function. The imbalanced production of these two adipokines (decreased adiponectin and elevated A-FABP) in obesity is an important contributor to the pathogenesis of endothelial dysfunction, hypertension, and atherosclerosis. Both adipokines have been proposed as useful therapeutic targets for treating obesity-related vascular diseases. Indeed, pharmacological agents that increase adiponectin production or inhibit A-FABP activity are effective in treating vascular disease in rodent models.

Despite these promising advances, it is important to note that most of the findings on adiponectin and A-FABP were obtained from rodent models. Whether the two adipokines possess the same effects in large animals and humans awaits further investigation. Although A-FABP is present in the bloodstream, whether or not the circulating form of A-FABP plays an endocrine role in modulating vascular function *in vivo* in humans remain to be defined. Further studies on these two major adipokines may not only provide an important insight on the crosstalk between adipose tissue and vasculature but also bring new hope to develop novel therapeutics for treating vascular diseases associated with obesity and diabetes.

Acknowledgments

A. X. is supported by a grant from the NSFC/RGC Joint Research Scheme sponsored by the Research Grants Council of Hong Kong and the National Natural Science Foundation of China (Project No: N_HKU 735/08 and 30831160518) and the Collaborative Research Fund (HKU 2/07C) from the Research Grants Council of Hong Kong.

Conflict of Interest: The authors have no conflicts of interest to declare.

Abbreviations

A-FABP adipocyte fatty acid binding protein
ADRF adipose tissue derived relaxation factor
ADCF adipose tissue derived constriction factor
AMPK AMP-activated protein kinase

AP-1	activated protein-1
BMI	body weight index
EPC	endothelial progenitor cells
ER	endoplasmic reticulum
HMW	high molecular weight
JNK	C-Jun N-terminal kinases
PVAT	perivascular adipose tissue
PPARγ	peroxisome proliferator-activated receptor γ
PKA	protein kinase A
ROS	reactive oxygen species
TZDs	thiazolidinediones
TNFα	tumor necrosis factor α

References

Berg, A. H., Combs, T. P., Du, X., Brownlee, M., & Scherer, P. E. (2001). The adipocyte-secreted protein Acrp30 enhances hepatic insulin action. *Natural Medicines*, *7*, 947–953.

Boord, J. B., Maeda, K., Makowski, L., Babaev, V. R., Fazio, S., Linton, M. F., & Hotamisligil, G. S. (2002). Adipocyte fatty acid-binding protein, aP2, alters late atherosclerotic lesion formation in severe hypercholesterolemia. *Arteriosclerosis, Thrombosis, and Vascular Biology*, *22*, 1686–1691.

Boord, J. B., Maeda, K., Makowski, L., Babaev, V. R., Fazio, S., Linton, M. F., & Hotamisligil, G. S. (2004). Combined adipocyte-macrophage fatty acid-binding protein deficiency improves metabolism, atherosclerosis, and survival in apolipoprotein E-deficient mice. *Circulation*, *110*, 1492–1498. Epub 2004 Sep 1497.

Cao, Y. (2010). Adipose tissue angiogenesis as a therapeutic target for obesity and metabolic diseases. *Nature Reviews*, *9*, 107–115.

Chatterjee, T. K., Stoll, L. L., Denning, G. M., Harrelson, A., Blomkalns, A. L., Idelman, G., Rothenberg, F. G., Neltner, B., Romig-Martin, S. A., Dickson, E. W., et al. (2009). Proinflammatory phenotype of perivascular adipocytes: Influence of high-fat feeding. *Circulation Research*, *104*, 541–549.

Chen, H., Montagnani, M., Funahashi, T., Shimomura, I., & Quon, M. J. (2003). Adiponectin stimulates production of nitric oxide in vascular endothelial cells. *The Journal of Biological Chemistry*, *278*, 45021–45026.

Cheng, K. K., Iglesias, M. A., Lam, K. S., Wang, Y., Sweeney, G., Zhu, W., Vanhoutte, P. M., Kraegen, E. W., & Xu, A. (2009). APPL1 potentiates insulin-mediated inhibition of hepatic glucose production and alleviates diabetes via Akt activation in mice. *Cell Metabolism*, *9*, 417–427.

Cheng, K. K., Lam, K. S., Wang, Y., Huang, Y., Carling, D., Wu, D., Wong, C., & Xu, A. (2007). Adiponectin-induced endothelial nitric oxide synthase activation and nitric oxide production are mediated by APPL1 in endothelial cells. *Diabetes*, *56*, 1387–1394.

Chow, W. S., Cheung, B. M., Tso, A. W., Xu, A., Wat, N. M., Fong, C. H., Ong, L. H., Tam, S., Tan, K. C., Janus, E. D., et al. (2007). Hypoadiponectinemia as a predictor for the development of hypertension: A 5-year prospective study. *Hypertension*, *49*, 1455–1461.

Clasen, R., Schupp, M., Foryst-Ludwig, A., Sprang, C., Clemenz, M., Krikov, M., Thone-Reineke, C., Unger, T., & Kintscher, U. (2005). PPARgamma-activating angiotensin type-1 receptor blockers induce adiponectin. *Hypertension*, *46*, 137–143. Epub 2005 Jun 2006.

Coe, N. R., Simpson, M. A., & Bernlohr, D. A. (1999). Targeted disruption of the adipocyte lipid-binding protein (aP2 protein) gene impairs fat cell lipolysis and increases cellular fatty acid levels. *Journal of Lipid Research*, *40*, 967–972.

Combs, T. P., Berg, A. H., Obici, S., Scherer, P. E., & Rossetti, L. (2001). Endogenous glucose production is inhibited by the adipose-derived protein Acrp30. *The Journal of Clinical Investigation*, *108*, 1875–1881.

Ding, X., Saxena, N. K., Lin, S., Xu, A., Srinivasan, S., & Anania, F. A. (2005). The roles of leptin and adiponectin: A novel paradigm in adipocytokine regulation of liver fibrosis and stellate cell biology. *The American Journal of Pathology*, *166*, 1655–1669.

Erbay, E., Babaev, V. R., Mayers, J. R., Makowski, L., Charles, K. N., Snitow, M. E., Fazio, S., Wiest, M. M., Watkins, S. M., Linton, M. F., et al. (2009). Reducing endoplasmic reticulum stress through a macrophage lipid chaperone alleviates atherosclerosis. *Nature Medicine*, *15*, 1383–1391.

Fadini, G. P., Sartore, S., Agostini, C., & Avogaro, A. (2007). Significance of endothelial progenitor cells in subjects with diabetes. *Diabetes Care*, *30*, 1305–1313.

Félétou, M., & Vanhoutte, P. M. (2006). Endothelial dysfunction: A multifaceted disorder (The Wiggers Award Lecture). *American Journal of Physiology. Heart and Circulatory Physiology*, *291*, H985–H1002.

Fesus, G., Dubrovska, G., Gorzelniak, K., Kluge, R., Huang, Y., Luft, F. C., & Gollasch, M. (2007). Adiponectin is a novel humoral vasodilator. *Cardiovascular Research*, *75*, 719–727.

Frizzell, N., Rajesh, M., Jepson, M. J., Nagai, R., Carson, J. A., Thorpe, S. R., & Baynes, J. W. (2009). Succination of thiol groups in adipose tissue proteins in diabetes: Succination inhibits polymerization and secretion of adiponectin. *The Journal of Biological Chemistry*, *284*, 25772–25781.

Fruebis, J., Tsao, T. S., Javorschi, S., Ebbets-Reed, D., Erickson, M. R., Yen, F. T., Bihain, B. E., & Lodish, H. F. (2001). Proteolytic cleavage product of 30-kDa adipocyte complement-related protein increases fatty acid oxidation in muscle and causes weight loss in mice. *Proceedings of the National Academy of Sciences of the United States of America*, *98*, 2005–2010.

Fu, Y., Luo, N., & Lopes-Virella, M. F. (2000). Oxidized LDL induces the expression of ALBP/aP2 mRNA and protein in human THP-1 macrophages. *Journal of Lipid Research*, *41*, 2017–2023.

Fu, Y., Luo, N., Lopes-Virella, M. F., & Garvey, W. T. (2002). The adipocyte lipid binding protein (ALBP/aP2) gene facilitates foam cell formation in human THP-1 macrophages. *Atherosclerosis*, *165*, 259–269.

Furuhashi, M., & Hotamisligil, G. S. (2008). Fatty acid-binding proteins: Role in metabolic diseases and potential as drug targets. *Nature Reviews*, *7*, 489–503.

Furuhashi, M., Tuncman, G., Gorgun, C. Z., Makowski, L., Atsumi, G., Vaillancourt, E., Kono, K., Babaev, V. R., Fazio, S., Linton, M. F., et al. (2007). Treatment of diabetes and atherosclerosis by inhibiting fatty-acid-binding protein aP2. *Nature*, *447*, 959–965.

Furuhashi, M., Ura, N., Higashiura, K., Murakami, H., Tanaka, M., Moniwa, N., Yoshida, D., & Shimamoto, K. (2003). Blockade of the renin-angiotensin system increases adiponectin concentrations in patients with essential hypertension. *Hypertension*, *42*, 76–81.

Galvez, B., de Castro, J., Herold, D., Dubrovska, G., Arribas, S., Gonzalez, M. C., Aranguez, I., Luft, F. C., Ramos, M. P., Gollasch, M., et al. (2006). Perivascular adipose tissue and mesenteric vascular function in spontaneously hypertensive rats. *Arteriosclerosis, Thrombosis, and Vascular biology*, *26*, 1297–1302.

Gao, Y. J. (2007). Dual modulation of vascular function by perivascular adipose tissue and its potential correlation with adiposity/lipoatrophy-related vascular dysfunction. *Current Pharmaceutical Design*, *13*, 2185–2192.

Goldstein, B. J., & Scalia, R. (2007). Adipokines and vascular disease in diabetes. *Current Diabetes Reports*, *7*, 25–33.
Greenstein, A. S., Khavandi, K., Withers, S. B., Sonoyama, K., Clancy, O., Jeziorska, M., Laing, I., Yates, A. P., Pemberton, P. W., Malik, R. A., et al. (2009). Local inflammation and hypoxia abolish the protective anticontractile properties of perivascular fat in obese patients. *Circulation*, *119*, 1661–1670.
Gustafsson, S., Lind, L., Soderberg, S., & Ingelsson, E. (2010). Associations of circulating adiponectin with measures of vascular function and morphology. *The Journal of Clinical Endocrinology and Metabolism*, *95*, 2927–2934.
Haider, D. G., Schindler, K., Bohdjalian, A., Prager, G., Luger, A., Wolzt, M., & Ludvik, B. (2007). Plasma adipocyte and epidermal fatty acid binding protein is reduced after weight loss in obesity. *Diabetes, Obesity and Metabolism*, *9*, 761–763.
Hattori, Y., Suzuki, M., Hattori, S., & Kasai, K. (2003). Globular adiponectin upregulates nitric oxide production in vascular endothelial cells. *Diabetologia*, *46*, 1543–1549.
Higuchi, A., Ohashi, K., Kihara, S., Walsh, K., & Ouchi, N. (2009). Adiponectin suppresses pathological microvessel formation in retina through modulation of tumor necrosis factor-alpha expression. *Circulation Research*, *104*, 1058–1065.
Hirosumi, J., Tuncman, G., Chang, L., Gorgun, C. Z., Uysal, K. T., Maeda, K., Karin, M., & Hotamisligil, G. S. (2002). A central role for JNK in obesity and insulin resistance. *Nature*, *420*, 333–336.
Hoo, R. C., Yeung, C. Y., Lam, K. S., & Xu, A. (2008). Inflammatory biomarkers associated with obesity and insulin resistance: A focus on lipocalin-2 and adipocyte fatty acid-binding protein. *Expert Review of Endocrinology and Metabolism*, *3*, 29–41.
Hotamisligil, G. S. (2010). Endoplasmic reticulum stress and the inflammatory basis of metabolic disease. *Cell*, *140*, 900–917.
Hsu, B. G., Chen, Y. C., Lee, R. P., Lee, C. C., Lee, C. J., & Wang, J. H. (2010). Fasting serum level of fatty-acid-binding protein 4 positively correlates with metabolic syndrome in patients with coronary artery disease. *Circulation Journal*, *74*, 327–331.
Hui, X., Li, H., Zhou, Z., Lam, K. S., Xiao, Y., Wu, D., Ding, K., Wang, Y., Vanhoutte, P. M., & Xu, A. (2010). Adipocyte fatty acid-binding protein modulates inflammatory responses in macrophages through a positive feedback loop involving c-Jun NH2-terminal kinases and activator protein-1. *The Journal of Biological Chemistry*, *285*, 10273–10280.
Iwaki, M., Matsuda, M., Maeda, N., Funahashi, T., Matsuzawa, Y., Makishima, M., & Shimomura, I. (2003). Induction of adiponectin, a fat-derived antidiabetic and antiatherogenic factor, by nuclear receptors. *Diabetes*, *52*, 1655–1663.
Kawamoto, A., & Losordo, D. W. (2008). Endothelial progenitor cells for cardiovascular regeneration. *Trends in Cardiovascular Medicine*, *18*, 33–37.
Kazemi, M. R., McDonald, C. M., Shigenaga, J. K., Grunfeld, C., & Feingold, K. R. (2005). Adipocyte fatty acid-binding protein expression and lipid accumulation are increased during activation of murine macrophages by toll-like receptor agonists. *Arteriosclerosis, Thrombosis, and Vascular Biology*, *25*, 1220–1224.
Kobashi, C., Urakaze, M., Kishida, M., Kibayashi, E., Kobayashi, H., Kihara, S., Funahashi, T., Takata, M., Temaru, R., Sato, A., et al. (2005). Adiponectin inhibits endothelial synthesis of Interleukin-8. *Circulation Research*, *97*, 1245–1252.
Kobayashi, H., Ouchi, N., Kihara, S., Walsh, K., Kumada, M., Abe, Y., Funahashi, T., & Matsuzawa, Y. (2004). Selective suppression of endothelial cell apoptosis by the high molecular weight form of adiponectin. *Circulation Research*, *94*, e27–31.
Koh, K. K., Han, S. H., Quon, M. J., Yeal Ahn, J., & Shin, E. K. (2005). Beneficial effects of fenofibrate to improve endothelial dysfunction and raise adiponectin levels in patients with primary hypertriglyceridemia. *Diabetes Care*, *28*, 1419–1424.

Koh, K. K., Quon, M. J., Han, S. H., Chung, W. J., Ahn, J. Y., Seo, Y. H., Kang, M. H., Ahn, T. H., Choi, I. S., & Shin, E. K. (2004). Additive beneficial effects of losartan combined with simvastatin in the treatment of hypercholesterolemic, hypertensive patients. *Circulation, 110*, 3687–3692.

Kubota, N., Terauchi, Y., Yamauchi, T., Kubota, T., Moroi, M., Matsui, J., Eto, K., Yamashita, T., Kamon, J., Satoh, H., Yano, W., Froguel, P., et al (2002). Disruption of adiponectin causes insulin resistance and neointimal formation. *Journal of Biological Chemistry*, 25863–25866.

Kubota, N., Terauchi, Y., Kubota, T., Kumagai, H., Itoh, S., Satoh, H., Yano, W., Ogata, H., Tokuyama, K., Takamoto, I., et al. (2006). Pioglitazone ameliorates insulin resistance and diabetes by both adiponectin-dependent and -independent pathways. *The Journal of Biological Chemistry*, *281*, 8748–8755.

Lam, K. S., & Xu, A. (2005). Adiponectin: Protection of the endothelium. *Current Diabetes Reports*, *5*, 254–259.

Li, C. J., Sun, H. W., Zhu, F. L., Chen, L., Rong, Y. Y., Zhang, Y., & Zhang, M. (2007). Local adiponectin treatment reduces atherosclerotic plaque size in rabbits. *The Journal of Endocrinology*, *193*, 137–145.

Lin, L. Y., Lin, C. Y., Su, T. C., & Liau, C. S. (2004). Angiotensin II-induced apoptosis in human endothelial cells is inhibited by adiponectin through restoration of the association between endothelial nitric oxide synthase and heat shock protein 90. *FEBS Letters*, *574*, 106–110.

Llaverias, G., Noe, V., Penuelas, S., Vazquez-Carrera, M., Sanchez, R. M., Laguna, J. C., Ciudad, C. J., & Alegret, M. (2004). Atorvastatin reduces CD68, FABP4, and HBP expression in oxLDL-treated human macrophages. *Biochemical and Biophysical Research Communications*, *318*, 265–274.

Makowski, L., Boord, J. B., Maeda, K., Babaev, V. R., Uysal, K. T., Morgan, M. A., Parker, R. A., Suttles, J., Fazio, S., Hotamisligil, G. S., et al. (2001). Lack of macrophage fatty-acid-binding protein aP2 protects mice deficient in apolipoprotein E against atherosclerosis. *Nature Medicine*, *7*, 699–705.

Makowski, L., Brittingham, K. C., Reynolds, J. M., Suttles, J., & Hotamisligil, G. S. (2005). The fatty acid-binding protein, aP2, coordinates macrophage cholesterol trafficking and inflammatory activity. Macrophage expression of aP2 impacts peroxisome proliferator-activated receptor gamma and IkappaB kinase activities. *The Journal of Biological Chemistry*, *280*, 12888–12895.

Makowski, L., & Hotamisligil, G. S. (2004). Fatty acid binding proteins–the evolutionary crossroads of inflammatory and metabolic responses. *The Journal of Nutrition*, *134*, 2464S–2468S.

Mallamaci, F., Zoccali, C., Cuzzola, F., Tripepi, G., Cutrupi, S., Parlongo, S., Tanaka, S., Ouchi, N., Kihara, S., Funahashi, T., et al. (2002). Adiponectin in essential hypertension. *Journal of Nephrology*, *15*, 507–511.

Matsuda, M., Shimomura, I., Sata, M., Arita, Y., Nishida, M., Maeda, N., Kumada, M., Okamoto, Y., Nagaretani, H., Nishizawa, H., et al. (2002). Role of adiponectin in preventing vascular stenosis. The missing link of adipo-vascular axis. *The Journal of Biological Chemistry*, *277*, 37487–37491.

Matsuo, Y., Imanishi, T., Kuroi, A., Kitabata, H., Kubo, T., Hayashi, Y., Tomobuchi, Y., & Akasaka, T. (2007). Effects of plasma adiponectin levels on the number and function of endothelial progenitor cells in patients with coronary artery disease. *Circulation Journal*, *71*, 1376–1382.

Mazurek, T., Zhang, L., Zalewski, A., Mannion, J. D., Diehl, J. T., Arafat, H., Sarov-Blat, L., O'Brien, S., Keiper, E. A., Johnson, A. G., et al. (2003). Human epicardial adipose tissue is a source of inflammatory mediators. *Circulation*, *108*, 2460–2466.

Milner, K. L., van der Poorten, D., Xu, A., Bugianesi, E., Kench, J. G., Lam, K. S., Chisholm, D. J., & George, J. (2009). Adipocyte fatty acid binding protein levels relate to inflammation and fibrosis in nonalcoholic fatty liver disease. *Hepatology*, *49*, 1926–1934.

Miyoshi, T., Onoue, G., Hirohata, A., Hirohata, S., Usui, S., Hina, K., Kawamura, H., Doi, M., Kusano, K. F., Kusachi, S., et al. (2010). Serum adipocyte fatty acid-binding protein is independently associated with coronary atherosclerotic burden measured by intravascular ultrasound. *Atherosclerosis*, *211*, 164–169 [Epub ahead of print].

Mohlig, M., Weickert, M. O., Ghadamgadai, E., Machlitt, A., Pfuller, B., Arafat, A. M., Pfeiffer, A. F., & Schofl, C. (2007). Adipocyte fatty acid-binding protein is associated with markers of obesity, but is an unlikely link between obesity, insulin resistance, and hyperandrogenism in polycystic ovary syndrome women. *European Journal of Endocrinology*, *157*, 195–200.

Motoshima, H., Wu, X., Mahadev, K., & Goldstein, B. J. (2004). Adiponectin suppresses proliferation and superoxide generation and enhances eNOS activity in endothelial cells treated with oxidized LDL. *Biochemical and Biophysical Research Communications*, *315*, 264–271.

Nakamura, N., Naruse, K., Matsuki, T., Hamada, Y., Nakashima, E., Kamiya, H., Matsubara, T., Enomoto, A., Takahashi, M., Oiso, Y., et al. (2009). Adiponectin promotes migration activities of endothelial progenitor cells via Cdc42/Rac1. *FEBS Letters*, *583*, 2457–2463.

Nishimura, M., Izumiya, Y., Higuchi, A., Shibata, R., Qiu, J., Kudo, C., Shin, H. K., Moskowitz, M. A., & Ouchi, N. (2008). Adiponectin prevents cerebral ischemic injury through endothelial nitric oxide synthase dependent mechanisms. *Circulation*, *117*, 216–223.

Ohashi, K., Kihara, S., Ouchi, N., Kumada, M., Fujita, K., Hiuge, A., Hibuse, T., Ryo, M., Nishizawa, H., Maeda, N., et al. (2006). Adiponectin replenishment ameliorates obesity-related hypertension. *Hypertension*, *47*, 1108–1116.

Okamoto, E., Couse, T., De Leon, H., Vinten-Johansen, J., Goodman, R. B., Scott, N. A., & Wilcox, J. N. (2001). Perivascular inflammation after balloon angioplasty of porcine coronary arteries. *Circulation*, *104*, 2228–2235.

Okamoto, Y., Kihara, S., Ouchi, N., Nishida, M., Arita, Y., Kumada, M., Ohashi, K., Sakai, N., Shimomura, I., Kobayashi, H., et al. (2002). Adiponectin reduces atherosclerosis in apolipoprotein E-deficient mice. *Circulation*, *106*, 2767–2770.

Osto, E., Matter, C. M., Kouroedov, A., Malinski, T., Bachschmid, M., Camici, G. G., Kilic, U., Stallmach, T., Boren, J., Iliceto, S., et al. (2008). c-Jun N-terminal kinase 2 deficiency protects against hypercholesterolemia-induced endothelial dysfunction and oxidative stress. *Circulation*, *118*, 2073–2080.

Ouchi, N., Kihara, S., Arita, Y., Okamoto, Y., Maeda, K., Kuriyama, H., Hotta, K., Nishida, M., Takahashi, M., Muraguchi, M., et al. (2000). Adiponectin, an adipocyte-derived plasma protein, inhibits endothelial NF-kappaB signaling through a cAMP-dependent pathway. *Circulation*, *102*, 1296–1301.

Ouchi, N., Kobayashi, H., Kihara, S., Kumada, M., Sato, K., Inoue, T., Funahashi, T., & Walsh, K. (2003a). Adiponectin stimulates angiogenesis by promoting cross-talk between AMP-activated protein kinase and Akt signaling in endothelial cells. *The Journal of Biological Chemistry*, *13*, 13.

Ouchi, N., Ohishi, M., Kihara, S., Funahashi, T., Nakamura, T., Nagaretani, H., Kumada, M., Ohashi, K., Okamoto, Y., Nishizawa, H., et al. (2003b). Association of hypoadiponectinemia with impaired vasoreactivity. *Hypertension*, *42*, 231–234.

Ouedraogo, R., Wu, X., Xu, S. Q., Fuchsel, L., Motoshima, H., Mahadev, K., Hough, K., Scalia, R., & Goldstein, B. J. (2006). Adiponectin suppression of high-glucose-induced reactive oxygen species in vascular endothelial cells: Evidence for involvement of a cAMP signaling pathway. *Diabetes*, *55*, 1840–1846.

Ozcan, U., Cao, Q., Yilmaz, E., Lee, A. H., Iwakoshi, N. N., Ozdelen, E., Tuncman, G., Gorgun, C., Glimcher, L. H., & Hotamisligil, G. S. (2004). Endoplasmic reticulum stress links obesity, insulin action, and type 2 diabetes. *Science*, *306*, 457–461.

Ozcan, U., Yilmaz, E., Ozcan, L., Furuhashi, M., Vaillancourt, E., Smith, R. O., Gorgun, C. Z., & Hotamisligil, G. S. (2006). Chemical chaperones reduce ER stress and restore glucose homeostasis in a mouse model of type 2 diabetes. *Science*, *313*, 1137–1140.

Pajvani, U. B., Du, X., Combs, T. P., Berg, A. H., Rajala, M. W., Schulthess, T., Engel, J., Brownlee, M., & Scherer, P. E. (2003). Structure-function studies of the adipocyte-secreted hormone Acrp30/adiponectin. Implications fpr metabolic regulation and bioactivity. *The Journal of Biological Chemistry*, *278*, 9073–9085.

Pajvani, U. B., Hawkins, M., Combs, T. P., Rajala, M. W., Doebber, T., Berger, J. P., Wagner, J. A., Wu, M., Knopps, A., Xiang, A. H., et al. (2004). Complex distribution, not absolute amount of adiponectin, correlates with thiazolidinedione-mediated improvement in insulin sensitivity. *The Journal of Biological Chemistry*, *279*, 12152–12162.

Pelton, P. D., Zhou, L., Demarest, K. T., & Burris, T. P. (1999). PPARgamma activation induces the expression of the adipocyte fatty acid binding protein gene in human monocytes. *Biochemical and Biophysical Research Communications*, *261*, 456–458.

Pischon, T., Girman, C. J., Hotamisligil, G. S., Rifai, N., Hu, F. B., & Rimm, E. B. (2004). Plasma adiponectin levels and risk of myocardial infarction in men. *JAMA*, *291*, 1730–1737.

Rajsheker, S., Manka, D., Blomkalns, A. L., Chatterjee, T. K., Stoll, L. L., & Weintraub, N. L. (2010). Crosstalk between perivascular adipose tissue and blood vessels. *Current Opinion in Pharmacology*, *10*, 191–196.

Rhee, E. J., Lee, W. Y., Park, C. Y., Oh, K. W., Kim, B. J., Sung, K. C., & Kim, B. S. (2009). The association of serum adipocyte fatty acid-binding protein with coronary artery disease in Korean adults. *European Journal of Endocrinology*, *160*, 165–172.

Ricci, R., Sumara, G., Sumara, I., Rozenberg, I., Kurrer, M., Akhmedov, A., Hersberger, M., Eriksson, U., Eberli, F. R., Becher, B., et al. (2004). Requirement of JNK2 for scavenger receptor A-mediated foam cell formation in atherogenesis. *Science*, *306*, 1558–1561.

Richards, A. A., Colgrave, M. L., Zhang, J., Webster, J., Simpson, F., Preston, E., Wilks, D., Hoehn, K. L., Stephenson, M., Macdonald, G. A., et al. (2010). Sialic acid modification of adiponectin is not required for multimerization or secretion but determines half-life in circulation. *Molecular Endocrinology*, *24*, 229–239.

Ryo, M., Nakamura, T., Kihara, S., Kumada, M., Shibazaki, S., Takahashi, M., Nagai, M., Matsuzawa, Y., & Funahashi, T. (2004). Adiponectin as a biomarker of the metabolic syndrome. *Circulation Journal*, *68*, 975–981.

Sabio, G., Das, M., Mora, A., Zhang, Z., Jun, J. Y., Ko, H. J., Barrett, T., Kim, J. K., & Davis, R. J. (2008). A stress signaling pathway in adipose tissue regulates hepatic insulin resistance. *Science*, *322*, 1539–1543.

Sambuceti, G., Morbelli, S., Vanella, L., Kusmic, C., Marini, C., Massollo, M., Augeri, C., Corselli, M., Ghersi, C., Chiavarina, B., et al. (2009). Diabetes impairs the vascular recruitment of normal stem cells by oxidant damage, reversed by increases in pAMPK, heme oxygenase-1, and adiponectin. *Stem Cells*, *27*, 399–407.

Sato, C., Yasukawa, Z., Honda, N., Matsuda, T., & Kitajima, K. (2001). Identification and adipocyte differentiation-dependent expression of the unique disialic acid residue in an adipose tissue-specific glycoprotein, adipo Q. *The Journal of Biological Chemistry*, *276*, 28849–28856.

Scheja, L., Makowski, L., Uysal, K. T., Wiesbrock, S. M., Shimshek, D. R., Meyers, D. S., Morgan, M., Parker, R. A., & Hotamisligil, G. S. (1999). Altered insulin secretion associated with reduced lipolytic efficiency in aP2-/- mice. *Diabetes*, *48*, 1987–1994.

Shibata, R., Sato, K., Pimentel, D. R., Takemura, Y., Kihara, S., Ohashi, K., Funahashi, T., Ouchi, N., & Walsh, K. (2005). Adiponectin protects against myocardial ischemia-reperfusion injury through AMPK- and COX-2-dependent mechanisms. *Nature Medicine*, *11*, 1096–1103.

Shibata, R., Skurk, C., Ouchi, N., Galasso, G., Kondo, K., Ohashi, T., Shimano, M., Kihara, S., Murohara, T., & Walsh, K. (2008). Adiponectin promotes endothelial progenitor cell number and function. *FEBS Letters*, *582*, 1607–1612.

Shimabukuro, M., Higa, N., Asahi, T., Oshiro, Y., Takasu, N., Tagawa, T., Ueda, S., Shimomura, I., Funahashi, T., & Matsuzawa, Y. (2003). Hypoadiponectinemia is closely linked to endothelial dysfunction in man. *The Journal of Clinical Endocrinology and Metabolism*, *88*, 3236–3240.

Shoelson, S. E., Herrero, L., & Naaz, A. (2007). Obesity, inflammation, and insulin resistance. *Gastroenterology*, *132*, 2169–2180.

Shore, S. A., Terry, R. D., Flynt, L., Xu, A., & Hug, C. (2006). Adiponectin attenuates allergen-induced airway inflammation and hyperresponsiveness in mice. *The Journal of Allergy and Clinical Immunology*, *118*, 389–395.

Stejskal, D., & Karpisek, M. (2006). Adipocyte fatty acid binding protein in a Caucasian population: A new marker of metabolic syndrome? *European Journal of Clinical Investigation*, *36*, 621–625.

Sulsky, R., Magnin, D. R., Huang, Y., Simpkins, L., Taunk, P., Patel, M., Zhu, Y., Stouch, T. R., Bassolino-Klimas, D., Parker, R., et al. (2007). Potent and selective biphenyl azole inhibitors of adipocyte fatty acid binding protein (aFABP). *Bioorganic and Medicinal Chemistry Letters*, *17*, 3511–3515.

Summer, R., Fiack, C. A., Ikeda, Y., Sato, K., Dwyer, D., Ouchi, N., Fine, A., Farber, H. W., & Walsh, K. (2009). Adiponectin deficiency: A model of pulmonary hypertension associated with pulmonary vascular disease. *American Journal of Physiology. Lung Cellular and Molecular Physiology*, *297*, L432–438.

Szmitko, P. E., Teoh, H., Stewart, D. J., & Verma, S. (2007). Adiponectin and cardiovascular disease: State of the art? *American Journal of Physiology. Heart and Circulatory Physiology*, *292*, H1655–H1663.

Tan, K. C., Xu, A., Chow, W. S., Lam, M. C., Ai, V. H., Tam, S. C., & Lam, K. S. (2004). Hypoadiponectinemia is associated with impaired endothelium-dependent vasodilation. *The Journal of Clinical Endocrinology and Metabolism*, *89*, 765–769.

Tao, L., Wang, Y., Gao, E., Zhang, H., Yuan, Y., Lau, W. B., Chan, L., Koch, W. J., & Ma, X. L. (2010). Adiponectin: an indispensable molecule in rosiglitazone cardioprotection following myocardial infarction. *Circulation Research*, *106*, 409–417.

Thakur, V., Pritchard, M. T., McMullen, M. R., & Nagy, L. E. (2006). Adiponectin normalizes LPS-stimulated TNF-alpha production by rat Kupffer cells after chronic ethanol feeding. *The American Journal of Physiology*, *290*, G998–G1007.

Tilg, H., & Moschen, A. R. (2006). Adipocytokines: Mediators linking adipose tissue, inflammation and immunity. *Nature Reviews. Immunology*, *6*, 772–783.

Torigoe, M., Matsui, H., Ogawa, Y., Murakami, H., Murakami, R., Cheng, X. W., Numaguchi, Y., Murohara, T., & Okumura, K. (2007). Impact of the high-molecular-weight form of adiponectin on endothelial function in healthy young men. *Clinical Endocrinology*, *67*, 276–281.

Trujillo, M. E., & Scherer, P. E. (2005). Adiponectin–journey from an adipocyte secretory protein to biomarker of the metabolic syndrome. *Journal of Internal Medicine*, *257*, 167–175.

Tsao, T. S., Tomas, E., Murrey, H. E., Hug, C., Lee, D. H., Ruderman, N. B., Heuser, J. E., & Lodish, H. F. (2003). Role of disulfide bonds in Acrp30/Adiponectin structure and signaling specificity: Different oligomers activate different signal transduction pathways. *The Journal of Biological Chemistry*, *278*, 50810–50817.

Tso, A. W., Xu, A., Sham, P. C., Wat, N. M., Wang, Y., Fong, C. H., Cheung, B. M., Janus, E. D., & Lam, K. S. (2007). Serum adipocyte fatty acid binding protein as a new biomarker predicting the development of type 2 diabetes: A 10-year prospective study in a Chinese cohort. *Diabetes Care*, *30*, 2667–2672.

Vanhoutte, P. M. (1997). Endothelial dysfunction and atherosclerosis. *European Heart Journal*, *18*, E19–E29.

Vanhoutte, P. M., Tang, E., Félétou, M., & Shimokawa, H. (2009). Endothelial dysfunction and vascular disease. *Acta Physiologica*, *196*, 193–222.

Verlohren, S., Dubrovska, G., Tsang, S. Y., Essin, K., Luft, F. C., Huang, Y., & Gollasch, M. (2004). Visceral periadventitial adipose tissue regulates arterial tone of mesenteric arteries. *Hypertension*, *44*, 271–276.

Waki, H., Yamauchi, T., Kamon, J., Ito, Y., Uchida, S., Kita, S., Hara, K., Hada, Y., Vasseur, F., Froguel, P., et al. (2003). Impaired multimerization of human adiponectin mutants associated with diabetes. Molecular structure and multimer formation of adiponectin. *The Journal of Biological Chemistry*, *278*, 40352–40363.

Wang, Y., Lam, K. S., Chan, L., Chan, K. W., Lam, J. B., Lam, M. C., Hoo, R. C., Mak, W. W., Cooper, G. J., & Xu, A. (2006b). Post-translational modifications of the four conserved lysine residues within the collagenous domain of adiponectin are required for the formation of its high molecular weight oligomeric complex. *The Journal of Biological Chemistry*, *281*, 16391–16400.

Wang, Y., Lam, J. B., Lam, K. S., Liu, J., Lam, M. C., Hoo, R. L., Wu, D., Cooper, G. J., & Xu, A. (2006a). Adiponectin modulates the glycogen synthase kinase-3beta/beta-catenin signaling pathway and attenuates mammary tumorigenesis of MDA-MB-231 cells in nude mice. *Cancer Research*, *66*, 11462–11470.

Wang, Y., Lam, K. S., & Xu, A. (2007). Adiponectin as a negative regulator in obesity-related mammary carcinogenesis. *Cell Research*, *17*, 280–282.

Wang, Y., Lam, K. S., Xu, J. Y., Lu, G., Xu, L. Y., Cooper, G. J., & Xu, A. (2005). Adiponectin inhibits cell proliferation by interacting with several growth factors in an oligomerization-dependent manner. *The Journal of Biological Chemistry*, *280*, 18341–18347.

Wang, Y., Lam, K. S., Yau, M. H., & Xu, A. (2008). Post-translational modifications of adiponectin: Mechanisms and functional implications. *The Biochemical Journal*, *409*, 623–633.

Wang, Y., Lu, G., Wong, W. P., Vliegenthart, J. F., Gerwig, G. J., Lam, K. S., Cooper, G. J., & Xu, A. (2004). Proteomic and functional characterization of endogenous adiponectin purified from fetal bovine serum. *Proteomics*, *12*, 3933–3942.

Wang, Y., Xu, A., Knight, C., Xu, L. Y., & Cooper, G. J. (2002). Hydroxylation and glycosylation of the four conserved lysine residues in the collagenous domain of adiponectin. Potential role in the modulation of its insulin-sensitizing activity. *The Journal of Biological Chemistry*, *277*, 19521–19529.

Werner, C., Kamani, C. H., Gensch, C., Bohm, M., & Laufs, U. (2007). The peroxisome proliferator-activated receptor-gamma agonist pioglitazone increases number and function of endothelial progenitor cells in patients with coronary artery disease and normal glucose tolerance. *Diabetes*, *56*, 2609–2615.

Whitlock, G., Lewington, S., Sherliker, P., Clarke, R., Emberson, J., Halsey, J., Qizilbash, N., Collins, R., & Peto, R. (2009). Body-mass index and cause-specific mortality in 900 000 adults: Collaborative analyses of 57 prospective studies. *Lancet*, *373*, 1083–1096.

Xiao, Y., Yao, L., Li, X., Zhong, H., Chen, X. Y., Tang, W. L., Liu, S. P., Xu, A. M., & Zhou, Z. G. (2010). Relationship of adipocyte fatty acid-binding protein to adiponectin ratio with femoral intima-media thickness and endothelium-dependent vasodilation in patients with newly-diagnosed type 2 diabetes mellitus. *Zhonghua Yi Xue Za Zhi*, *90*, 231–235.

Xu, A., Chan, K. W., Hoo, R. L., Wang, Y., Tan, K. C., Zhang, J., Chen, B., Lam, M. C., Tse, C., Cooper, G. J., et al. (2005). Testosterone selectively reduces the high molecular weight form of adiponectin by inhibiting its secretion from adipocytes. *The Journal of Biological Chemistry*, *280*, 18073–18080.

Xu, A., Tso, A. W., Cheung, B. M., Wang, Y., Wat, N. M., Fong, C. H., Yeung, D. C., Janus, E. D., Sham, P. C., & Lam, K. S. (2007). Circulating adipocyte-fatty acid binding protein levels predict the development of the metabolic syndrome: A 5-year prospective study. *Circulation*, *115*, 1537–1543.

Xu, A., Wang, H., Hoo, R. L., Sweeney, G., Vanhoutte, P. M., Wang, Y., Wu, D., Chu, W., Qin, G., & Lam, K. S. (2009). Selective elevation of adiponectin production by the natural compounds derived from a medicinal herb alleviates insulin resistance and glucose intolerance in obese mice. *Endocrinology*, *150*, 625–633.

Xu, A., Wang, Y., Keshaw, H., Xu, L. Y., Lam, K. S., & Cooper, G. J. (2003). The fat-derived hormone adiponectin alleviates alcoholic and nonalcoholic fatty liver diseases in mice. *The Journal of Clinical Investigation*, *112*, 91–100.

Xu, A., Wang, Y., Xu, J. Y., Stejskal, D., Tam, S., Zhang, J., Wat, N. M., Wong, W. K., & Lam, K. S. (2006). Adipocyte fatty acid binding protein is a plasma biomarker closely associated with obesity and metabolic syndrome. *Clinical Chemistry*, *52*, 405–413.

Xu, A., Yin, S., Wong, L., Chan, K. W., & Lam, K. S. (2004). Adiponectin ameliorates Dyslipidemia induced by the HIV protease inhibitor ritonavir in Mice. *Endocrinology*, *145*, 487–494.

Yamauchi, T., Kamon, J., Waki, H., Terauchi, Y., Kubota, N., Hara, K., Mori, Y., Ide, T., Murakami, K., Tsuboyama-Kasaoka, N., et al. (2001). The fat-derived hormone adiponectin reverses insulin resistance associated with both lipoatrophy and obesity. *Nature Medicine*, *7*, 941–946.

Yeung, D. C., Xu, A., Cheung, C. W., Wat, N. M., Yau, M. H., Fong, C. H., Chau, M. T., & Lam, K. S. (2007). Serum adipocyte fatty acid-binding protein levels were independently associated with carotid atherosclerosis. *Arteriosclerosis, Thrombosis, and Vascular Biology*, *27*, 1796–1802.

Yeung, D. C., Xu, A., Tso, A. W., Chow, W. S., Wat, N. M., Fong, C. H., Tam, S., Sham, P. C., & Lam, K. S. (2009). Circulating levels of adipocyte and epidermal fatty acid-binding proteins in relation to nephropathy staging and macrovascular complications in type 2 diabetic patients. *Diabetes Care*, *32*, 132–134.

Zhou, L., Deepa, S. S., Etzler, J. C., Ryu, J., Mao, X., Fang, Q., Liu, D. D., Torres, J. M., Jia, W., Lechleiter, J. D., et al. (2009). Adiponectin activates AMP-activated protein kinase in muscle cells via APPL1/LKB1-dependent and phospholipase C/Ca2+/Ca2+/calmodulin-dependent protein kinase kinase-dependent pathways. *The Journal of Biological Chemistry*, *284*, 22426–22435.

Zhu, W., Cheng, K. K., Vanhoutte, P. M., Lam, K. S., & Xu, A. (2008). Vascular effects of adiponectin: Molecular mechanisms and potential therapeutic intervention. *Clinical Science (London)*, *114*, 361–374.

Anantha Vijay R. Santhanam, Livius V. d'Uscio, and Zvonimir S. Katusic

Departments of Anesthesiology and Molecular Pharmacology and Experimental Therapeutics, Mayo Clinic College of Medicine, Rochester, Minnesota, USA

Cardiovascular Effects of Erythropoietin: An Update

Abstract

Erythropoietin (EPO) is a therapeutic product of recombinant DNA technology and it has been in clinical use as stimulator of erythropoiesis over the last two decades. Identification of EPO and its receptor (EPOR) in the cardiovascular system expanded understanding of physiological and pathophysiological role of EPO. In experimental models of cardiovascular and cerebrovascular disorders, EPO exerts protection either by preventing apoptosis of cardiac myocytes, smooth muscle cells, and endothelial cells, or by increasing endothelial production of nitric oxide. In addition, EPO stimulates mobilization of progenitor cells from bone marrow thereby accelerating

Advances in Pharmacology, Volume 60

1054-3589/10 $35.00
10.1016/S1054-3589(10)60001-4

repair of injured endothelium and neovascularization. A novel signal transduction pathway involving EPOR—β-common heteroreceptor is postulated to enhance EPO-mediated tissue protection. A better understanding of the role of β-common receptor signaling as well as development of novel analogs of EPO with enhanced nonhematopoietic protective effects may expand clinical application of EPO in prevention and treatment of cardiovascular and cerebrovascular disorders.

I. Introduction

The humoral regulation of erythropoiesis was identified more than 50 years ago (Bonsdorff & Jalavisto, 1948). Studies of erythropoietin (EPO) had been severely restricted, due to its circulating levels being only picomolar, until EPO was purified for the first time in 1977 from human urine (Miyake et al., 1977) and the amino acid sequence of human urinary EPO was characterized (Lai et al., 1986). EPO is a member of the class I family of cytokines and is mainly synthesized and secreted by the kidney (Jelkmann, 1992). The ability of EPO to stimulate erythropoiesis expanded clinical application of recombinant EPO to patients suffering from anemia and chronic kidney disease. To date, recombinant EPO is the most successful biotechnology drug and indeed is the world's largest selling biopharmaceutical. EPO mediates erythropoiesis by binding to its specific receptor, EPO-receptor (EPOR), expressed on the surface of immature erythroblasts (Noguchi et al., 1991). Studies over the last decade have demonstrated that EPO and EPOR are expressed in a number of cell types including those within the cardiovascular and nervous system, suggesting that the effects of EPO extend beyond regulation of erythropoiesis.

In the cardiovascular system, EPO exerts its effects on cardiac as well as on the vascular tissues. EPO and EPOR are expressed in cardiomyocytes, vascular endothelial cells, and smooth muscle cells (Ammarguellat et al., 1996; Anagnostou et al., 1994; Brines & Cerami, 2005). It is important to note that high doses are required to observe EPO-induced tissue protection, in comparison with doses of EPO required for hematopoietic effects (Ammarguellat et al., 1996; Anagnostou et al., 1994; Brines & Cerami, 2005). Based on the existing studies with these high doses, a tangential paradigm evolved. On one hand, protective effects of EPO and the underlying mechanisms were elucidated in cardiovascular and neurological system, prompting initiation of clinical trials in patients with myocardial infarction, aneurysmal subarachnoid hemorrhage, and acute stroke. In contrast, adverse events associated with EPO therapy were identified, due to its pleiotropic effects mainly in the cardiovascular system including hypertension, thrombosis, and augmented tumor angiogenesis. In this review, the present understanding of the beneficial and detrimental cardiovascular effects of EPO will be discussed.

II. Pharmacokinetics of Erythropoietin

A. Recombinant Human Erythropoietin

EPO was produced in Chinese hamster ovary cells by recombinant DNA technology (Sasaki et al., 1987, 1988). Following isolation of the human erythropoietin gene (Jacobs et al., 1985; Lin et al., 1985), it was inserted into and expressed by cultured mammalian cells, which are capable of synthesizing unlimited quantities of the hormone. Epoetin α, epoetin β, and epoetin γ are analogs of recombinant human EPO (rhEPO) derived from a cloned human erythropoietin gene, licensed by international regulatory bodies to stimulate erythropoiesis. All have the same 165 amino acid sequences with a molecular weight of 30,400 Da and have the same pharmacological actions as native EPO (Jelkmann, 1992). After intravenous administration, epoetin is distributed in a volume comparable to the plasma volume. Epoetin plasma concentrations decay at a much lower rate after subcutaneous than intravenous administration (Table I). Epoetin α and epoetin β exhibit some differences in their pharmacokinetic profiles, such as elimination half-life, due to differences in glycosylation pattern (Table I; Storring et al., 1998). Biological activity of EPO *in vivo* is abolished when EPO is deglycosylated, suggesting that the carbohydrate moiety is essential to prevent degradation and to delay clearance of EPO from the circulation (Takeuchi & Kobata, 1991).

Normal serum concentrations of EPO for individuals with normal hematocrit range from 4 to 27 mU/mL but EPO levels can be increased 100–1000 times the normal serum EPO concentration in response to hypoxia and anemia (Jelkmann, 1992). Subcutaneous administration of a single 600 U/kg dose of epoetin α to healthy volunteers produced a peak serum concentration of over 1000 mU/mL after 24 h (Ramakrishnan et al., 2004). Furthermore, it is important to consider differences in pharmacokinetic profile of rhEPO between human, dog, rat, and mouse for achieving the desired therapeutic effects with EPO. Indeed, the total body clearance of EPO increases in the order human = dog < rat ≪ mouse (Bleuel et al., 1996; Egrie et al., 1986).

B. New Generations of Erythropoietin-Related Drugs

1. Novel Erythropoiesis Stimulating Protein

The first-generation erythropoiesis stimulating agents were succeeded by the development and production of longer-acting EPO analogs. EPO is known to be desialylated *in vivo*, cleared from plasma, and is bound to galactose receptors in the liver (Egrie et al., 1993). Thus, there is a direct relationship between the amount of sialic acid-containing carbohydrates, plasma half-life, and *in vivo* biological activity, and an inverse relationship with receptor affinity (Egrie & Browne, 2001; Egrie et al., 1993; Jelkmann, 1992).

TABLE I Molecular Characterization and Elimination Half-Life of EPO and Its Analogs

Drug	*Molecule structure*			*Route of administration*		*References*
	Molecular weight (Da)	*Glycosylation type*	*Number of sialic acid*	*Intravenous (h)*	*Subcutaneous (h)*	
Epoetin α	30,400	3N-linked oligosaccharide chains	10–14	4–8	24	Halstenson et al. (1991), Macdougall et al. (1991), McMahon et al. (1990)
Epoetin β	30,400	3N-linked oligosaccharide chains	10–14	4–10	13–28	Halstenson et al. (1991), Macdougall et al. (1991), McMahon et al. (1990)
Darbepoetin α (NESP)	37,100	5N-linked oligosaccharide chains	22	~25	~50	Egrie and Browne (2001), Egrie et al. (2003), Macdougall et al. (1999, 2007)
CERA	~60,000	30 kDa methoxy-polyethylene glycol polymer chain	10–14	~134	~139	Locatelli et al. (2007), Macdougall and Eckardt (2006), Provenzano et al. (2007)

A second-generation NESP (darbepoetin) was created to test the hypothesis that an analog with more sialic acid-containing oligosaccharides than EPO would have an extended circulating half-life and thereby an increased *in vivo* biological activity (Egrie et al., 2003; Macdougall et al., 1999). When compared with rhEPO, darbepoetin is a hyperglycosylated rhEPO analog with two extra carbohydrate chains with more sialic acid (Table I) and has a molecular weight of 37,100 Da (Macdougall, 2001). The amino acid sequence differs from that of native human EPO at five positions (Macdougall, 2001). Darbepoetin has a threefold longer circulating half-life than rhEPO (Table I), and due to the pharmacokinetic differences, the relative potency of the two molecules varies as a function of the dosing frequency. Darbepoetin α is 3.6-fold more potent than rhEPO in increasing the hematocrit of normal mice when each is administered thrice weekly, but when the administration frequency is reduced to once weekly, darbepoetin α has approximately 13-fold higher *in vivo* potency than rhEPO (Egrie et al., 2003; Macdougall, 2001).

Epoetin and darbepoetin bind to the EPOR to induce signal transduction by the same intracellular molecules as native EPO. However, differences in the glycosylation pattern lead to variations in the pharmacodynamic profiles. Darbepoetin has approximately fourfold lower EPOR binding activity than rhEPO despite higher potency. The apparent paradox is explained by the counteracting effects of sialic acid containing carbohydrate on clearance (Egrie et al., 2003; Elliott et al., 2004).

2. *Continuous Erythropoietin Receptor Activator*

More recently, a third-generation EPO-related molecule has been manufactured called continuous EPOR activator (CERA; methoxy polyethylene glycol–epoetin β), which was created by inserting a single 30 kDa polymer chain into the EPO molecule (Macdougall & Eckardt, 2006). The elimination half-life of CERA in humans is considerably increased to about 130 h and is comparable after intravenous or subcutaneous administrations (Table I; Macdougall et al., 2006). Furthermore, CERA at up to once monthly intervals maintains sustained and stable control of hemoglobin levels (Locatelli et al., 2007; Provenzano et al., 2007).

III. Signal Transduction by Erythropoietin

Survival, proliferation, and differentiation of red blood cells by EPO are mediated by activation of the homodimeric EPOR, which belongs to the cytokine receptor superfamily (Richmond et al., 2005). Binding of EPO to the erythrocytic EPOR induces a conformational change of the homodimeric-EPOR and triggers Janus protein tyrosine kinase 2 (Jak2) phosphorylation and activation (Miura et al., 1994b; Witthuhn et al., 1993).

Phosphorylation of Jak2, in turn, phosphorylates several tyrosine residues on EPOR providing docking sites for binding of several intracellular proteins and activation of multiple signaling cascades.

The tissue-protective effects of EPO, beyond the hematopoietic system, is mediated but not restricted to activation of homodimeric EPOR, but is also believed to be mediated by its actions on the heterotrimeric complex consisting of EPOR and the β-common receptor (βCR, also known as CD131). The β-common receptor is a shared receptor subunit of interleukin 3, interleukin 5, and granulocyte macrophage-colony-stimulating factor (Brines & Cerami, 2005; Brines et al., 2004). This heterotrimeric complex is postulated to mediate EPO-induced cytoprotection in the nervous system, while the involvement of EPOR–βCR complex in EPO-mediated protection in the cardiovascular system has not yet been demonstrated. Nevertheless, following binding of EPO to either homodimeric or heterotrimeric receptor complex activates Jak2. Signal transduction pathways mediating tissue protection in the cardiovascular system activated by EPO-induced phosphorylation by Jak2 are shown in Fig. 1.

Stimulation with EPO, on one hand, activates the phosphatidylinositol 3-kinase (PI3K) signaling by recruiting its p85 regulatory subunit to EPOR (Miura et al., 1994a). Phosphorylation and activation of PI3K is coupled to activation of protein kinase B/Akt, a pathway known to stimulate antiapoptotic signals that facilitate the inhibition of mitochondrial cytochrome *c* release, in part by translocating proapoptotic Bad to mitochondria, and help maintain mitochondrial membrane potential (Fig. 1). EPO-induced Akt activation prevents nuclear translocation of the proapoptotic forkhead transcription factor FOXO3a by facilitating its binding with the protein 14-3-3 (Chong & Maiese, 2007). In addition, EPO stimulation in the myocardium may lead to phosphorylation and inhibition of glycogen synthase kinase-3β (GSK-3β) in either an Akt-dependent or an independent mechanism (Miki et al., 2009; Nishihara et al., 2006). In peripheral and cerebral vasculature, activation of Akt by EPO phosphorylates endothelial nitric oxide synthase (eNOS) and increases nitric oxide (NO) production (d'Uscio et al., 2007; Santhanam et al., 2005, 2006). Akt activation may also prevent activation of caspase-9 to prevent apoptosis in the cerebral vasculature (Chong et al., 2003).

Phosphorylation of Jak2 may directly activate inhibitor of nuclear factor κB kinase (IκB kinase or IKK), or indirectly activate Akt to stimulate nuclear translocation of nuclear factor κB (NFκB) for nuclear transcription (Fig. 1) and exerts cardioprotection (Xu et al., 2005). In isolated perfused rabbit hearts, EPO has been shown to protect the heart against ischemia by activating signal transducer and activator of transcription (STAT-5) and/or mitogen-activated protein kinase (MAPK) pathway (Rafiee et al., 2005).

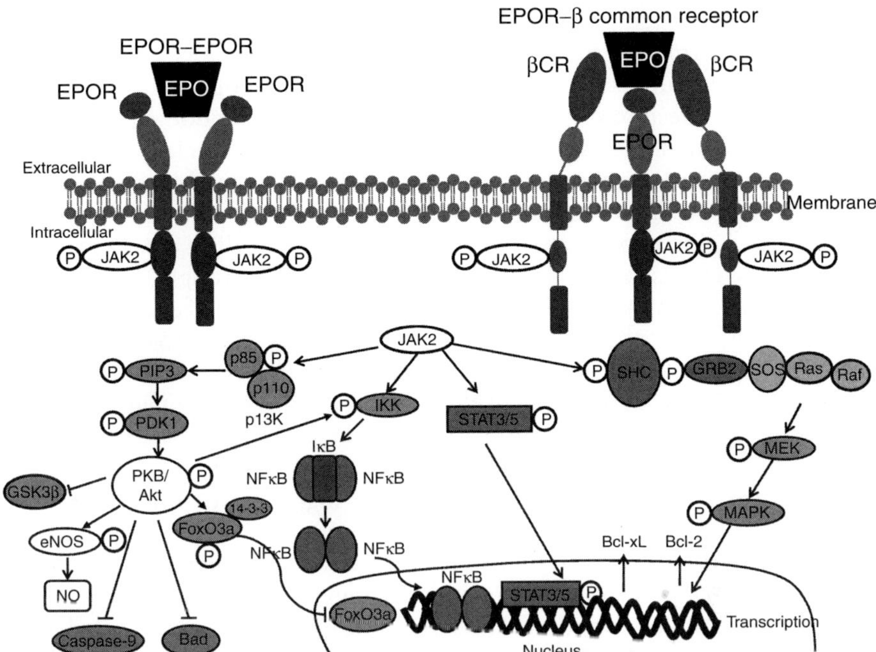

FIGURE 1 Signaling pathways mediating EPO-induced protection in the cardiovascular system. Binding of EPO to either the homodimeric EPO-receptor (EPOR) complex or the heterotrimeric EPOR–βCR complex first activates Janus tyrosine kinase 2 (Jak2). Phosphorylation and activation of Jak2 recruits secondary messengers and activates secondary signaling pathways including phosphatidylinositol 3-kinase (PI3K), mitogen-activated protein kinase (MAPK), and nuclear factor κB (NFκB). Some of the pathways mediating EPO-mediated protection are similar to those mediating erythropoiesis or neuroprotection. In particular, EPO-mediated inhibition of apoptosis involves activation of protein kinase B/Akt to inhibit caspase-9 or activate NFκB or inhibit Bad, or stimulate signal transducer and activator of transcription (STAT), or MAPK pathway to increase antiapoptotic messengers in the mitochondria including Bcl-2 and Bcl-xL. The most common pathway for EPO-mediated cardioprotection as well as vascular protection is EPO-induced augmentation of nitric oxide (NO) production by PI3K/Akt phosphorylation and activation of endothelial nitric oxide synthase (eNOS). EPO is also shown to regulate intracellular Ca^{2+} by activating phospholipase C (PLC).

IV. Mechanisms of Vascular Protective Effects of Erythropoietin

A. Erythropoietin and Endothelial Nitric Oxide Synthase

NO is a potent vasodilator and plays a key role in control of the cardiovascular system (Lüscher & Vanhoutte, 1990). NO is formed in endothelial cells from L-arginine by oxidation of its terminal guanidino-nitrogen, requiring the cofactors (6*R*)-5,6,7,8-tetrahydrobiopterin (BH_4), nicotinamide adenine dinucleotide phosphate-oxidase, flavin adenine dinucleotide, heme, and Zn^{2+} (Ignarro, 1990; Palmer et al., 1987, 1988; Raman et al., 1998). The formation of NO occurs by activation of eNOS, which is expressed constitutively (Förstermann et al., 1991; Pollock et al., 1991). Relaxations in response to the abluminal release of endothelium-derived NO are associated with stimulation of soluble guanylyl cyclase and in turn formation of cyclic guanosine 3′,5′-monophosphate (cGMP) in vascular smooth muscle cells (Rapoport et al., 1983).

Accumulating experimental evidence suggests that EPO can exert nonerythropoietic effects in vascular endothelium and is increasingly regarded as a potent tissue-protective cytokine. Indeed, EPO decreases tissue damage by inhibition of apoptosis and reduction of inflammatory cytokines (Burger et al., 2006; Iversen et al., 1999; Li et al., 2006; Rui et al., 2005). Indeed, *in vitro* treatment with low dose of rhEPO increased eNOS protein expression and NO_2^-/NO_3^- levels in cultured endothelial cells (Table II; Banerjee et al., 2000; Beleslin-Cokic et al., 2004; Wu et al., 1999). However, incubation of human coronary artery endothelial cells with high dose of rhEPO for 24 h inhibited eNOS expression and NO production (Wang & Vaziri, 1999). Moreover, asymmetric dimethylarginine concentrations were increased after high dose of rhEPO or NESP over longer period leading to a significant reduction of NO synthesis in cultured endothelial cells (Table II) suggesting that high dose of EPO may have detrimental effects on endothelial function (Scalera et al., 2005).

As illustrated in Table III, *in vivo* treatment with rhEPO has been shown to increase urinary NO_2^-/NO_3^- levels in normotensive rats (del Castillo et al., 1995; Tsukahara et al., 1997), increase vascular eNOS phosphorylation and eNOS protein expressions (d'Uscio et al., 2007; Kanagy et al., 2003; Ruschitzka et al., 2000), improve endothelium-dependent relaxations in isolated aortas of rats and mice (Iversen et al., 1999; Kanagy et al., 2003; Ruschitzka et al., 2000; Tsukahara et al., 1997), and improve endothelial function in predialysis patients (Kuriyama et al., 1996). Furthermore, in transgenic mice overexpression of human EPO markedly increased aortic eNOS protein expression, NO-mediated endothelium-dependent relaxation, and circulating and vascular tissue NO levels (Table III). These mice do not develop hypertension, stroke, myocardial infarction, or thromboembolic complications despite excessive erythrocytosis exhibiting very high hematocrit levels of 80% (Ruschitzka et al., 2000). The increased NO production in these animals appears to counteract increased expression of potent vasoconstrictor endothelin-1 (ET-1) (Quaschning et al., 2003). Indeed, EPO transgenic mice treated with the NO synthase inhibitor exhibited high systolic blood pressure and showed increased mortality, whereas wild-type siblings developed only hypertension (Ruschitzka et al., 2000). Despite concomitant activation of the ET-1 system observed in transgenic mice, elevated NO levels led to a pronounced vasodilation, thereby protecting the transgenic animals from cardiovascular complications. Consistent with this concept, studies on cultured endothelial cells demonstrated that inactivation of NO synthesis

TABLE II *In Vitro* Effects of Recombinant Human EPO on eNOS Expression in Cultured Endothelial Cells

Cell type	*Dose of rhEPO (U/mL)*	*Treatment duration*	*Reguation*	*Method of detection*	*References*
Human umbilical vein endothelial cells	0.1, 1, 20, and 40	8 h	Upregulation	NO_2^-/NO_3^- bioassay	Wu et al. (1999)
Human umbilical vein, dermis, and pulmonary artery endothelial cells	4	1–6 days	Upregulation	RT-PCR, L-arginine-to-L-citrulline conversion assay	Banerjee et al. (2000)
Human bone marrow microvascular endothelial cells	5	1 h	Upregulation	Western blot, NO_3^- bioassay	Beleslin-Cokic et al. (2004)
Human umbilical vein endothelial cells	5	0.5 h	Upregulation	NO_3^- bioassay	Beleslin-Cokic et al. (2004)
Bovine aortic endothelial cells	0.1–10	1–24 h	No change	Western blot	Lopez Ongil et al. (1996)
Human artery endothelial cells	5	1 h	No change	Western blot, NO_3^- bioassay	Beleslin-Cokic et al.(2004)
Human coronary artery endothelial cells	5 and 20	24 h	Downregulation	Western blot, NO_2^-/NO_3^- bioassay	Wang and Vaziri (1999)
Human umbilical vein endothelial cells	10, 50, 100, and 200	24 h	Downregulation	NO_2^-/NO_3^- bioassay	Scalera et al. (2005)

rhEPO, recombinant human erythropoietin; eNOS, endothelial nitric oxide synthase; NO_2^-, nitrite; NO_3^-, nitrate.

TABLE III *In Vivo* Effects of Recombinant Human EPO on Vascular eNOS in Animal Studies

Animal species	*Dose of rhEPO*	*Treatment duration (week)*	*Cell tissue*	*Result*	*Method of detection*	*References*
Nephrectomized Sprague–Dawley rat	150 U/kg, intraperitoneal, twice a week	6	Thoracic aorta	No change	Western blot, NOx bioassay	Ni et al. (1998)
Rabbit	400 U/kg, intravenous, each other day	1	Carotid artery	No change	Endothelium-dependent vasodilation studies in isolated artery	Noguchi et al. (2001)
Sprague–Dawley rat	100 or 300 U/kg, subcutaneous, each other day	2	Urine	Upregulation	NOx bioassay	Tsukahara et al. (1997)
Sprague–Dawley rat	150 U/kg, subcutaneous, three times per week	3	Urine	Upregulation	NOx bioassay	del Castillo et al. (1995)
Sprague–Dawley rat	~160 U/kg/day, subcutaneous	2	Thoracic aorta	Upregulation	Western blot, NOx bioassay, endothelium-dependent vasodilation studies in isolated aorta	Kanagy et al. (2003)
EPO–transgenic mice	n/a	n/a	Aorta	Upregulation	Western blot, NOx bioassay, endothelium-dependent vasodilation studies in isolated aorta	Ruschitzka et al. (2000)
C57BL/6 mice	1000 U/kg, subcutaneous, twice a week	2	Aorta	Upregulation	Western blot	d'Uscio et al. (2007)
C57BL/6 mice	1000 U/kg, intraperitoneal, for the initial 3 days	2	Endothelial progenitor cells	Upregulation	Immunohistochemistry, NOx bioassay	Urao et al. (2006)

caused increased production of ET-1 (Boulanger & Lüscher, 1990). In addition, EPO increased systolic blood pressure, accelerated thrombus formation, and exacerbated medial thickening of injured carotid arteries in eNOS-deficient mice (Fig. 2; d'Uscio et al., 2007; Lindenblatt et al., 2007). These observations underscore the importance of eNOS during vascular adaptation to increased circulating levels of EPO.

Because of an increased number of circulating red blood cells by EPO, subsequent increase in shear stress is a powerful stimulus for upregulation of eNOS in endothelial cells (Berk et al., 1995; Boo et al., 2002; Davis et al., 2001; Lam et al., 2006; Ruschitzka et al., 2000). However, administration of EPO for 3 days, although not affecting the number of circulating red blood cells, stimulated phosphorylation of eNOS to a similar degree as did treatment with EPO for 14 days. These findings suggest that EPO has a direct stimulatory effect on phosphorylation of eNOS in vascular endothelium and that this effect is independent of hematopoietic effects of EPO (d'Uscio et al., 2007). Most recently, it was demonstrated that activation of eNOS by hypoxia was abolished in EPOR-deficient mice, the latter mutants also

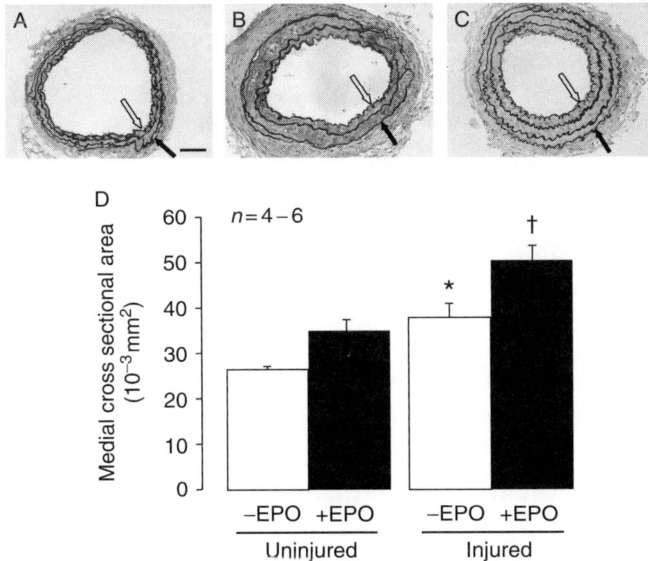

FIGURE 2 Morphological studies of carotid arteries of eNOS-deficient mice undertaken 14 days after injury. Carotid arteries were stained with standard Verhoeff van-Giessen. Representative photomicrographs of uninjured carotid arteries (A), carotid arteries after injury (B), and injured carotid arteries of eNOS-deficient mice treated with EPO for 14 days (C). The media is demarcated by internal elastic lamina (open arrow) and external elastic lamina (black arrow). Original magnification 200×. Size bar = 50 μm. (D) Quantitative histomorphometric analyses of medial CSA in carotid arteries of eNOS-deficient mice without (−) and with (+) EPO treatment. Data are shown as means ± SEM ($n = 4$–6). $^*P < 0.05$ versus control uninjured; $^{\dagger}P < 0.05$ versus uninjured + EPO (ANOVA with Bonferroni's) (d'Uscio et al., 2007).

exhibiting an exacerbation of pulmonary hypertension and vascular injury. These findings suggest that the vascular protective effects of EPO are dependent on NO production via activation of the EPOR in endothelial cells (Beleslin-Cokic et al., 2004; Satoh et al., 2006).

B. Erythropoietin and Protein Kinase B/Akt

As discussed earlier, EPO elicits phosphorylation of Akt and this is dependent upon the activation of Jak2 and PI3K. Numerous studies demonstrated the tissue-protective effects of both rhEPO- and darbepoetin α-mediated by phosphorylation of Akt and a subsequent increase in the production of NO (Bahlmann et al., 2004b; Chong & Maiese, 2007; Chong et al., 2002; d'Uscio & Katusic, 2008; Lindenblatt et al., 2007; Santhanam et al., 2006). To date, the cardiovascular effects of third generation of EPO analogs are largely unknown.

C. Erythropoietin and Tetrahydrobiopterin

Tetrahydrobiopterin (BH_4) is an essential cofactor required for enzymatic activity of eNOS (Raman et al., 1998). The biosynthesis of BH_4 is dependent on activity of the rate-limiting enzyme GTP-cyclohydrolase I (GTPCH I) (Nichol et al., 1985). Recent study showed that EPO causes an increase in intracellular levels of BH_4 via activation of GTPCH I (d'Uscio & Katusic, 2008). Pharmacological inhibition of Jak2 with AG490 abolished EPO-induced BH_4 biosynthesis, suggesting that increased phosphorylation and activation of Jak2 are the molecular mechanisms underlying the observed effect of EPO on BH_4 synthesis. Further analysis revealed that PI3K activity is an upstream activator of Akt1, because pharmacological and genetic inactivation of PI3K/Akt1 abolishes the stimulatory effects of EPO on GTPCH I activity and biosynthesis of BH_4 in mouse aorta (d'Uscio & Katusic, 2008). The ability of EPO to upregulate GTPCH I activity and eNOS phosphorylation in a coordinated fashion is most likely designed to optimize the production of NO in vascular endothelium.

D. Erythropoietin and Antioxidant Enzymes

The local concentrations of NO in arterial wall are not only dependent on enzymatic activity of eNOS but are also determined by concentrations of superoxide anions (Harrison, 1994). Indeed, recent study showed that in wild-type mice, treatment with EPO increases vascular CuZn superoxide dismutase (SOD1) expression and effectively prevents vascular remodeling after carotid artery injury (d'Uscio et al., 2010). In contrast, genetic inactivation of SOD1 abolished ability of EPO to reduce concentrations of superoxide anions thereby suggesting that EPO exerts antioxidant effect in blood vessel wall by regulating expression and activity of SOD1 protein.

E. Adverse Effects of Erythropoietin

Normally, endothelial cells contribute to the regulation of blood pressure and blood flow by releasing vasodilators such as NO and prostacyclin, as well as vasoconstrictors including ET-1 and prostanoids (Lüscher & Vanhoutte, 1990). Long-term administration of rhEPO has been associated with hypertension (Krapf & Hulter, 2009). The mechanisms of hypertension induced by long-term administration of EPO include increased cytoplasmic Ca^{2+} concentrations and increased ET-1 production leading to a blunted response to the vasodilator NO in rabbits and rats (Bode-Boger et al., 1992; Vaziri et al., 1995). Chronic use of EPO raises cytosolic Ca^{2+} concentration in smooth muscle cells, endothelial cells, or platelets (Neusser et al., 1993; Van Geet et al., 1990; Vaziri et al., 1996) and in turn, may contribute to impaired responses to NO donors. It is also suggested that EPO, at doses exceeding 200 U/mL, activates tyrosine phosphorylation of phospholipase Cγ, which hydrolyzes phosphatidyinositol bisphosphate to inositol 1,4,5-triphosphate and diacylglycerol. These second messengers may, in turn, elevate Ca^{2+} concentration as well as activate protein kinase C in vascular smooth muscle cells (Akimoto et al., 1999, 2001). Similar doses of EPO (200 U/mL) released ET-1 and increased constrictor prostanoids over dilator prostanoids *in vitro* in endothelial cells (Bode-Boger et al., 1996; Krapf & Hulter, 2009) and these observations may help to explain EPO-induced hypertension observed in animals and in humans. One implication of these results is that endothelial dysfunction predisposes to EPO-dependent hypertension. In a recent study, EPO treatment caused hypertension in rats treated in combination with a NOS inhibitor but not in rats treated with EPO alone (Moreau et al., 2000). In addition, a study using polycythemic mice overexpressing rhEPO observed that NOS blockade caused the normotensive polycythemic mice to develop hypertension (Ruschitzka et al., 2000). Therefore, erythropoiesis with a raised hematocrit is not likely associated with an increased risk for hypertension and thrombosis as long as endothelial NO production serves as compensatory mechanism (d'Uscio & Katusic, 2008; d'Uscio et al., 2007; Lindenblatt et al., 2007; Moreau et al., 2000). Impaired endothelial-dependent dilation in hypertensives (Spieker et al., 2000) as well as in hemodialyzed patients (Joannides et al., 1997) and the impaired ability to synthesize endothelial NO may increase susceptibility to EPO-induced hypertension.

V. Erythropoietin and Cardioprotection

Endogenous EPO system facilitates cardiomyocyte survival after ischemia-reperfusion injury and accelerates left ventricular remodeling (Tada et al., 2006). Deficiency of endogenous EPO–EPOR system resulted in acceleration of pressure overload-induced cardiac dysfunction by accelerating left ventricular hypertrophy, dysfunction, and reduced survival (Asaumi et al., 2007).

Therapeutic potential of EPO to exert cardioprotection was demonstrated in models of myocardial ischemia wherein EPO inhibited apoptosis and augmented survival of cardiac myocytes (Calvillo et al., 2003; Parsa et al., 2003), mediated in part, by activating EPOR expressed on the cardiac myocytes (Wright et al., 2004). Favorable results from numerous studies offered promise for administering EPO to treat myocardial infarction (Cai et al., 2003; Calvillo et al., 2003; Moon et al., 2003; Parsa et al., 2003; Wright et al., 2004).

However, subsequent *in vitro* studies identified activation of multiple signaling pathways in EPO-induced cardioprotection (Rafiee et al., 2005; Shi et al., 2004). In particular, EPO-mediated cardioprotection involves activation of one or more of the following pathways: Jak/STAT, Jak2/PI3K/Akt/GSK3β, Jak2/MAPK as well as activation of protein kinase Cε. Cardioprotection by EPO also seems to be mediated by activation of potassium channels, in particular, K_{ATP} and mitochondrial calcium-activated potassium channels (Shi et al., 2004). Administration of EPO before ischemia, at the onset of ischemia, or after reperfusion triggered response mimicking preconditioning in the ischemic myocardium (Baker, 2005; Gross & Gross, 2006; Parsa et al., 2003). However, the role of EPOR in this supposed preconditioning response remains to be clarified. In addition, the ability of EPO to stimulate eNOS activation may also facilitate cardiomyocyte survival, as observed during hypoxia-induced apoptosis of cardiac myocytes (Burger et al., 2006). As mentioned earlier, the doses of EPO for cardioprotection far-exceeded the doses of EPO used in treatment of anemia and clinical translations of these results may be hampered by the onset of widely reported adverse events. In this regard, understanding of the role of EPOR–βCR receptor complex may help expand clinical applications of nonhematopoietic analogs of EPO to cardioprotection. Recent success with carbamylated EPO, which selectively binds to EPOR-βCR heteromer, in a mouse model of ischemia–reperfusion injury (Xu et al., 2009) provides a promising outlook toward expanding tissue protection by EPO in the clinic to cardioprotection.

Occurrence of anemia as a risk factor for morbidity and mortality in patients with chronic heart failure necessitated evaluation of EPO in clinical trials. Randomized, single-center studies successfully demonstrated treatment of anemia in patients with mild, moderate, or severe heart failure with associated improvement in exercise capacity, cardiac and renal function, and reduced use of diuretics (Mancini et al., 2003; Namiuchi et al., 2005; Palazzuoli et al., 2006; Silverberg et al., 2001). Recently, three trials evaluated the safety and efficacy of darbepoetin α (0.75 μg/kg once every 2 weeks) in symptomatic heart failure patients (Ghali et al., 2008; Ponikowski et al., 2007; van Veldhuisen et al., 2007). Incidence of adverse events in these trials was similar between placebo- and darbepoetin α-treated patients (Klapholz et al., 2009), while an increase in hemoglobin with darbepoetin α treatment tended to correlate with improved health-related quality of life. Results obtained in these clinical trials extend support as a proof of concept and encourage the need for larger outcome trials for treatment of anemia with erythropoiesis-stimulating agents.

In addition to treatment of anemia, results from preclinical studies attribute multiple mechanisms of protection by EPO against myocardial disorders. Treatment with EPO may decrease apoptosis of myocytes, induce neovascularization by promoting myocardial angiogenesis, reduce collagen deposition in ischemic myocardium, and improve left ventricular function. Employing either nonhematopoietic analogs of EPO or a novel EPO delivery system (Kobayashi et al., 2008) may expand therapeutic boundaries for EPO-mediated cardioprotection. Success from these studies will lay the frame work for future clinical evaluation in the treatment of myocardial infarction.

VI. Erythropoietin and Cerebrovascular Disorders

Identification of EPO and EPOR in the brain (Masuda et al., 1994; Tan et al., 1992) expanded investigations into the tissue-protective effects of EPO, beyond the hematopoietic system. Initial research ascribed EPO-mediated protection to the cytokine's ability to inhibit apoptosis in tissues adjacent to a pathological insult in the brain. The current understanding of the tissue-protective effects of EPO in the nervous system involves interaction of the nonhematopoietic βCR with the classical EPOR and subsequent activation of multiple signaling cascades, as reviewed by Brines and Cerami (2005). In the nervous system, EPO mediates neuroprotection following ischemic, hypoxic, metabolic, neurotoxic, and excitotoxic stress (Genc et al., 2004). EPO-mediated protective effects in the brain may involve one or more of the following mechanisms: (a) prevention of excitatory aminoacid release, (b) inhibition of apoptosis, (c) antioxidant effects, (d) anti-inflammatory effects, and (e) stimulation of neurogenesis and angiogenesis.

EPO exhibits higher affinity for endothelium of cerebral arteries as compared to neuronal cells (Brines & Cerami, 2006). Prior studies have demonstrated that EPO activates cerebrovascular protective mechanisms (Grasso, 2004; Grasso et al., 2002; Santhanam et al., 2005, 2006). In cerebral arteries exposed to recombinant EPO, the expressions of eNOS and its phosphorylated (S1177) form were increased. Basal levels of cGMP were also significantly elevated consistent with increased NO production. Overexpression of EPO in cerebral arteries reversed vasospasm in rabbits induced by injection of autologous blood into cisterna magna. Arteries-transduced with recombinant EPO demonstrated significant augmentation of the endothelium-dependent relaxations to acetylcholine. Overexpression of EPO further increased the expression of phosphorylated Akt and eNOS and elevated basal levels of cGMP in the spastic arteries (Santhanam et al., 2005). Cerebrovascular protective effects of EPO against cerebral vasospasm following experimental subarachnoid hemorrhage may appear to be mediated in part by phosphorylation and activation of endothelial Akt/eNOS pathway.

Administration of EPO into the brain reduced neurological dysfunction in rodent models of stroke (Bernaudin et al., 1999; Brines et al., 2000; Sadamoto et al., 1998; Sakanaka et al., 1998; Siren et al., 2001). Following the success of EPO in preclinical models of stroke and cerebral ischemia, a successful proof-of-concept clinical trial demonstrated that intravenously injected EPO was well tolerated in patients with acute ischemic stroke (Ehrenreich et al., 2002). In addition, outcomes from a recent Phase II trial report beneficial effect of EPO, as observed from a reduction in delayed ischemic deficits after aneurysmal subarachnoid hemorrhage (Tseng et al., 2009).

However, results from the German Multicenter EPO Stroke Phase II/III trial warrant caution in administering EPO with thrombolytics in patients with acute stroke (Ehrenreich et al., 2009). In this trial, patients receiving EPO alone demonstrated a reduction in National Institute of Health Stroke Scale (NIHSS) score from day 1 to day 90, an index of improved neurological outcome. On the contrary, when combined with thrombolytic recombinant tissue plasminogen activator, patients receiving EPO demonstrated increased risk of complications including death, intracerebral hemorrhage, brain edema, and thromboembolic events compared to patients receiving placebo. Further understanding of the complex interactions between different components of the cardiovascular system initiated by EPO is necessary for designing better strategies to maximize therapeutic potential of this cytokine.

With new evidence pointing toward the role of nonhematopoietic βCR in EPO-mediated tissue-protective effects (Grasso et al., 2004), use of nonerythrocytic analogs of EPO currently under development offers promise of therapy as adjunct for patients with stroke and other cerebral vascular disorders with enhanced safety and reduced adverse effects. Another novel strategy currently investigated for acute stroke has employed the neurotrophic ability of EPO to differentiate neural progenitor cells into neurons. The Phase IIb prospective randomized, double-blind study of NTx^{TM}-265, comprising human chorionic gonadotrophin (for proliferation of endogenous neural stem cells) and EPO (to differentiate these neural stem cells) aims to improve neurological outcome in acute stroke patients, and was based on the success of preclinical studies adopting similar strategy in rats (Belayev et al., 2009).

VII. Erythropoietin and Progenitor Cells

Seminal studies by Heeschen et al. (2003) demonstrated that circulating levels of EPO in humans significantly correlated with the number of stem and progenitor cells in the bone marrow as well as to the number and function of circulating progenitor cells. In addition, treatment of mice with EPO increased the number of stem and progenitor cells in the bone marrow as well as an increase in the number of peripheral blood endothelial progenitor cells (EPCs), suggestive of stimulation of mobilization by EPO. Indeed, mobilization of progenitor cells by EPO-stimulated postnatal neovascularization in a model of hind limb ischemia (Heeschen et al., 2003; Kato et al., 2010; Li et al., 2009). Numerous studies, subsequently, have established a crucial proangiogenic role of EPO, as indicated by enhanced mobilization of

TABLE IV Contribution to Vascular Repair by Progenitor Cells Mobilized by EPO

Species	*Dose of EPO*	*Functional outcome*	*Phenotype of mobilized progenitor cells*	*References*
Mice, humans	1000 U/kg	Increased angiogenesis in models of disk neovascularization and hind limb ischemia	Lin-1$^+$/Sca-1$^+$ stem cells Sca-1$^+$/Flk-1$^+$ or CD34$^+$/Flk-1$^+$ cells	Heeschen et al. (2003)
Humans	5000±674 U	Increased angiogenesis in EPCs of patients exposed to EPO	CD34$^+$/CD45$^+$ cells UEA-1$^+$/acLDL-DiI$^+$ cells	Bahlmann et al. (2004a)
Mice	1000 U/kg	Inhibition of neointimal hyperplasia after wire injury of femoral artery	CD45dim/Flk-1$^+$ cells Sca-1$^+$/Flk-1$^+$ cells	Urao et al. (2006)
Mice	100 U	Accelerated revascularization in a model of hind limb ischemia	CXCR4$^+$/VEGFR1$^+$-hemangiocytes	Jin et al. (2006)
Dogs	1000 U/kg	Augmented neovascularization in a model of myocardial infarction	CD34$^+$-mononuclear cells Di-acLDL$^+$/UEA-I cells	Hirata et al. (2006)
Rats	40 μg/kg	Increased functional neovascularization following acute myocardial infarction	Ac-LDL$^+$/Lectin$^+$ cells per high powered field	Westenbrink et al. (2007)
Mice	1000 U/kg	Epo-induced mobilization impaired in eNOS$^{-/-}$ mice	Lin-1$^-$/Sca-1$^+$/c-Kit$^+$ cells CD34$^+$/Flk-1$^+$ cells	Santhanam et al. (2008)
Mice	500 U/Kkg	Acceleration of smooth muscle lesion formation by EPO in a model of carotid artery ligation	Ter-119$^-$/CD45lo/c-Kit$^+$/Sca-1$^+$ cells	Janmaat et al. (2010)

This table summarizes the results from studies wherein EPO was administered to stimulate mobilization of progenitor cells in a model of vascular injury. Studies relating increased endogenous levels of EPO with enhanced mobilization of progenitor cells and their contribution to vascular repair have not been discussed.

progenitor cells, to elicit vascular repair. As illustrated in Table IV, treatment with EPO significantly augments mobilization of diverse population of progenitor cells from bone marrow distinguished by their distinct phenotype and, in turn, the mobilized progenitor cells contributed to either neovascularization or repair of denuded endothelium.

Multiple mechanisms are likely to mediate mobilization of progenitor cells by EPO (Aicher et al., 2005; Heeschen et al., 2003; Urao et al., 2006). Prior studies from our laboratory have demonstrated that activation of endothelial nitric oxide synthase is critical to EPO-induced mobilization of hematopoietic stem and progenitor cells and $CD34^+$/Flk-1^+ EPCs (Santhanam et al., 2008). The ability of EPO to upregulate the antioxidant capacity of EPCs, in particular, activation of SOD1 may also help to explain the pronounced vascular protection observed with this pleiotropic cytokine (He et al., 2005).

Endogenous EPO–EPOR system in the vasculature also dictates mobilization of progenitor cells and facilitates vascular repair. Satoh et al. (2006) successfully demonstrated this crucial role of endogenous EPO–EPOR using EPOR null mutant mice that expresses EPO-R exclusively in the erythroid lineage ($EPOR^{-/-}$ rescued mice). Lack of EPOR signaling in the hematopoietic system resulted in attenuated mobilization of Flk-1^+/$CD133^+$ EPCs (hematopoietic progenitor cell population; Satoh et al., 2006), as well as alteration in VEGF/VEGFR signaling and impaired recovery after hind limb ischemia in mice (Nakano et al., 2007). During hypoxia-induced pulmonary hypertension, in addition to impaired endogenous mobilization, recruitment of administered progenitor cells was impaired in $EPOR^{-/-}$ rescued mice (Satoh et al., 2006). It is likely that endogenous EPO-mediated effects are not restricted to its effects on mature vascular cells, but also contributes to mobilization, recruitment, and activation of progenitor cells as demonstrated either in hypoxia-induced pulmonary hypertension (Satoh et al., 2006) or in a model of myocardial ischemia and reperfusion (Tada et al., 2006). Elevated plasma levels of EPO in patients with acute myocardial infarction correlated with mobilization of $CD34^+$/$CD133^+$/$VEGFR2^+$ EPCs (Ferrario et al., 2007). It is therefore likely that the extent of alteration of endogenous EPO levels following a cardiovascular insult may dictate the proportion of endogenous progenitor cells mobilized in the mononuclear cell population, which in turn, may contribute to the extent of damage or repair.

VIII. Conclusion

Extensive research has demonstrated that not only does EPO affect the hematopoietic system but it also plays an important role in control of cardiovascular system. EPO has been shown to be vascular protective by exerting its effects on the endothelial cells and vascular smooth muscle cells. Tissue-protective effects of EPO on the vasculature are mediated, in part, by prevention of apoptosis or stimulation of endothelial nitric oxide production and augmented vasodilatation. In addition, EPO regulates mobilization of proangiogenic cells, including EPCs, from the bone marrow and stimulates neovascularization. EPO may exert its beneficial effects on the vasculature by acting through its homodimeric EPOR or through EPOR–βCR and activation of Jak2 to stimulate PI3K/Akt, NFκB, and MAPK signaling pathways. Further studies are however needed to clarify the receptor(s) involved in protective effects of EPO in different vascular cell types and injury models. In addition, exacerbation of injury in EPO-treated mice deficient in either eNOS or CuZn superoxide dismutase highlights the critical role of endogenous NO or antioxidant defense mechanisms in triggering adaptation of vascular wall to elevated doses of EPO, commonly used in preclinical and clinical settings. These studies may help to explain the adverse events precluding the clinical use of EPO in treatment of cardiovascular and neurological disorders.

Results obtained from clinical trials with EPO on cerebrovascular disorders, conducted over the last decade, suggest EPO as a promising lead toward designing and developing novel strategies for these disorders. Future research on better understanding of protective effects of the nonhematopoietic analogs of EPO as well as harnessing the EPO-induced ability to stimulate progenitor cell function may favor therapeutic benefits with minimal side effects associated with this pleiotropic cytokine.

Acknowledgments

This work was supported by National Institutes of Health Grants HL53524, HL91867 (to Z. S. K.), Roche Foundation for Anemia Research, American Heart Association Scientist Development Grants #07301333N (to L. V. D.), #0835436N (to A. V. R. S.), and the Mayo Foundation.

Conflict of interest: The authors have no conflicts of interest to declare.

Abbreviations

EPO	erythropoietin
EPOR	erythropoietin receptor
βCR	β-common receptor
rhEPO	recombinant human erythropoietin
NESP	novel erythropoiesis stimulating protein
CERA	continuous erythropoietin receptor activator
Jak2	Janus kinase 2
NFκB	nuclear factor κ B
STAT-5	signal transducer and activator of transcription 5
MAPK	mitogen-activated protein kinase

PI3K phosphatidylinositol 3-kinase
NO nitric oxide
eNOS endothelial nitric oxide synthase
cGMP cyclic guanosine 3′,5′-monophosphate
ET-1 endothelin-1

References

Aicher, A., Zeiher, A. M., & Dimmeler, S. (2005). Mobilizing endothelial progenitor cells. *Hypertension*, *45*, 321–325.

Akimoto, T., Kusano, E., Ito, C., Yanagiba, S., Inoue, M., Amemiya, M., Ando, Y., & Asano, Y. (2001). Involvement of erythropoietin-induced cytosolic free calcium mobilization in activation of mitogen-activated protein kinase and DNA synthesis in vascular smooth muscle cells. *Journal of Hypertension*, *19*, 193–202.

Akimoto, T., Kusano, E., Muto, S., Fujita, N., Okada, K., Saito, T., Komatsu, N., Ono, S., Ebata, S., Ando, Y., Homma, S., & Asano, Y. (1999). The effect of erythropoietin on interleukin-1beta mediated increase in nitric oxide synthesis in vascular smooth muscle cells. *Journal of Hypertension*, *17*, 1249–1256.

Ammarguellat, F., Gogusev, J., & Drueke, T. B. (1996). Direct effect of erythropoietin on rat vascular smooth-muscle cell via a putative erythropoietin receptor. *Nephrology, Dialysis, Transplantation*, *11*, 687–692.

Anagnostou, A., Liu, Z., Steiner, M., Chin, K., Lee, E. S., Kessimian, N., & Noguchi, C. T. (1994). Erythropoietin receptor mRNA expression in human endothelial cells. *Proceedings of the National Academy of Sciences of the United States of America*, *91*, 3974–3978.

Asaumi, Y., Kagaya, Y., Takeda, M., Yamaguchi, N., Tada, H., Ito, K., Ohta, J., Shiroto, T., Shirato, K., Minegishi, N., & Shimokawa, H. (2007). Protective role of endogenous erythropoietin system in nonhematopoietic cells against pressure overload-induced left ventricular dysfunction in mice. *Circulation*, *115*, 2022–2032.

Bahlmann, F. H., De Groot, K., Spandau, J. M., Landry, A. L., Hertel, B., Duckert, T., Boehm, S. M., Menne, J., Haller, H., & Fliser, D. (2004a). Erythropoietin regulates endothelial progenitor cells. *Blood*, *103*, 921–926.

Bahlmann, F. H., Song, R., Boehm, S. M., Mengel, M., von Wasielewski, R., Lindschau, C., Kirsch, T., de Groot, K., Laudeley, R., Niemczyk, E., Guler, F., Menne, J., Haller, H., & Fliser, D. (2004b). Low-dose therapy with the long-acting erythropoietin analogue darbewpoetin alpha persistently activates endothelial Akt and attenuates progressive organ failure. *Circulation*, *110*, 1006–1012.

Baker, J. E. (2005). Erythropoietin mimics ischemic preconditioning. *Vascular Pharmacology*, *42*, 233–241.

Banerjee, D., Rodriguez, M., Nag, M., & Adamson, J. W. (2000). Exposure of endothelial cells to recombinant human erythropoietin induces nitric oxide synthase activity. *Kidney International*, *57*, 1895–1904.

Belayev, L., Khoutorova, L., Zhao, K. L., Davidoff, A. W., Moore, A. F., & Cramer, S. C. (2009). A novel neurotrophic therapeutic strategy for experimental stroke. *Brain Research*, *1280*, 117–123.

Beleslin-Cokic, B. B., Cokic, V. P., Yu, X., Weksler, B. B., Schechter, A. N., & Noguchi, C. T. (2004). Erythropoietin and hypoxia stimulate erythropoietin receptor and nitric oxide production by endothelial cells. *Blood*, *104*, 2073–2080.

Berk, B. C., Corson, M. A., Peterson, T. E., & Tseng, H. (1995). Protein kinases as mediators of fluid shear stress stimulated signal transduction in endothelial cells: A hypothesis for calcium-dependent and calcium-independent events activated by flow. *Journal of Biomechanics*, *28*, 1439–1450.

Bernaudin, M., Marti, H. H., Roussel, S., Divoux, D., Nouvelot, A., MacKenzie, E. T., & Petit, E. (1999). A potential role for erythropoietin in focal permanent cerebral ischemia in mice. *Journal of Cerebral Blood Flow and Metabolism*, *19*, 643–651.

Bleuel, H., Hoffmann, R., Kaufmann, B., Neubert, P., Ochlich, P. P., & Schaumann, W. (1996). Kinetics of subcutaneous versus intravenous epoetin-beta in dogs, rats and mice. *Pharmacology*, *52*, 329–338.

Bode-Boger, S. M., Boger, R. H., Kuhn, M., Radermacher, J., & Frolich, J. C. (1992). Endothelin release and shift in prostaglandin balance are involved in the modulation of vascular tone by recombinant erythropoietin. *Journal of Cardiovascular Pharmacology*, *20*(Suppl. 12), S25–S28.

Bode-Boger, S. M., Boger, R. H., Kuhn, M., Radermacher, J., & Frolich, J. C. (1996). Recombinant human erythropoietin enhances vasoconstrictor tone via endothelin-1 and constrictor prostanoids. *Kidney International*, *50*, 1255–1261.

Bonsdorff, E., & Jalavisto, E. (1948). A humoral mechanism in anoxic erythrocytosis. *Acta Physiologica Scandinavica*, *16*, 150–170.

Boo, Y. C., Sorescu, G., Boyd, N., Shiojima, I., Walsh, K., Du, J., & Jo, H. (2002). Shear stress stimulates phosphorylation of endothelial nitric-oxide synthase at Ser1179 by Akt-independent mechanisms: role of protein kinase A. *Journal of Cerebral Blood Flow and Metabolism*, *277*, 3388–3396.

Boulanger, C., & Luscher, T. F. (1990). Release of endothelin from the porcine aorta: Inhibition by endothelium-derived nitric oxide. *Journal of Clinical Investigation*, *85*, 587–590.

Brines, M., & Cerami, A. (2005). Emerging biological roles for erythropoietin in the nervous system. *Nature Reviews. Neuroscience*, *6*, 484–494.

Brines, M., & Cerami, A. (2006). Discovering erythropoietin's extra-hematopoietic functions: Biology and clinical promise. *Kidney International*, *70*, 246–250.

Brines, M. L., Ghezzi, P., Keenan, S., Agnello, D., de Lanerolle, N. C., Cerami, C., Itri, L. M., & Cerami, A. (2000). Erythropoietin crosses the blood-brain barrier to protect against experimental brain injury. *Proceedings of the National Academy of Sciences of the United States of America*, *97*, 10526–10531.

Brines, M., Grasso, G., Fiordaliso, F., Sfacteria, A., Ghezzi, P., Fratelli, M., Latini, R., Xie, Q. W., Smart, J., Su-Rick, C. J., Pobre, E., Diaz, D., Gomez, D., Hand, C., Coleman, T., & Cerami, A. (2004). Erythropoietin mediates tissue protection through an erythropoietin and common beta-subunit heteroreceptor. *Proceedings of the National Academy of Sciences of the United States of America*, *101*, 14907–14912.

Burger, D., Lei, M., Geoghegan-Morphet, N., Lu, X., Xenocostas, A., & Feng, Q. (2006). Erythropoietin protects cardiomyocytes from apoptosis via up-regulation of endothelial nitric oxide synthase. *Cardiovascular Research*, *72*, 51–59.

Cai, Z., Manalo, D. J., Wei, G., Rodriguez, E. R., Fox-Talbot, K., Lu, H., Zweier, J. L., & Semenza, G. L. (2003). Hearts from rodents exposed to intermittent hypoxia or erythropoietin are protected against ischemia-reperfusion injury. *Circulation*, *108*, 79–85.

Calvillo, L., Latini, R., Kajstura, J., Leri, A., Anversa, P., Ghezzi, P., Salio, M., Cerami, A., & Brines, M. (2003). Recombinant human erythropoietin protects the myocardium from ischemia-reperfusion injury and promotes beneficial remodeling. *Proceedings of the National Academy of Sciences of the United States of America*, *100*, 4802–4806.

Chong, Z. Z., Kang, J. Q., & Maiese, K. (2002). Erythropoietin is a novel vascular protectant through activation of Akt1 and mitochondrial modulation of cysteine proteases. *Circulation*, *106*, 2973–2979.

Chong, Z. Z., Kang, J. Q., & Maiese, K. (2003). Apaf-1, Bcl-xL, cytochrome c, and caspase-9 form the critical elements for cerebral vascular protection by erythropoietin. *Journal of Cerebral Blood Flow and Metabolism*, *23*, 320–330.

Chong, Z. Z., & Maiese, K. (2007). Erythropoietin involves the phosphatidylinositol 3-kinase pathway, 14-3-3 protein and FOXO3a nuclear trafficking to preserve endothelial cell integrity. *British Journal of Pharmacology*, *150*, 839–850.

Davis, M. E., Cai, H., Drummond, G. R., & Harrison, D. G. (2001). Shear stress regulates endothelial nitric oxide synthase expression through c-Src by divergent signaling pathways. *Circulation Research*, *89*, 1073–1080.

del Castillo, D., Raij, L., Shultz, P. J., & Tolins, J. P. (1995). The pressor effect of recombinant human erythropoietin is not due to decreased activity of the endogenous nitric oxide system. *Nephrology, Dialysis, Transplantation*, *10*, 505–508.

d'Uscio, L. V., & Katusic, Z. S. (2008). Erythropoietin increases endothelial biosynthesis of tetrahydrobiopterin by activation of protein kinase B alpha/Akt1. *Hypertension*, *52*, 93–99.

d'Uscio, L. V., Smith, L. A., & Katusic, Z. S. (2010). Erythropoietin increases expression and function of vascular copper- and zinc-containing superoxide dismutase. *Hypertension*, *55*, 998–1004.

d'Uscio, L. V., Smith, L. A., Santhanam, A. V., Richardson, D., Nath, K. A., & Katusic, Z. S. (2007). Essential role of endothelial nitric oxide synthase in vascular effects of erythropoietin. *Hypertension*, *49*, 1142–1148.

Egrie, J. C., & Browne, J. K. (2001). Development and characterization of novel erythropoiesis stimulating protein (NESP). *British Journal of Cancer*, *84*(Suppl. 1), 3–10.

Egrie, J. C., Dwyer, E., Browne, J. K., Hitz, A., & Lykos, M. A. (2003). Darbepoetin alfa has a longer circulating half-life and greater *in vivo* potency than recombinant human erythropoietin. *Experimental Hematology*, *31*, 290–299.

Egrie, J. C., Grant, J. R., Gillies, D. K., Aoki, K. H., & Strickland, T. W. (1993). The role of carbohydrate on the biological activity of erythropoietin. *Glycoconjugate Journal*, *10*, 263.

Egrie, J. C., Strickland, T. W., Lane, J., Aoki, K., Cohen, A. M., Smalling, R., Trail, G., Lin, F. K., Browne, J. K., & Hines, D. K. (1986). Characterization and biological effects of recombinant human erythropoietin. *Immunobiology*, *172*, 213–224.

Ehrenreich, H., Hasselblatt, M., Dembowski, C., Cepek, L., Lewczuk, P., Stiefel, M., Rustenbeck, H. H., Breiter, N., Jacob, S., Knerlich, F., Bohn, M., Poser, W., Ruther, E., Kochen, M., Gefeller, O., Gleiter, C., Wessel, T. C., De Ryck, M., Itri, L., Prange, H., Cerami, A., Brines, M., & Siren, A. L. (2002). Erythropoietin therapy for acute stroke is both safe and beneficial. *Molecular Medicine*, *8*, 495–505.

Ehrenreich, H., Weissenborn, K., Prange, H., Schneider, D., Weimar, C., Wartenberg, K., Schellinger, P. D., Bohn, M., Becker, H., Wegrzyn, M., Jahnig, P., Herrmann, M., Knauth, M., Bahr, M., Heide, W., Wagner, A., Schwab, S., Reichmann, H., Schwendemann, G., Dengler, R., Kastrup, A., & Bartels, C. (2009). Recombinant human erythropoietin in the treatment of acute ischemic stroke. *Stroke*, *40*, e647–656.

Elliott, S., Egrie, J., Browne, J., Lorenzini, T., Busse, L., Rogers, N., & Ponting, I. (2004). Control of rHuEPO biological activity: The role of carbohydrate. *Experimental Hematology*, *32*, 1146–1155.

Ferrario, M., Massa, M., Rosti, V., Campanelli, R., Ferlini, M., Marinoni, B., De Ferrari, G. M., Meli, V., De Amici, M., Repetto, A., Verri, A., Bramucci, E., & Tavazzi, L. (2007). Early haemoglobin-independent increase of plasma erythropoietin levels in patients with acute myocardial infarction. *European Heart Journal*, *28*, 1805–1813.

Förstermann, U., Schmitt, H. H. H. W., Pollock, J. S., Sheng, H., Mitchell, J. A., Warner, T. D., Nakane, M., & Murrad, F. (1991). Isoforms of nitric oxide synthase: Characterization and purification from different cell types. *Biochemical Pharmacology*, *42*, 1849–1857.

Genc, S., Koroglu, T. F., & Genc, K. (2004). Erythropoietin and the nervous system. *Brain Research*, *1000*, 19–31.

Ghali, J. K., Anand, I. S., Abraham, W. T., Fonarow, G. C., Greenberg, B., Krum, H., Massie, B. M., Wasserman, S. M., Trotman, M. L., Sun, Y., Knusel, B., & Armstrong, P. (2008). Randomized double-blind trial of darbepoetin alfa in patients with symptomatic heart failure and anemia. *Circulation*, *117*, 526–535.

Grasso, G. (2004). An overview of new pharmacological treatments for cerebrovascular dysfunction after experimental subarachnoid hemorrhage. *Brain Research. Brain Research Reviews*, *44*, 49–63.

Grasso, G., Buemi, M., Alafaci, C., Sfacteria, A., Passalacqua, M., Sturiale, A., Calapai, G., De Vico, G., Piedimonte, G., Salpietro, F. M., & Tomasello, F. (2002). Beneficial effects of systemic administration of recombinant human erythropoietin in rabbits subjected to subarachnoid hemorrhage. *Proceedings of the National Academy of Sciences of the United States of America*, *99*, 5627–5631.

Grasso, G., Sfacteria, A., Cerami, A., & Brines, M. (2004). Erythropoietin as a tissue-protective cytokine in brain injury: What do we know and where do we go? *The Neuroscientist*, *10*, 93–98.

Gross, E. R., & Gross, G. J. (2006). Ligand triggers of classical preconditioning and postconditioning. *Cardiovascular Research*, *70*, 212–221.

Halstenson, C. E., Macres, M., Katz, S. A., Schnieders, J. R., Watanabe, M., Sobota, J. T., & Abraham, P. A. (1991). Comparative pharmacokinetics and pharmacodynamics of epoetin alfa and epoetin beta. *Clinical Pharmacology and Therapeutics*, *50*, 702–712.

Harrison, D. G. (1994). Endothelial dysfunction in atherosclerosis. *Basic Research in Cardiology*, *89*(Suppl. 1), 87–102.

He, T., Richardson, D., Oberley, L., & Katusic, Z. (2005). Erythropoietin increases expression of CuZn superoxide dismutase in endothelial progenitor cells. *Stroke*, *36*, 426.

Heeschen, C., Aicher, A., Lehmann, R., Fichtlscherer, S., Vasa, M., Urbich, C., Mildner-Rihm, C., Martin, H., Zeiher, A. M., & Dimmeler, S. (2003). Erythropoietin is a potent physiologic stimulus for endothelial progenitor cell mobilization. *Blood*, *102*, 1340–1346.

Hirata, A., Minamino, T., Asanuma, H., Fujita, M., Wakeno, M., Myoishi, M., Tsukamoto, O., Okada, K., Koyama, H., Komamura, K., Takashima, S., Shinozaki, Y., Mori, H., Shiraga, M., Kitakaze, M., & Hori, M. (2006). Erythropoietin enhances neovascularization of ischemic myocardium and improves left ventricular dysfunction after myocardial infarction in dogs. *Journal of the American College of Cardiology*, *48*, 176–184.

Ignarro, L. J. (1990). Biosynthesis and metabolism of endothelium-derived nitric oxide. *Annual Review of Pharmacology and Toxicology*, *30*, 535–560.

Iversen, P. O., Nicolaysen, A., Kvernebo, K., Benestad, H. B., & Nicolaysen, G. (1999). Human cytokines modulate arterial vascular tone via endothelial receptors. *Pflugers Archiv*, *439*, 93–100.

Jacobs, K., Shoemaker, C., Rudersdorf, R., Neill, S. D., Kaufman, R. J., Mufson, A., Seehra, J., Jones, S. S., Hewick, R., Fritsch, E. F., et al. (1985). Isolation and characterization of genomic and cDNA clones of human erythropoietin. *Nature*, *313*, 806–810.

Janmaat, M. L., Heerkens, J. L., de Bruin, A. M., Klous, A., de Waard, V., & de Vries, C. J. (2010). Erythropoietin accelerates smooth muscle cell-rich vascular lesion formation in mice through endothelial cell activation involving enhanced PDGF-BB release. *Blood*, *115*, 1453–1460.

Jelkmann, W. (1992). Erythropoietin: Structure, control of production, and function. *Physiological Reviews*, *72*, 449–489.

Jin, D. K., Shido, K., Kopp, H. G., Petit, I., Shmelkov, S. V., Young, L. M., Hooper, A. T., Amano, H., Avecilla, S. T., Heissig, B., Hattori, K., Zhang, F., Hicklin, D. J., Wu, Y., Zhu, Z., Dunn, A., Salari, H., Werb, Z., Hackett, N. R., Crystal, R. G., Lyden, D., & Rafii, S. (2006). Cytokine-mediated deployment of SDF-1 induces revascularization through recruitment of CXCR4+ hemangiocytes. *Nature Medicine*, *12*, 557–567.

Joannides, R., Bakkali, E. H., Le Roy, F., Rivault, O., Godin, M., Moore, N., Fillastre, J. P., & Thuillez, C. (1997). Altered flow-dependent vasodilatation of conduit arteries in maintenance haemodialysis. *Nephrology, Dialysis, Transplantation*, *12*, 2623–2628.

Kanagy, N. L., Perrine, M. F., Cheung, D. K., & Walker, B. R. (2003). Erythropoietin administration *in vivo* increases vascular nitric oxide synthase expression. *Journal of Cardiovascular Pharmacology*, *42*, 527–533.

Kato, S., Amano, H., Ito, Y., Eshima, K., Aoyama, N., Tamaki, H., Sakagami, H., Satoh, Y., Izumi, T., & Majima, M. (2010). Effect of erythropoietin on angiogenesis with the increased adhesion of platelets to the microvessels in the hind-limb ischemia model in mice. *Journal of Pharmacological Sciences*, *112*, 167–175.

Klapholz, M., Abraham, W. T., Ghali, J. K., Ponikowski, P., Anker, S. D., Knusel, B., Sun, Y., Wasserman, S. M., & van Veldhuisen, D. J. (2009). The safety and tolerability of darbepoetin alfa in patients with anaemia and symptomatic heart failure. *European Journal of Heart Failure*, *11*, 1071–1077.

Kobayashi, H., Minatoguchi, S., Yasuda, S., Bao, N., Kawamura, I., Iwasa, M., Yamaki, T., Sumi, S., Misao, Y., Ushikoshi, H., Nishigaki, K., Takemura, G., Fujiwara, T., Tabata, Y., & Fujiwara, H. (2008). Post-infarct treatment with an erythropoietin-gelatin hydrogel drug delivery system for cardiac repair. *Cardiovascular Research*, *79*, 611–620.

Krapf, R., & Hulter, H. N. (2009). Arterial hypertension induced by erythropoietin and erythropoiesis-stimulating agents (ESA). *Clinical Journal of the American Society of Nephrology*, *4*, 470–480.

Kuriyama, S., Hopp, L., Yoshida, H., Hikita, M., Tomonari, H., Hashimoto, T., & Sakai, O. (1996). Evidence for amelioration of endothelial cell dysfunction by erythropoietin therapy in predialysis patients. *American Journal of Hypertension*, *9*, 426–431.

Lai, P. H., Everett, R., Wang, F. F., Arakawa, T., & Goldwasser, E. (1986). Structural characterization of human erythropoietin. *The Journal of Biological Chemistry*, *261*, 3116–3121.

Lam, C. F., Peterson, T. E., Richardson, D. M., Croatt, A. J., d'Uscio, L. V., Nath, K. A., & Katusic, Z. S. (2006). Increased blood flow causes coordinated upregulation of arterial eNOS and biosynthesis of tetrahydrobiopterin. *American Journal of Physiology. Heart and Circulatory Physiology*, *290*, H786–793.

Li, L., Okada, H., Takemura, G., Esaki, M., Kobayashi, H., Kanamori, H., Kawamura, I., Maruyama, R., Fujiwara, T., Fujiwara, H., Tabata, Y., & Minatoguchi, S. (2009). Sustained release of erythropoietin using biodegradable gelatin hydrogel microspheres persistently improves lower leg ischemia. *Journal of the American College of Cardiology*, *53*, 2378–2388.

Li, Y., Takemura, G., Okada, H., Miyata, S., Maruyama, R., Li, L., Higuchi, M., Minatoguchi, S., Fujiwara, T., & Fujiwara, H. (2006). Reduction of inflammatory cytokine expression and oxidative damage by erythropoietin in chronic heart failure. *Cardiovascular Research*, *71*, 684–694.

Lin, F. K., Suggs, S., Lin, C. H., Browne, J. K., Smalling, R., Egrie, J. C., Chen, K. K., Fox, G. M., Martin, F., Stabinsky, Z., et al. (1985). Cloning and expression of the human erythropoietin gene. *Proceedings of the National Academy of Sciences of the United States of America*, *82*, 7580–7584.

Lindenblatt, N., Menger, M. D., Klar, E., & Vollmar, B. (2007). Darbepoetin-alpha does not promote microvascular thrombus formation in mice: Role of eNOS-dependent protection through platelet and endothelial cell deactivation. *Arteriosclerosis, Thrombosis, and Vascular Biology*, *27*, 1191–1198.

Locatelli, F., Villa, G., de Francisco, A. L., Albertazzi, A., Adrogue, H. J., Dougherty, F. C., & Beyer, U. (2007). Effect of a continuous erythropoietin receptor activator (C.E.R.A.) on stable haemoglobin in patients with CKD on dialysis: Once monthly administration. *Current Medical Research and Opinion*, *23*, 969–979.

Lopez Ongil, S. L., Saura, M., Lamas, S., Rodriguez Puyol, M., & Rodriguez Puyol, D. (1996). Recombinant human erythropoietin does not regulate the expression of endothelin-1 and constitutive nitric oxide synthase in vascular endothelial cells. *Experimental Nephrology*, *4*, 37–42.

Lüscher, T. F., & Vanhoutte, P. M. (1990). *The Endothelium: Modulator of Cardiovascular Function*. Boca Raton, Fla: CRC Press.

Macdougall, I. (2001). An overview of the efficacy and safety of novel erythropoiesis stimulating protein (NESP). *Nephrology, Dialysis, Transplantation*, *16*(Suppl. 3), 14–21.

Macdougall, I. C., & Eckardt, K. U. (2006). Novel strategies for stimulating erythropoiesis and potential new treatments for anaemia. *Lancet*, *368*, 947–953.

Macdougall, I. C., Gray, S. J., Elston, O., Breen, C., Jenkins, B., Browne, J., & Egrie, J. (1999). Pharmacokinetics of novel erythropoiesis stimulating protein compared with epoetin alfa in dialysis patients. *Journal of the American Society of Nephrology*, *10*, 2392–2395.

Macdougall, I. C., Padhi, D., & Jang, G. (2007). Pharmacology of darbepoetin alfa. *Nephrology, Dialysis, Transplantation*, *22*(Suppl. 4), iv2–iv9.

Macdougall, I. C., Roberts, D. E., Coles, G. A., & Williams, J. D. (1991). Clinical pharmacokinetics of epoetin (recombinant human erythropoietin). *Clinical Pharmacokinetics*, *20*, 99–113.

Macdougall, I. C., Robson, R., Opatrna, S., Liogier, X., Pannier, A., Jordan, P., Dougherty, F. C., & Reigner, B. (2006). Pharmacokinetics and pharmacodynamics of intravenous and subcutaneous Continuous Erythropoietin Receptor Activator (C.E.R.A.) in patients with chronic kidney disease. *Clinical Journal of the American Society of Nephrology*, *1*, 1211–1215.

Mancini, D. M., Katz, S. D., Lang, C. C., LaManca, J., Hudaihed, A., & Androne, A. S. (2003). Effect of erythropoietin on exercise capacity in patients with moderate to severe chronic heart failure. *Circulation*, *107*, 294–299.

Masuda, S., Okano, M., Yamagishi, K., Nagao, M., Ueda, M., & Sasaki, R. (1994). A novel site of erythropoietin production. Oxygen-dependent production in cultured rat astrocytes. *The Journal of Biological Chemistry*, *269*, 19488–19493.

McMahon, F., Vargas, R., Ryan, M., Jain, A., Abels, R., Perry, B., et al. (1990). Pharmacokinetics and effects of recombinant human erythropoietin after intravenous and subcutaneous injections in healthy volunteers. *Blood*, *76*, 1718–1722.

Miki, T., Miura, T., Hotta, H., Tanno, M., Yano, T., Sato, T., Terashima, Y., Takada, A., Ishikawa, S., & Shimamoto, K. (2009). Endoplasmic reticulum stress in diabetic hearts abolishes erythropoietin-induced myocardial protection by impairment of phospho-glycogen synthase kinase-3beta-mediated suppression of mitochondrial permeability transition. *Diabetes*, *58*, 2863–2872.

Miura, O., Nakamura, N., Ihle, J. N., & Aoki, N. (1994a). Erythropoietin-dependent association of phosphatidylinositol 3-kinase with tyrosine-phosphorylated erythropoietin receptor. *The Journal of Biological Chemistry*, *269*, 614–620.

Miura, O., Nakamura, N., Quelle, F. W., Witthuhn, B. A., Ihle, J. N., & Aoki, N. (1994b). Erythropoietin induces association of the JAK2 protein tyrosine kinase with the erythropoietin receptor *in vivo*. *Blood*, *84*, 1501–1507.

Miyake, T., Kung, C. K., & Goldwasser, E. (1977). Purification of human erythropoietin. *The Journal of Biological Chemistry*, *252*, 5558–5564.

Moon, C., Krawczyk, M., Ahn, D., Ahmet, I., Paik, D., Lakatta, E. G., & Talan, M. I. (2003). Erythropoietin reduces myocardial infarction and left ventricular functional decline after coronary artery ligation in rats. *Proceedings of the National Academy of Sciences of the United States of America*, *100*, 11612–11617.

Moreau, C., Lariviere, R., Kingma, I., Grose, J. H., & Lebel, M. (2000). Chronic nitric oxide inhibition aggravates hypertension in erythropoietin-treated renal failure rats. *Clinical and Experimental Hypertension*, *22*, 663–674.

Nakano, M., Satoh, K., Fukumoto, Y., Ito, Y., Kagaya, Y., Ishii, N., Sugamura, K., & Shimokawa, H. (2007). Important role of erythropoietin receptor to promote VEGF expression and angiogenesis in peripheral ischemia in mice. *Circulation Research*, *100*, 662–669.

Namiuchi, S., Kagaya, Y., Ohta, J., Shiba, N., Sugi, M., Oikawa, M., Kunii, H., Yamao, H., Komatsu, N., Yui, M., Tada, H., Sakuma, M., Watanabe, J., Ichihara, T., & Shirato, K. (2005). High serum erythropoietin level is associated with smaller infarct size in patients with acute myocardial infarction who undergo successful primary percutaneous coronary intervention. *Journal of the American College of Cardiology*, *45*, 1406–1412.

Neusser, M., Tepel, M., & Zidek, W. (1993). Erythropoietin increases cytosolic free calcium concentration in vascular smooth muscle cells. *Cardiovascular Research*, *27*, 1233–1236.

Ni, Z., Wang, X. Q., & Vaziri, N. D. (1998). Nitric oxide metabolism in erythropoietin-induced hypertension: Effect of calcium channel blockade. *Hypertension*, *32*, 724–729.

Nichol, C. A., Smith, G. K., & Duch, D. S. (1985). Biosynthesis and metabolism of tetrahydrobiopterin and molybdopterin. *Annual Review of Biochemistry*, *54*, 729–764.

Nishihara, M., Miura, T., Miki, T., Sakamoto, J., Tanno, M., Kobayashi, H., Ikeda, Y., Ohori, K., Takahashi, A., & Shimamoto, K. (2006). Erythropoietin affords additional cardioprotection to preconditioned hearts by enhanced phosphorylation of glycogen synthase kinase-3 beta. *American Journal of Physiology. Heart and Circulatory Physiology*, *291*, H748–755.

Noguchi, C. T., Bae, K. S., Chin, K., Wada, Y., Schechter, A. N., & Hankins, W. D. (1991). Cloning of the human erythropoietin receptor gene. *Blood*, *78*, 2548–2556.

Noguchi, K., Yamashiro, S., Matsuzaki, T., Sakanashi, M., Nakasone, J., & Miyagi, K. (2001). Effect of 1-week treatment with erythropoietin on the vascular endothelial function in anaesthetized rabbits. *British Journal of Pharmacology*, *133*, 395–405.

Palazzuoli, A., Silverberg, D., Iovine, F., Capobianco, S., Giannotti, G., Calabro, A., Campagna, S. M., & Nuti, R. (2006). Erythropoietin improves anemia exercise tolerance and renal function and reduces B-type natriuretic peptide and hospitalization in patients with heart failure and anemia. *American Heart Journal*, *152*(1096), e1099–1015.

Palmer, R. M. J., Ashton, D. S., & Moncada, S. (1988). Vascular endothelial cells synthesize nitric oxide from L-arginine. *Nature*, *333*, 664–666.

Palmer, R. M. J., Ferrige, A. G., & Moncada, S. (1987). Nitric oxide release accounts for the biological activity of endothelium-derived relaxing factor. *Nature*, *327*, 524–526.

Parsa, C. J., Matsumoto, A., Kim, J., Riel, R. U., Pascal, L. S., Walton, G. B., Thompson, R. B., Petrofski, J. A., Annex, B. H., Stamler, J. S., & Koch, W. J. (2003). A novel protective effect of erythropoietin in the infarcted heart. *The Journal of Clinical Investigation*, *112*, 999–1007.

Pollock, J. S., Förstermann, U., Mitchell, J. A., Warner, T. D., Schmidt, H. H., Nakane, M., & Murad, F. (1991). Purification and characterization of particulate endothelium-derived relaxing factor synthase from cultured and native bovine aortic endothelial cells. *Proceedings of the National Academy of Sciences of the United States of America*, *88*, 10480–10484.

Ponikowski, P., Anker, S. D., Szachniewicz, J., Okonko, D., Ledwidge, M., Zymlinski, R., Ryan, E., Wasserman, S. M., Baker, N., Rosser, D., Rosen, S. D., Poole-Wilson, P. A., Banasiak, W., Coats, A. J., & McDonald, K. (2007). Effect of darbepoetin alfa on exercise tolerance in anemic patients with symptomatic chronic heart failure: A randomized, double-blind, placebo-controlled trial. *Journal of the American College of Cardiology*, *49*, 753–762.

Provenzano, R., Besarab, A., Macdougall, I. C., Ellison, D. H., Maxwell, A. P., Sulowicz, W., Klinger, M., Rutkowski, B., Correa-Rotter, R., & Dougherty, F. C. (2007). The continuous erythropoietin receptor activator (C.E.R.A.) corrects anemia at extended administration intervals in patients with chronic kidney disease not on dialysis: Results of a phase II study. *Clinical Nephrology*, *67*, 306–317.

Quaschning, T., Ruschitzka, F., Stallmach, T., Shaw, S., Morawietz, H., Goettsch, W., Hermann, M., Slowinski, T., Theuring, F., Hocher, B., Lüscher, T. F., & Gassmann, M. (2003). Erythropoietin-induced excessive erythrocytosis activates the tissue endothelin system in mice. *FASEB Journal*, *17*, 259–261.

Rafiee, P., Shi, Y., Su, J., Pritchard, K. A., Jr., Tweddell, J. S., & Baker, J. E. (2005). Erythropoietin protects the infant heart against ischemia-reperfusion injury by triggering multiple signaling pathways. *Basic Research in Cardiology*, *100*, 187–197.

Ramakrishnan, R., Cheung, W. K., Wacholtz, M. C., Minton, N., & Jusko, W. J. (2004). Pharmacokinetic and pharmacodynamic modeling of recombinant human erythropoietin after single and multiple doses in healthy volunteers. *Journal of Clinical Pharmacology*, *44*, 991–1002.

Raman, C. S., Li, H., Martasek, P., Kral, V., Masters, B. S., & Poulos, T. L. (1998). Crystal structure of constitutive endothelial nitric oxide synthase: A paradigm for pterin function involving a novel metal center. *Cell*, *95*, 939–950.

Rapoport, R. M., Draznin, M. B., & Murad, F. (1983). Endothelium-dependent relaxation in rat aorta may be mediated through cyclic GMP-dependent protein phosphorylation. *Nature*, *306*, 174–176.

Richmond, T. D., Chohan, M., & Barber, D. L. (2005). Turning cells red: Signal transduction mediated by erythropoietin. *Trends in Cell Biology*, *15*, 146–155.

Rui, T., Feng, Q., Lei, M., Peng, T., Zhang, J., Xu, M., Abel, E. D., Xenocostas, A., & Kvietys, P. R. (2005). Erythropoietin prevents the acute myocardial inflammatory response induced by ischemia/reperfusion via induction of AP-1. *Cardiovascular Research*, *65*, 719–727.

Ruschitzka, F. T., Wenger, R. H., Stallmach, T., Quaschning, T., de Wit, C., Wagner, K., Labugger, R., Kelm, M., Noll, G., Rulicke, T., Shaw, S., Lindberg, R. L., Rodenwaldt, B., Lutz, H., Bauer, C., Lüscher, T. F., & Gassmann, M. (2000). Nitric oxide prevents cardiovascular disease and determines survival in polyglobulic mice overexpressing erythropoietin. *Proceedings of the National Academy of Sciences of the United States of America*, *97*, 11609–11613.

Sadamoto, Y., Igase, K., Sakanaka, M., Sato, K., Otsuka, H., Sakaki, S., Masuda, S., & Sasaki, R. (1998). Erythropoietin prevents place navigation disability and cortical infarction in rats with permanent occlusion of the middle cerebral artery. *Biochemical and Biophysical Research Communications*, *253*, 26–32.

Sakanaka, M., Wen, T. C., Matsuda, S., Masuda, S., Morishita, E., Nagao, M., & Sasaki, R. (1998). *In vivo* evidence that erythropoietin protects neurons from ischemic damage. *Proceedings of the National Academy of Sciences of the United States of America*, *95*, 4635–4640.

Santhanam, A. V., d'Uscio, L. V., Peterson, T. E., & Katusic, Z. S. (2008). Activation of endothelial nitric oxide synthase is critical for erythropoietin-induced mobilization of progenitor cells. *Peptides*, *29*, 1451–1455.

Santhanam, A. V., Smith, L. A., Akiyama, M., Rosales, A. G., Bailey, K. R., & Katusic, Z. S. (2005). Role of endothelial NO synthase phosphorylation in cerebrovascular protective effect of recombinant erythropoietin during subarachnoid hemorrhage-induced cerebral vasospasm. *Stroke*, *36*, 2731–2737.

Santhanam, A. V., Smith, L. A., Nath, K. A., & Katusic, Z. S. (2006). *In vivo* stimulatory effect of erythropoietin on endothelial nitric oxide synthase in cerebral arteries. *American Journal of Physiology. Heart and Circulatory Physiology*, *291*, H781–H786.

Sasaki, H., Bothner, B., Dell, A., & Fukuda, M. (1987). Carbohydrate structure of erythropoietin expressed in Chinese hamster ovary cells by a human erythropoietin cDNA. *The Journal of Biological Chemistry*, *262*, 12059–12076.

Sasaki, H., Ochi, N., Dell, A., & Fukuda, M. (1988). Site-specific glycosylation of human recombinant erythropoietin: Analysis of glycopeptides or peptides at each glycosylation site by fast atom bombardment mass spectrometry. *Biochemistry*, *27*, 8618–8626.

Satoh, K., Kagaya, Y., Nakano, M., Ito, Y., Ohta, J., Tada, H., Karibe, A., Minegishi, N., Suzuki, N., Yamamoto, M., Ono, M., Watanabe, J., Shirato, K., Ishii, N., Sugamura, K., & Shimokawa, H. (2006). Important role of endogenous erythropoietin system in recruitment of endothelial progenitor cells in hypoxia-induced pulmonary hypertension in mice. *Circulation*, *113*, 1442–1450.

Scalera, F., Kielstein, J. T., Martens-Lobenhoffer, J., Postel, S. C., Tager, M., & Bode-Boger, S. M. (2005). Erythropoietin increases asymmetric dimethylarginine in endothelial cells: Role of dimethylarginine dimethylaminohydrolase. *Journal of the American Society of Nephrology*, *16*, 892–898.

Shi, Y., Rafiee, P., Su, J., Pritchard, K. A. Jr., Tweddell, J. S., & Baker, J. E. (2004). Acute cardioprotective effects of erythropoietin in infant rabbits are mediated by activation of protein kinases and potassium channels. *Basic Research in Cardiology*, *99*, 173–182.

Silverberg, D. S., Wexler, D., Sheps, D., Blum, M., Keren, G., Baruch, R., Schwartz, D., Yachnin, T., Steinbruch, S., Shapira, I., Laniado, S., & Iaina, A. (2001). The effect of correction of mild anemia in severe, resistant congestive heart failure using subcutaneous erythropoietin and intravenous iron: A randomized controlled study. *Journal of the American College of Cardiology*, *37*, 1775–1780.

Siren, A. L., Fratelli, M., Brines, M., Goemans, C., Casagrande, S., Lewczuk, P., Keenan, S., Gleiter, C., Pasquali, C., Capobianco, A., Mennini, T., Heumann, R., Cerami, A., Ehrenreich, H., & Ghezzi, P. (2001). Erythropoietin prevents neuronal apoptosis after cerebral ischemia and metabolic stress. *Proceedings of the National Academy of Sciences of the United States of America*, *98*, 4044–4049.

Spieker, L. E., Noll, G., Ruschitzka, F. T., Maier, W., & Luscher, T. F. (2000). Working under pressure: The vascular endothelium in arterial hypertension. *Journal of Human Hypertension*, *14*, 617–630.

Storring, P. L., Tiplady, R. J., Gaines Das, R. E., Stenning, B. E., Lamikanra, A., Rafferty, B., & Lee, J. (1998). Epoetin alfa and beta differ in their erythropoietin isoform compositions and biological properties. *British Journal of Haematology*, *100*, 79–89.

Tada, H., Kagaya, Y., Takeda, M., Ohta, J., Asaumi, Y., Satoh, K., Ito, K., Karibe, A., Shirato, K., Minegishi, N., & Shimokawa, H. (2006). Endogenous erythropoietin system in non-hematopoietic lineage cells plays a protective role in myocardial ischemia/ reperfusion. *Cardiovascular Research*, *71*, 466–477.

Takeuchi, M., & Kobata, A. (1991). Structures and functional roles of the sugar chains of human erythropoietins. *Glycobiology*, *1*, 337–346.

Tan, C. C., Eckardt, K. U., Firth, J. D., & Ratcliffe, P. J. (1992). Feedback modulation of renal and hepatic erythropoietin mRNA in response to graded anemia and hypoxia. *The American Journal of Physiology*, *263*, F474–F481.

Tseng, M. Y., Hutchinson, P. J., Richards, H. K., Czosnyka, M., Pickard, J. D., Erber, W. N., Brown, S., & Kirkpatrick, P. J. (2009). Acute systemic erythropoietin therapy to reduce delayed ischemic deficits following aneurysmal subarachnoid hemorrhage: A Phase II randomized, double-blind, placebo-controlled trial. Clinical article. *Journal of Neurosurgery*, *111*, 171–180.

Tsukahara, H., Hiraoka, M., Hori, C., Hata, I., Okada, T., Gejyo, F., et al. (1997). Chronic erythropoietin treatment enhances endogenous nitric oxide production in rats. *Scandinavian Journal of Clinical and Laboratory Investigation*, *57*, 487–493.

Urao, N., Okigaki, M., Yamada, H., Aadachi, Y., Matsuno, K., Matsui, A., Matsunaga, S., Tateishi, K., Nomura, T., Takahashi, T., Tatsumi, T., & Matsubara, H. (2006). Erythropoietin-mobilized endothelial progenitors enhance reendothelialization via Akt-endothelial nitric oxide synthase activation and prevent neointimal hyperplasia. *Circulation Research*, *98*, 1405–1413.

Van Geet, C., Van Damme-Lombaerts, R., Vanrusselt, M., de Mol, A., Proesmans, W., & Vermylen, J. (1990). Recombinant human erythropoietin increases blood pressure platelet aggregability and platelet free calcium mobilisation in uraemic children: A possible link? *Thrombosis and Haemostasis*, *64*, 7–10.

van Veldhuisen, D. J., Dickstein, K., Cohen-Solal, A., Lok, D. J., Wasserman, S. M., Baker, N., Rosser, D., Cleland, J. G., & Ponikowski, P. (2007). Randomized, double-blind, placebo-controlled study to evaluate the effect of two dosing regimens of darbepoetin alfa in patients with heart failure and anaemia. *European Heart Journal*, *28*, 2208–2216.

Vaziri, N. D., Zhou, X. J., Naqvi, F., Smith, J., Oveisi, F., Wang, Z. Q., & Purdy, R. E. (1996). Role of nitric oxide resistance in erythropoietin-induced hypertension in rats with chronic renal failure. *The American Journal of Physiology*, *271*, E113–122.

Vaziri, N. D., Zhou, X. J., Smith, J., Oveisi, F., Baldwin, K., & Purdy, R. E. (1995). *In vivo* and *in vitro* pressor effects of erythropoietin in rats. *The American Journal of Physiology*, *269*, F838–F845.

Wang, X. Q., & Vaziri, N. D. (1999). Erythropoietin depresses nitric oxide synthase expression by human endothelial cells. *Hypertension*, *33*, 894–899.

Westenbrink, B. D., Lipsic, E., van der Meer, P., van der Harst, P., Oeseburg, H., Du Marchie Sarvaas, G. J., Koster, J., Voors, A. A., van Veldhuisen, D. J., van Gilst, W. H., & Schoemaker, R. G. (2007). Erythropoietin improves cardiac function through endothelial progenitor cell and vascular endothelial growth factor mediated neovascularization. *European Heart Journal*, *28*, 2018–2027.

Witthuhn, B. A., Quelle, F. W., Silvennoinen, O., Yi, T., Tang, B., Miura, O., & Ihle, J. N. (1993). JAK2 associates with the erythropoietin receptor and is tyrosine phosphorylated and activated following stimulation with erythropoietin. *Cell*, *74*, 227–236.

Wright, G. L., Hanlon, P., Amin, K., Steenbergen, C., Murphy, E., & Arcasoy, M. O. (2004). Erythropoietin receptor expression in adult rat cardiomyocytes is associated with an acute cardioprotective effect for recombinant erythropoietin during ischemia-reperfusion injury. *The FASEB Journal*, *18*, 1031–1033.

Wu, X. C., Johns, E. J., & Richards, N. T. (1999). Relationship between erythropoietin and nitric oxide in the contraction of rat renal arcuate arteries and human umbilical vein endothelial cells. *Clinical Science (London)*, *97*, 413–419.

Xu, X., Cao, Z., Cao, B., Li, J., Guo, L., Que, L., Ha, T., Chen, Q., Li, C., & Li, Y. (2009). Carbamylated erythropoietin protects the myocardium from acute ischemia/reperfusion injury through a PI3K/Akt-dependent mechanism. *Surgery*, *146*, 506–514.

Xu, B., Dong, G. H., Liu, H., Wang, Y. Q., Wu, H. W., & Jing, H. (2005). Recombinant human erythropoietin pretreatment attenuates myocardial infarct size: A possible mechanism involves heat shock Protein 70 and attenuation of nuclear factor-kappaB. *Annals of Clinical and Laboratory Science*, *35*, 161–168.

Index

Contents of Previous Volumes

Volume 40

Volume 41

Volume 42

Catecholamine: Bridging Basic Science
Edited by David S. Goldstein, Graeme Eisenhofer, and Richard McCarty

Part A. Catecholamine Synthesis and Release

Part B. Catecholamine Reuptake and Storage

Part C. Catecholamine Metabolism

Part D. Catecholamine Receptors and Signal Transduction

Part E. Catecholamine in the Periphery

Part F. Catecholamine in the Central Nervous System

Part G. Novel Catecholaminergic Systems

Part H. Development and Plasticity

Part I. Drug Abuse and Alcoholism

Volume 43

Overview: Pharmacokinetic Drug–Drug Interactions
Albert P. Li and Malle Jurima-Romet

Role of Cytochrome P450 Enzymes in Drug–Drug Interactions
F. Peter Guengerich

The Liver as a Target for Chemical–Chemical Interactions
John-Michael Sauer, Eric R. Stine, Lhanoo Gunawardhana, Dwayne A. Hill, and I. Glenn Sipes

Application of Human Liver Microsomes in Metabolism-Based Drug–Drug Interactions: *In Vitro–in Vivo* Correlations and the Abbott Laboratories Experience
A. David Rodrigues and Shekman L. Wong

Primary Hepatocyte Cultures as an *in Vitro* Experimental Model for the Evaluation of Pharmacokinetic Drug–Drug Interactions
Albert P. Li

Volume 44

Volume 45

Volume 46

Volume 47

Hormones and Signaling
Edited by Bert W. O'Malley

Volume 48

HIV: Molecular Biology and Pathogenesis: Viral Mechanisms
Edited by Kuan-Teh Jeang

Volume 49

HIV: Molecular Biology and Pathogenesis: Clinical Applications
Edited by Kuan-Teh Jeang

Volume 50

Volume 51

Treatment of Leukemia and Lymphoma
Edited by David A. Scheinberg and Joseph G. Jurcic

Volume 52

Volume 53

Volume 54

Volume 55

Volume 56

Volume 57

Volume 58

Volume 59